ELEMENTS OF CONTROL SYSTEMS

Sudhir Gupta

Fanshawe College
London, Ontario, Canada

Prentice Hall

Upper Saddle River, New Jersey
Columbus, Ohio

Library of Congress Cataloging-in-Publication Data

Gupta, Sudhir (Sudhir K.)
Elements of control systems / Sudhir Gupta.
p. cm.
ISBN 0-13-011989-X
1. Automatic control. 2. Control theory. I. Title.

TJ213.G883 2002
629.8—dc21 00-051494

Vice President and Editor in Chief: Stephen Helba
Assistant Vice President and Publisher: Charles E. Stewart, Jr.
Production Editor: Tricia L. Rawnsley
Production Coordination: bookworks, Lisa Garboski
Design Coordinator: Robin G. Chukes
Text Designer: Carlisle Communications
Cover art: Corbis
Cover Designer: Linda Fares
Production Manager: Matthew Ottenweller
Electronic Text Management: Karen L. Bretz

This book was set in New Century Schoolbook by York Graphic Services, Inc. and was printed and bound by R. R. Donnelley & Sons Company. The cover was printed by Phoenix Color Corp.

Prentice Hall

10 9 8 7 6 5 4 3 2 1
ISBN 0-13-011989-X

I dedicate this book to my parents
Satyendra Kumar and Pushpa Kumar
in appreciation of all they taught me.

Sudhir Gupta

PREFACE

There are a number of excellent textbooks covering the subject of control engineering at the engineering level. These books are typically written for a two-semester course in control engineering or for an intense one-semester course in the senior year. As is typical of engineering textbooks, these treat control theory from a mathematical point of view.

Elements of Control Systems does not do that. This book is suitable for

- Two- or four-year college programs requiring an in-depth understanding of control systems
- A one-semester university course at the freshman level
- Industry personnel interested in developing a greater understanding of control principles

It has been very difficult to decide on the precise content of such a textbook. It is not a handbook, so it cannot contain every single aspect of it, and it is not a "how-to" recipe book focusing on very specific topics. An attempt has been made to cover the major topics in control system technology. It should permit the reader to develop sufficient understanding to operate, maintain, and regulate control systems. At the same time, it should permit the reader to design and develop basic control systems.

No attempt has been made to cover PLCs or their operation or programming. This topic is a field in itself, and a chapter in this text could not have done justice to it.

Even though a number of PLCs have begun to offer continuous control modules, their primary application is in digital I/O and not in process control.

Sensors, transducers, and actuators form an important segment of control technology, but so do microcontrollers, data converters, and instrumentation interface circuits. These again have not been covered in lieu of more relevant subject matter.

Where possible, an attempt has been made to point out the issues with circuit components and their impact on system operations. A chapter has been dedicated to digital control, still treating it from the classical control point of view. A chapter on fuzzy logic control provides a solid introduction to the topic. An attempt has been made to avoid excessive use of some of the cryptic terms commonly used with fuzzy control.

How to Use the Book

The book consists of two major sections: the first part covers control system theory and the second part covers controllers and their applications.

It is important that the first part be studied with some diligence. The reader should be fully comfortable with the notions of time and frequency response before tackling controllers and their applications. The first chapter introduces the concept of a control system. The second chapter covers block diagrams, followed by a chapter on Laplace transformation. Laplace transformation is covered with an application-oriented approach. It is not intended to provide a solid foundation in operational calculus but apply it to determine system transfer functions in the following chapter on mathematical modeling. Having an understanding of the origin of transfer functions, time and frequency response techniques are covered. The discussion on stability is the coverage on controllers. It is intended to develop awareness for the consequences of improper controller selection and adjustment.

The second part of the book deals with controllers, their operation, adjustment, and analog and digital implementation. Some of these topics can actually be studied independently of the first, but it is not recommended. The section on digital controllers requires some understanding of microcontrollers and their programming. Annotated software listings have been provided in each case.

I would like to thank a number of people who have provided valuable support for this book project: Charles Stewart, publisher at Prentice Hall for his acceptance of the initial manuscript and his encouraging words; my copyeditor, Linda Thompson and project editor, Lisa Garboski, for their skills and painstaking effort; Eric Sells and Al Lovrich of Microchip Technology; Mark Siegesmund of CCS, Inc.; and Naomi Fernandes of The MathWorks Inc.

Many thanks also should go to the reviewers: Jeff Slutsky, Rochester Institute of Technology; Richard L. Windley, ECPI College of Technology; C. Richard G. Helps, Brigham Young University; John Glozy, DeVry Institute of Technology; and Dr. Nebojsa Jaksic, DeVry Institute of Technology.

I would also like to thank my family: my wife Nita for putting up with my spending countless hours working on the manuscript, my son Sachin for very valuable help with computer work, and my daughter Avina for help in transforming the handwritten notes into a printed document. Above all I would like to thank them for their constant encouragement.

Feedback

This being the first edition, it is possible that mistakes may have been made, in spite of the best efforts of the author and the publisher. It would be greatly appreciated if you can point these out. Any kind of constructive criticism is welcome.

Sudhir K. Gupta
sgupta@mcjournal.com

CONTENTS

1

BASIC CONCEPTS

INTRODUCTION

Automatic control systems play a vital role in the (technological) progress of human civilization. In the early days of civilization, people used levers and linkages to control process, which required energy and power beyond human reach. Now, society tends to rely heavily on automatic control systems for its day-to-day operation. These control systems range from very simple to fairly complex in nature. Automatic washing machines, refrigerators, and ovens are examples of some of the simpler systems used in the home. Aircraft automatic pilots, welding and inspection robots used in manufacturing, and electric power generation and distribution systems represent complex control systems.

In this chapter some of the terms commonly used to describe the operation, analysis, and design of control systems are discussed. The concept of a control system is introduced and its operation is explained.

CONTROL TERMINOLOGY

Some of the terms commonly used in control systems are as follows.

System

A **system** is a collection, set, or arrangement of elements (subsystems).

The concept of a system is perhaps the most important—but also the most misunderstood—concept. A system is an entity that can be considered on its own. It may be very large and complex, such as a planetary system, or it may be small and simple, such as the thermostatic temperature control of a steam iron or the water-level control system in a toilet.

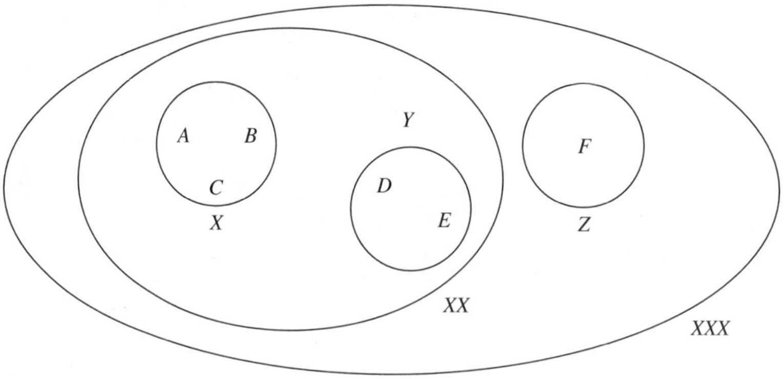

Figure 1.1 The concept of a system.

The concept of a system can perhaps be better understood through an illustration. Figure 1.1 shows various elements (objects) *A*, *B*, *C*, . . . , *F*. System *Z* contains only one element *F*. System *Y* contains *D*, and *E*, whereas system *X* contains elements *A*, *B*, and *C*. In other words, system *Y* is concerned only with the interaction of elements *D* and *E*. If the operation of a system is dependent upon elements *A*, *B*, *C*, *D*, and *E*, a larger system, *XX*, is formed. *X* and *Y*, formerly known as systems, are now better described as subsystems. Similarly, in case of system *XXX*, *XX* is termed a subsystem of *XXX*.

Subsystem

A **subsystem** is part of a system.

Any system may be considered as a subsystem of a larger system.

Element (Component)

An **element (component)** is the smallest part of a system that can be treated as a whole (entity).

It is well known that any physical object can be broken down to atomic (or, perhaps, subatomic) level. In the study of control systems, objects do not need to be broken down to the smallest level. An electrical control system contains several active and passive components, some of which may work together to form an active filter. It may be sufficient to deal with a filter network containing operational amplifiers, resistors, and capacitors network as a whole. Even though individual component properties dictate the filter characteristics, it is sufficient to treat the filter as a black box, which contains all the characteristics of an active filter network. It is then fair to say that the active filter can be treated as an element of the control system, without any reference to its components.

In a small socioeconomic system, a community may be treated as an element, even though it may incorporate several families, with every family consisting of one or more members.

Block

A **block** is a set of elements that can be grouped together, with overall characteristics described by an input/output relationship.

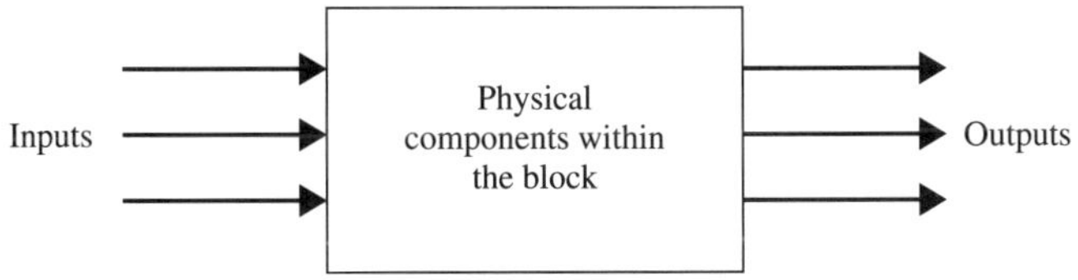

Figure 1.2 Block.

The block is an important concept in analysis and design of control systems (Figure 1.2). Input and output characteristics of entire groups of elements within the block can be described by suitable mathematical expressions.

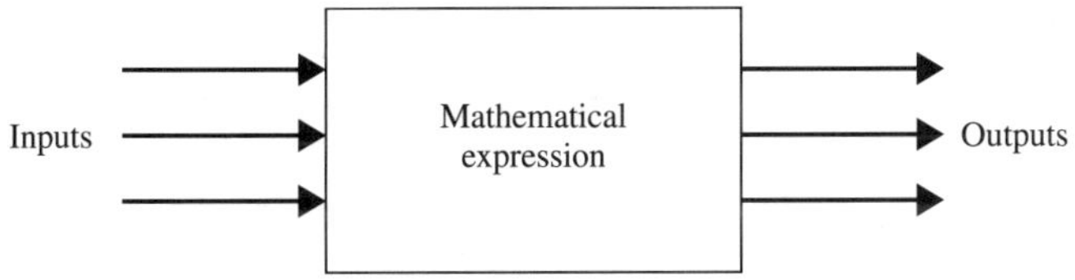

Figure 1.3 Block representation.

Block diagram representation contains only the input/output characteristics and leaves no indication about the physical arrangement of the elements within the block (Figure 1.3). It is nearly impossible to determine the true nature of elements from a block representation, such as whether it is electrical, pneumatic, mechanical, thermal, etc.

Control

Control means to regulate, direct, or govern.

Automatic

Automatic implies self-action without any human intervention.

Control System

A **control system** is a group of physical components arranged to control themselves or another system.

Automatic Control System

An **automatic control system** is a control system that is self-regulating, without any human intervention.

An automatic control system does not require human intervention for its operation. It monitors the output variable through a feedback loop and manipulates the plant or process to maintain the output at or near the desired value. Once the controller setpoint input has been entered, an automatic control system will attempt to maintain the output variable at or near the desired value.

EXAMPLE 1.1

An oven is an example of an automatic control system. Once the desired temperature is set on the dial, the oven-temperature controller takes over and maintains the oven temperature to the desired value.

Manual Control System

A **manual control system** is a control system regulated through human intervention.

Control systems range from very simple to very complex.

EXAMPLE 1.2

The following are examples of manual and automatic control.

- Opening or closing of a window in a room, thus regulating air temperature or air quality (*manual control*)
- Activation of a light switch to change illumination in a room (*manual control*)
- Porch-light activation through an optical sensor at dusk or dawn (*automatic control*)
- Thermostatic temperature control of home (*automatic control*)
- River water-flow control system (may be a very complex system involving several hydroelectric stations and reservoirs) (*combination of manual and automatic control*)

- Boiler temperature control in a nuclear reactor (*complex automatic control*)
- Autopilot control system in a commercial airliner (*complex automatic control*)

Open-Loop Control System

An **open-loop control system** is a control system in which the control (regulating) action is independent of the output.

Closed-Loop Control System

A **closed-loop control system** is a control system in which the control (regulating) action is influenced by the output.

Feedback

An important feature of a closed-loop control system is that a control system output signal is fed back to the input stage, providing information to the controller about the current output state and facilitating an appropriate control action. It is not possible to have a fully functional automatic or even manual control system without a feedback signal.

EXAMPLE 1.3

Consider a car being driven on a road. In this case, the driver acts as the controller; the car, its components, and the road become part of a closed-loop control system. Visual feedback provides information to the driver about the location of the car with respect to the surroundings. Through the help of this feedback, the driver can slow down or speed up and manipulate the steering in order to maintain a desired heading. If this (visual) feedback signal is removed (e.g., by a newspaper covering the windshield), the control system becomes an open-loop control system, rendering safe driving nearly impossible. Admittedly, if there is no traffic or curves on the road, it is still possible to maintain the car heading without a disaster. But the system will fail as soon as there is a change in the operating environment.

Regulator System

A **regulator system** is a control system where the reference (input) is normally fixed.

A home-heating system is an example of a regulator system. The reference temperature is normally set to a constant value, and the furnace is regulated to maintain the output temperature at a steady value. Variation in outside temperatures and wind velocity place varying heating loads and tend to vary the house temperature.

A temperature controller senses the variation in temperature and corrects it by changing the amount of heat added through the furnace. Regulator systems also experience variation in reference input. In a home-heating control system, the nighttime temperature setting may be different than the daytime temperature setting. However, these changes are less frequent, and so the control system has to cater mainly to load variation due to external disturbances.

Servomechanism (Follower System)

A **servomechanism (follower system)** is a control system where the reference (input) varies continuously and the system operates so that the output follows the input.

Servomechanism is (sometimes) used to describe control systems where the output variable is a mechanical position, velocity, or acceleration of an object. It originated from the Greek term *servus*, meaning servant. A solar power panel with continuous tracking of the sun (for optimum power generation) falls in this category. Other examples include robotics-movement, plotter-position, and hard-disk-drive control systems.

Single-Input, Single-Output (SISO) System

A **single-input, single-output (SISO) system** is a system where only one parameter is entered as input and only one parameter represents the output.

Figure 1.4 Single-input, single-output system.

A home-heating (or cooling) system involves a SISO system (see Figure 1.4). A single parameter, the desired room temperature, is entered as the reference variable and results in only one controlled (output) variable: actual room temperature. It is highly desirable to incorporate control of humidity, but that is left to a separate humidifier control system.

Multiple-Input, Multiple-Output (MIMO) System

A **multiple-input, multiple-output (MIMO) system** is a system where several parameters may be entered as input and output is represented by multiple variables.

A typical residential home-heating system is a SISO system where a single parameter, the room temperature, is monitored and controlled. There is a growing consensus that humidity control should be added to the home-heating system (environmental

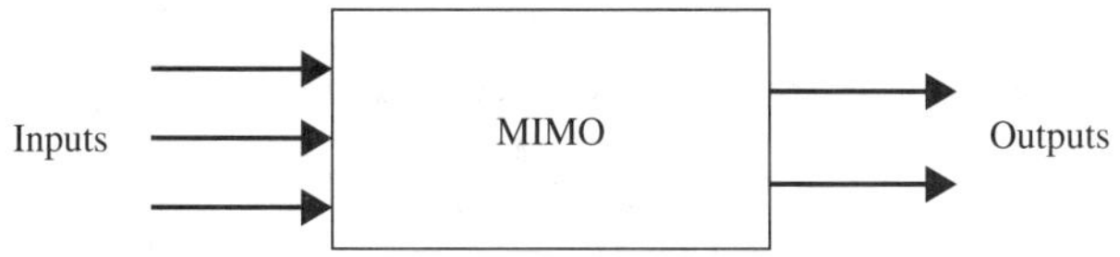

Figure 1.5 Multiple-input, multiple-output system.

control system). With this, both temperature and humidity are monitored and controlled, making the system an MIMO type (see Figure 1.5). A chemical process plant typically has several reference inputs and requires the control of several output variables.

Linear System

A **linear system** is a system where input/output relationships may be represented by a linear differential equation.

Ohm's law assumes a linear relationship between the voltage drop across a resistor and the current flow through it. It is certainly valid over a wide range of currents and voltages but does not hold for very high values of current flow through a metal wire.

Time-Invariant System

A **time-invariant system** is a system described by a differential equation with constant coefficients.

A spring mass and damper system represents a time-invariant system if the characteristics of all three components do not change with time.

Time-Variant System

A **time-variant system** is a system described by a differential equation with variable coefficients.

A rocket burning fuel (Figure 1.6) represents a time-variant system. The mass of the rocket does not remain constant during a flight but changes as fuel is burned.

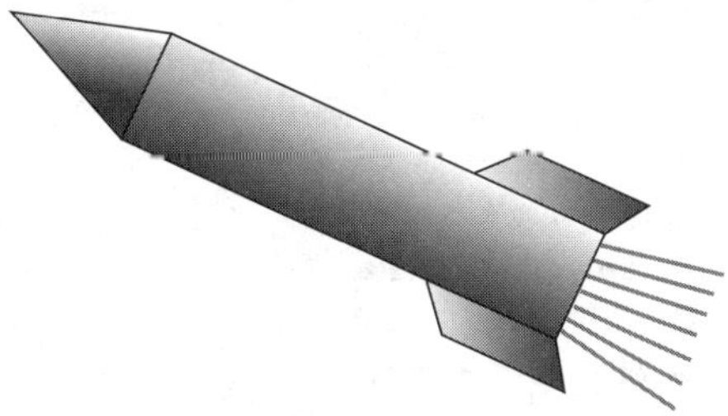

Figure 1.6 Time-variant system.

Thus the differential equation describing the rocket dynamics at the time of liftoff does not properly represent the system after a significant amount of fuel has been burned. Classical control techniques do not lend themselves to the treatment of such systems. Modern control theory, on the other hand, does not suffer from these limitations and has gained significant ground with the advent of microcontrollers and computers.

All the discussions in this book assume that all the control components and systems are linear and time-invariant. It is further assumed that the control systems are of the single-input, single-output (SISO) type. Furthermore, all systems considered are negative-feedback closed-loop control systems. Fuzzy control techniques provide a simple and effective way to deal with control systems without some of these restrictions. Fuzzy systems are discussed later in the book.

FUNCTIONING OF A CONTROL SYSTEM

All control systems, whether simple or complex, function in a very similar manner. The controller (brain of the control system) compares the actual value of the controlled variable to the desired value and tries to manipulate the process in such a fashion that the controlled variable approaches the desired value. Operation of a control system is no different than tasks routinely carried out by humans. Consider the task of reaching for a glass on a table (Figure 1.7).

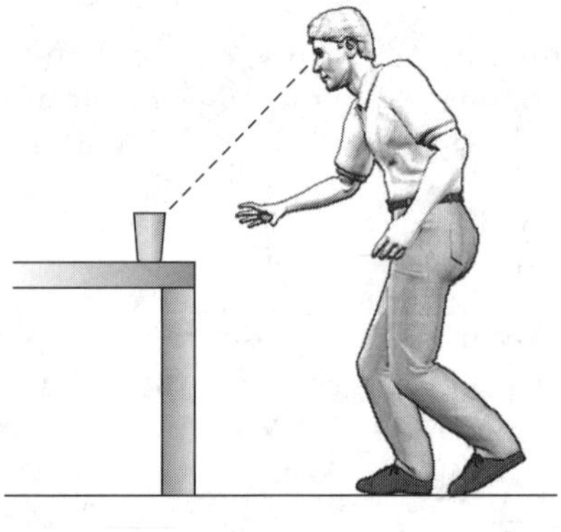

Figure 1.7 Reaching for a glass.

The objective of the task is fairly simple: the hand has to be moved so that it touches the glass. The entire process can be summarized as follows:

- **Reference input (objective)** Reach for the glass.
- **Controller (human brain)** Look at the glass (feedback) and determine (estimate) the distance between the hand and the glass.

- **Manipulate (final control element)** Move the arm toward the glass.
- **Feedback (eye)** Get a new estimate of distance between the hand and the glass.
- **Loop (repeat)** Move the arm until feedback (eye) indicates that the hand has reached the glass.

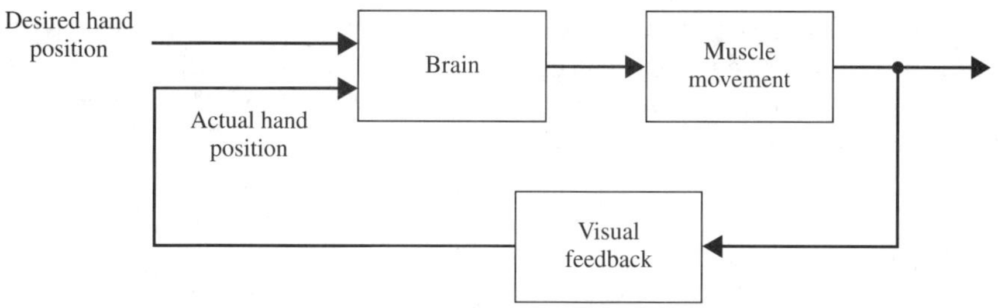

Figure 1.8 Information flow: reaching for a glass.

The visual feedback is the single most important part of the whole process (Figure 1.8). Without visual feedback (or auditory or any other kind), it is not possible to reach for the glass. Movement of the human hand is very imprecise. It cannot move through a precise distance on command (desire), but the use of (visual) feedback allows the hand movement to be very precisely controlled. Have you ever considered using a fancy robot to thread a needle? With feedback, the controller can modify its action, and even imprecise machinery can be made to work satisfactorily in demanding applications. *If it can be measured, it can be controlled.*

EXAMPLES OF CLOSED-LOOP CONTROL SYSTEMS

The following are some of the more common closed-loop control systems used today.

1. **Residential oven-temperature control** The oven contains a closed-loop temperature control system. A controlling relay located at the back panel of the oven is responsible for maintaining the oven at the desired temperature. The user turns the oven on and sets the temperature dial to the desired value. Actual oven temperature is measured by a thermocouple (temperature sensor) located at the back wall of the oven. It provides the crucial feedback of oven temperature to the controlling relay. If the actual temperature is below the desired temperature by a (factory-) set margin, the oven-heating element is turned on. The element is turned off when the oven temperature exceeds the desired value by the set margin. This is an on-off type of control and is not continuous in nature. These are being replaced by continuous-phase control systems.

2. **Automobile cruise (speed) control** The driver brings the automobile to the desired speed and activates the cruise control, which latches the speed data into the onboard computer memory. The computer gets the actual vehicle speed data from the speedometer (velocity transducer) and compares it with the set speed. It then adjusts the fuel intake into the engine manifold to bring the vehicle speed to the desired value.
3. **Electronic voltage regulator** A voltage regulator (Figure 1.9) provides a constant output voltage regardless of variation in supply voltage. These are found in almost all electronic devices. A voltage regulator can be either a fixed- or variable-output type.

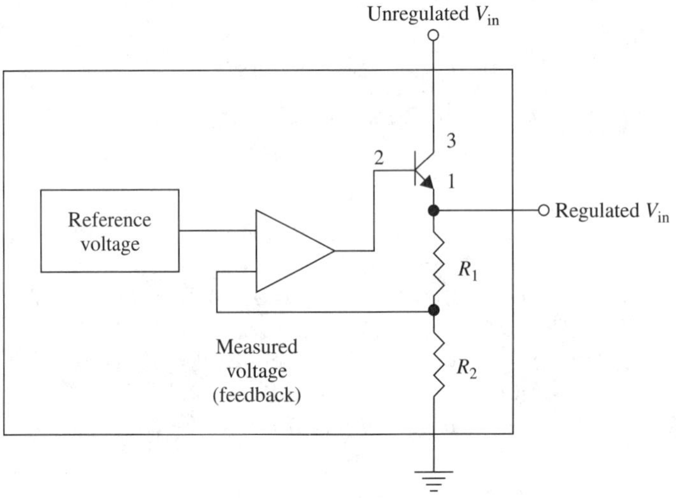

Figure 1.9 Voltage-regulator system.

Figure 1.9 shows a simplified view of a three-terminal, fixed-voltage regulator. The unregulated voltage-input terminal is connected to the source voltage and the output voltage is obtained from the regulated voltage-output terminal. In the case of a fixed-voltage regulator, the input is kept fixed (factory set) to a predefined voltage, e.g., 5 V. A buried voltage reference provides this fixed-input reference voltage. Two resistors, R_1 and R_2, provide feedback information about the actual output voltage.

An onboard op-amp acts as a controller. It compares the two input voltages and controls the amount of current passing through the series pass transistor, thus controlling the output voltage. An adjustable voltage regulator does not use an internal fixed reference but allows a user programmable voltage to be entered as the reference. It is generally set either using a potentiometer or as a combination of two precise resistors. A practical voltage regulator contains additional components to provide protection against excessive current and/or overheating.

4. **Aircraft instrument landing system** Instrument landing systems are well established and relied upon to land aircraft safely in adverse weather conditions. These systems are highly complex and very sophisticated. A computer aboard the aircraft acts as a controller. This controller is responsible for receiving real-time data from the airport computer and operating various flight controls. Ground-based instrumentation provides information about aircraft position, velocity, altitude, etc., to the aircraft controller. This controller, in turn, continuously adjusts the various controls on the aircraft to gently bring it down to the airport runway.
5. **Metal-rolling mill velocity control** In metal-rolling operations, multiple rolling stands are utilized. Material entering the first stand has the largest thickness, and the material leaving the last stand has the smallest thickness. As metal is rolled at a stand, its thickness is reduced, and as a result the linear velocity of material leaving the stand is larger than the velocity of material entering the stand. In order to make up for increased material velocity, the rolls at the next stand have to spin faster.

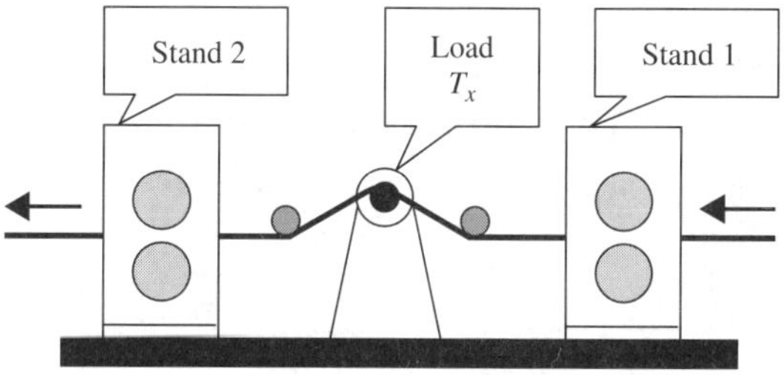

Figure 1.10 Twin-stand rolling mill.

Figure 1.10 shows a small twin-stand rolling mill. Each stand is driven by a dc servomotor. The rotational velocity of the servomotor at the second stand is slightly higher than the velocity of the motor at the first stand. The actual difference in velocity depends on several factors, including the material stiffness, reduction in thickness at the first stand, and entry velocity. A high-performance controller with fast response actuators and a precision tension load cell is required for successful implementation. The controller is responsible for controlling the velocity of the second stand. Instead of using the linear velocity as the input, the controller works on the sheet material tension in between the stands. If the rolls at the second stand match the velocity of incoming material (from the first stand), the tension in between the two stands will be minimal. The operator enters the desired tension value and the controller accepts the actual tension from the load cell as the feedback signal. It then manipulates the velocity of the second-stand servomotor, thus controlling the tension to the desired level. See Figure 1.11.

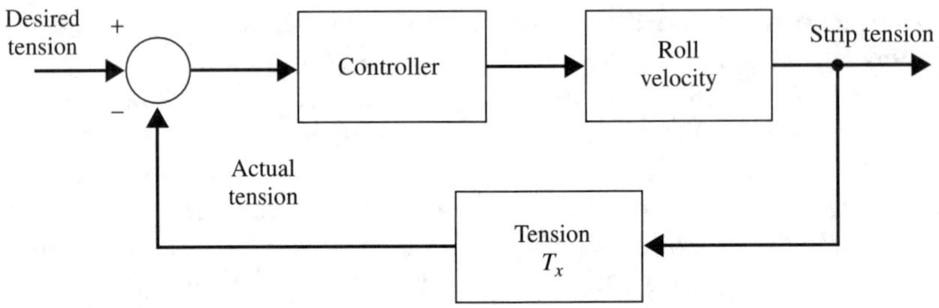

Figure 1.11 Information flow: rolling mill velocity control.

PROBLEMS

1. Provide an example of each of the following:
   a. Manual control system
   b. Automatic control system
   c. Biological control system
   d. Natural control system.
2. Provide an example of an open-loop control system that is manually controlled.
3. State whether the following systems are open-loop or closed-loop.
   a. Firing of a bullet
   b. Opening of a faucet
   c. Throwing a ball
   d. Turning on of a hot plate to a certain setting
   e. Operation of an oven (setting it to a specified temperature)
4. Consider the task of adjusting the water temperature in a shower. Identify the input, output, and control action.

2

BLOCK DIAGRAMS

INTRODUCTION

A control system consists of a number of components connected to perform a desired function. These components vary in nature and may be electrical, hydraulic, pneumatic, or mechanical. Control system theory is not concerned with the physical nature of components and/or the type of energy exchange that may be taking place. Once a component (system or subsystem) is reduced to a mathematical model, it can be represented as a *black box* (*block*), with the component operation described by the mathematical function. This process is repeated for all the parts of the system, so the system is reduced to a number of interconnected black boxes. These interconnected blocks can be combined into a single block (mathematical expression). This overall mathematical expression contains all the necessary information about the performance of the original control system. This greatly simplifies the analysis and design process, where the signal flow between these components is of most concern. Block diagrams also lend themselves to easier simulation using operational amplifiers as well as digital computers.

In this chapter the process of converting a pictorial diagram to a block diagram is illustrated, and the various techniques for reducing a complex block diagram are discussed. The use of operational amplifiers to simulate control systems is also demonstrated.

DEFINITION OF A BLOCK DIAGRAM

In a block diagram representation, every component (or subsystem) is represented as a rectangular block containing one input and one output. All the blocks are interconnected by arrows representing the signal-flow path. The operation of the component is described by the mathematical function enclosed within the block. See Figure 2.1.

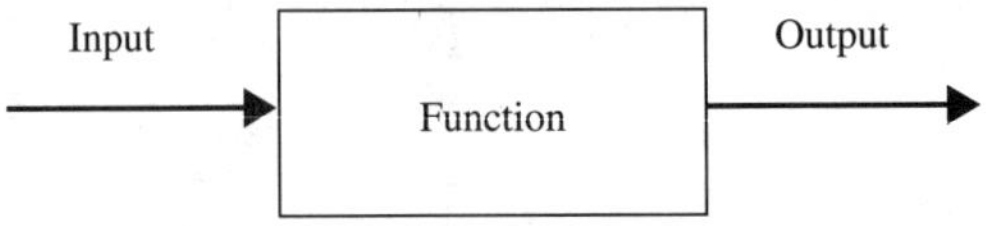

Figure 2.1 Block.

EXAMPLE 2.1

An operational amplifier has a voltage gain of 10. Figure 2.2 is a block diagram representation of this amplifier.

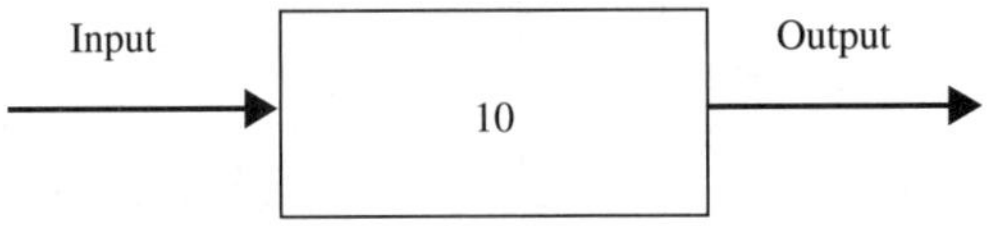

Figure 2.2 Example 2.1.

A complete amplifier has a number of passive components associated with it and it also requires a power supply for operation. The block diagram representation does not provide that information. Its sole purpose is to describe the relationship between input and output signals.

EXAMPLE 2.2

Figure 2.3 represents a differentiator function. This differentiator may be mechanical, fluidic, or electronic in nature, but this information is not available from the block diagram.

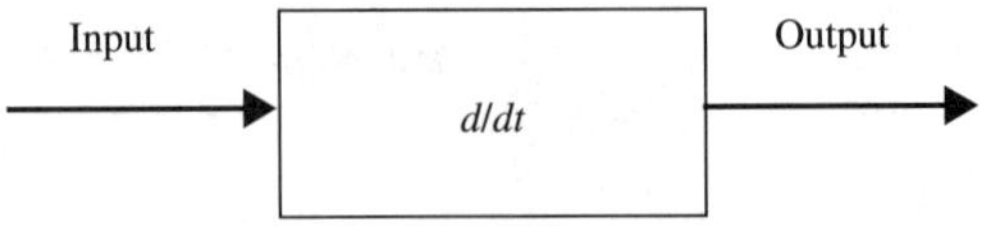

Figure 2.3 Differentiator function for Example 2.2.

EXAMPLE 2.3

A rotary potentiometer is being used as an angular position transducer. It has a rotational range of 0° to 300° and it is powered by a 12 V supply. The potentiometer is linear over the entire operating range. Determine its transfer function.

Solution

$$\text{potentiometer TF} = \frac{\text{output voltage}}{\text{input rotation}} = \frac{12\text{ V}}{300^\circ} = 0.04\text{ V/degree}$$

Figure 2.4 describes the transducer function.

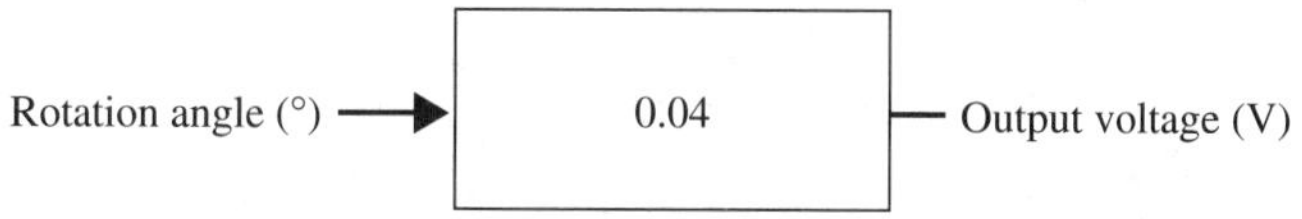

Figure 2.4 Potentiometer for Example 2.3.

Summing Point

The summing point is the block (Figure 2.5) used to represent the addition and/or subtraction of signals. It is shown as a small circle connected to arrows representing signal lines. Incoming signal lines are further identified with either a positive or negative sign, indicating the addition or subtraction process. Output of a summing point is again a signal, shown with an outgoing arrow. No sign is attached to this outgoing signal. It is also known as a *summing junction* or *summation point*.

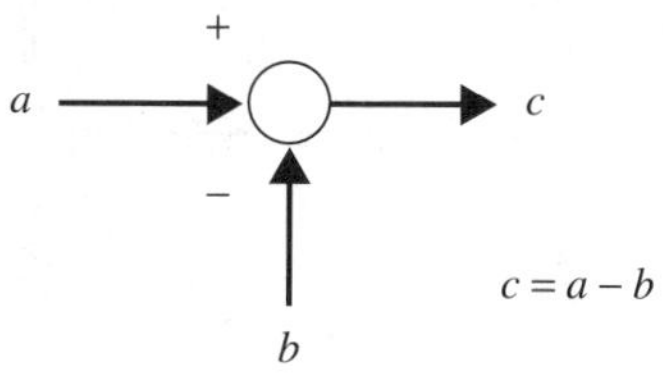

Figure 2.5 Summing point.

Takeoff Point

A takeoff point is used when a signal needs to be routed to two or more blocks. It can be thought of as an electrical junction with voltage being sent to all connected terminals. It is represented as a dot (solid circle) with arrows pointing away from it (Figure 2.6).

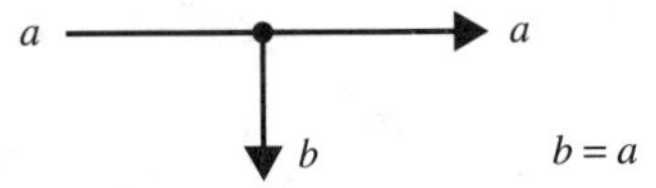

Figure 2.6 Takeoff point.

BLOCK DIAGRAM REPRESENTATION OF AN EQUATION

Algebraic equations and mathematical functions can be very conveniently represented using block diagrams.

EXAMPLE 2.4

a. $Y = mx + c$

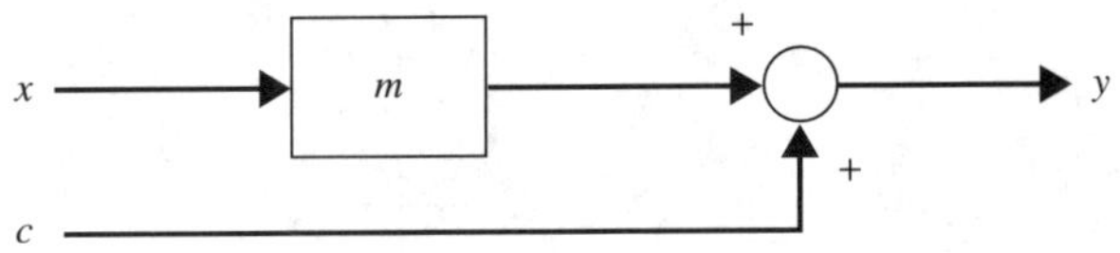

b. $z = 1.2x - 10y + 25$

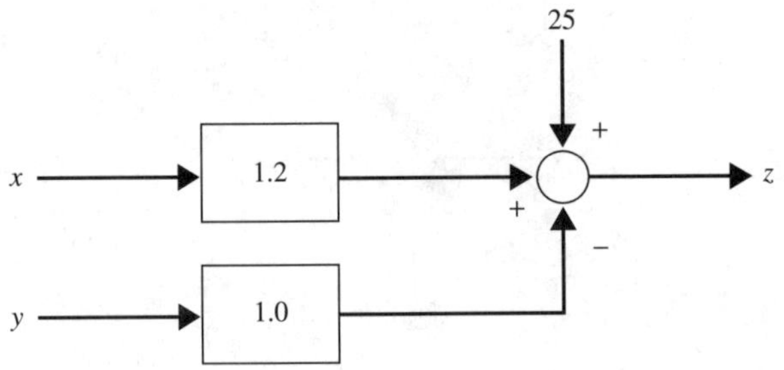

BLOCK DIAGRAM REPRESENTATION OF CONTROL SYSTEMS

A pictorial diagram of a process generally conveys information about the physical components used in the process and does not provide information about the interaction of control signals. For the purpose of analysis, it is vital to know the way control information is being transferred from one component to another. Once components are replaced by equivalent blocks and these blocks are connected through signal-flow lines, the pictorial view has been transformed into an equivalent block diagram.

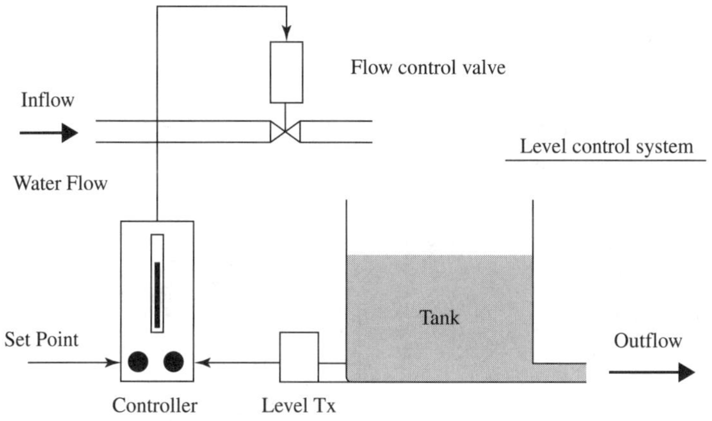

Figure 2.7 Level control system.

Figure 2.7 shows the main components of a level control system. Before this pictorial can be transformed into a block diagram, an understanding of system operation and components interaction need to be developed. Two basic questions can be asked.

1. *What is being controlled?* The objective of this control system is to maintain (control) the fluid level (*controlled variable*) in the tank (*process, or plant*) at the desired value (*set point, or reference*).
2. *How is it being controlled?* Fluid level in the tank cannot be directly controlled. It can be controlled only by changing (manipulating) the water flow (*manipulated variable*) into the tank.

Control System Operation

- The operator sets the desired level at the controller panel (*SP*).
- The level transducer provides a signal proportional to the actual level in the tank (*MV*).
- The difference between *SP* and *MV* generates an error signal ($E = SP - MV$).

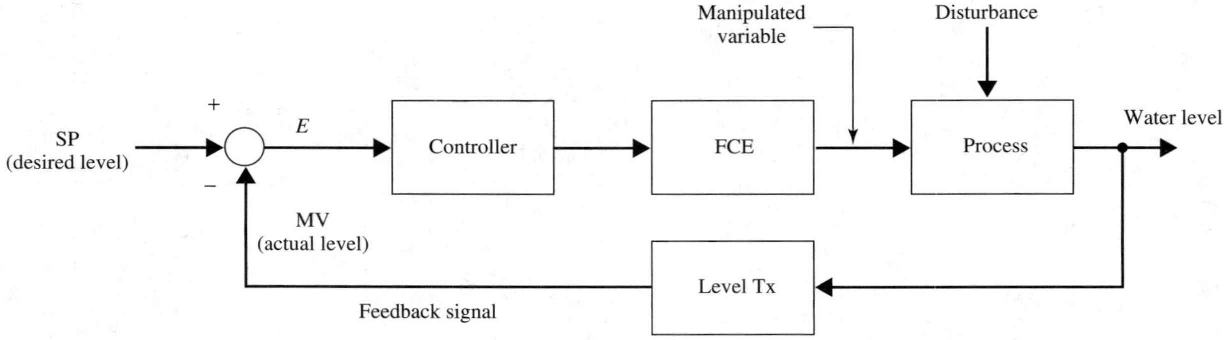

Figure 2.8 Block diagram representation of a level control system.

in and out not shown b/c not part of control/process

- If the error signal is positive, then it indicates to the controller that actual level is lower than desired level and an actuating signal is generated, which opens the flow control valve (FCE—final control element) to allow a higher flow rate into the tank. On the other hand, if the actual level is higher than the desired level, then the valve opening is reduced to reduce inflow rate.
- If the control system has been set up (tuned) properly, then the level will reach the desired value.
- The control cycle continues to monitor the fluid level, and if it changes due to variation in outflow rate (which is externally controlled), then the controller will manipulate the flow rate to maintain the fluid level.

Once a clear-cut understanding of control system operation is developed, a block diagram representation can be prepared (Figure 2.8).

EXERCISES

2.1 Identify the various components of an oven and draw a block diagram.

2.2 Draw a block diagram of a closed-loop home-heating system.

2.3 Modify the block diagram in Exercise 2.2 to reflect a different method of heating.

TRANSFER FUNCTIONS

General Transfer Function

The transfer function (TF) of a system (or a block) is defined as the ratio of output to input (Figure 2.9). It is normally expressed in Laplace domain (Laplace transformed expression). For simple systems, not containing integral and/or differential terms, the transfer function is a numerical value, as seen in an earlier section.

$$\text{transfer function (TF)} = \frac{\text{output}}{\text{input}}$$

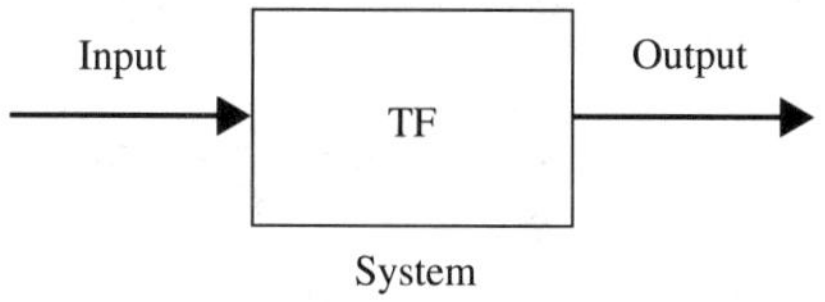

Figure 2.9 Transfer function.

Transfer Function of a Closed-Loop System

Closed-loop control systems can be divided into two categories, based upon the nature of the feedback signal.

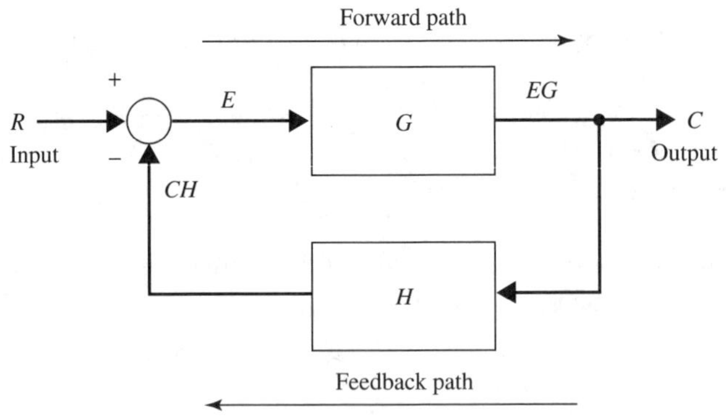

Figure 2.10 Negative-feedback closed-loop system.

Negative-Feedback Closed-Loop System Figure 2.10 shows one of the simplest forms of a negative-feedback closed-loop control system. It has one block in the forward path and one block in the feedback path. Practically, all control systems can be reduced to this simplified form. Block G represents the gain of all the blocks in the forward path, and block H represents the gain of all the blocks in the feedback path. Our objective is to reduce this control system to just a single block. (See Figure 2.11.)

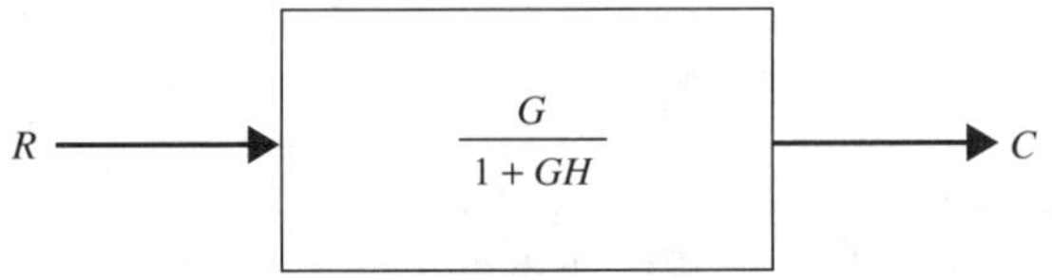

Figure 2.11 Reduced form of a closed-loop control system.

The following relationships can be determined from the block diagram:

$$\text{feedback signal} = \text{output} \times \text{feedback path gain}$$
$$= CH \tag{2.1}$$
$$\text{error signal } E = \text{input} - \text{feedback signal}$$
$$= R - CH \tag{2.2}$$
$$\text{output signal } C = \text{error signal} \times \text{forward path gain}$$
$$= EG \tag{2.3}$$

Substituting the value of the error signal from (2.2) to (2.3),

$$\text{output } C = (R - CH) \times G$$
$$C = RG - CGH$$
$$C + CGH = RG$$
$$C(1 + GH) = RG$$
$$\frac{C}{R} = \frac{G}{1 + GH}$$

$$\text{closed-loop transfer function} = \frac{\text{forward path gain}}{1 + \text{forward gain} \times \text{feedback gain}}$$

$$\text{output } C = \text{closed-loop transfer function} \times \text{input } R$$

$$\text{negative-feedback CLTF } \frac{C}{R} = \frac{G}{1 + GH}$$

Some observations can be drawn from this relationship:

- The control system transfer function is a property of the system.
- The transfer function is dependent only upon its internal structure and components.
- The transfer function is independent of the input applied to the system.

When an input signal is applied to a closed-loop control system, an output is generated, which is dependent not only upon the input, but also upon the system transfer function.

Positive-Feedback Closed-Loop System Figure 2.12 shows a simple positive-feedback closed-loop control system It again has one block in the forward path and one block in the feedback path.

The transfer function of the positive-feedback closed-loop control system can be similarly determined and is given as

$$\text{closed-loop transfer function} = \frac{\text{forward path gain}}{1 - \text{forward gain} \times \text{feedback gain}}$$

$$\text{CLTF } \frac{C}{R} = \frac{G}{1 - GH}$$

$$\text{output } C = \text{closed-loop transfer function} \times \text{input } R$$

$$\text{positive-feedback CLTF } \frac{C}{R} = \frac{G}{1 - GH}$$

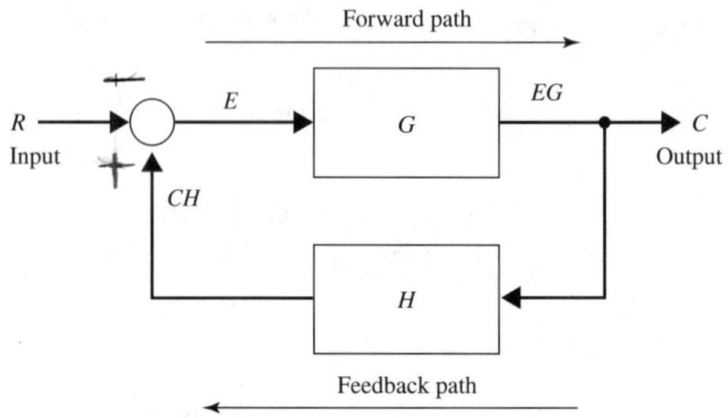

Figure 2.12 Positive-feedback closed-loop system.

Comparison of Positive-Feedback and Negative-Feedback Systems

Negative-Feedback Closed-Loop System

- The transfer function (CLTF) is generally less than 1.
- The CLTF remains a finite value as long as the denominator is greater than zero.
- It is commonly used in control systems.

Positive-Feedback Closed-Loop System

- The CLTF tends to become very large as GH (loop gain) approaches 1.
- It is generally restricted to specialized applications, such as oscillators.

EXAMPLE 2.5

Determine the overall gain of a negative-feedback closed-loop system with forward gain of 10 and feedback gain of 2.

Solution

Given $G = 10$ and $H = 2$. The overall gain of a closed-loop system is same as its closed-loop transfer function. For a negative-feedback closed-loop control system,

$$\text{CLTF} = \frac{G}{1 + GH}$$

Substituting the values of G and H:

$$\text{CLTF} = \frac{10}{1 + 10 \times 2} = 0.476$$

Note: For the sake of brevity, TF will be used to indicate a closed-loop transfer function (CLTF).

EXAMPLE 2.6

A negative-feedback closed-loop system is subjected to an input of 5 V. Determine the output voltage. The system has a forward gain of 1 and a feedback gain of 1.

Solution

Given $G = 1$, $H = 1$, and input $R = 5$ V.

$$\text{output } C = \text{transfer function} \times \text{input } R$$

The TF is given by

$$\frac{C}{R} = \frac{G}{1 + GH}$$

Substituting the values of G and H,

$$\frac{C}{R} = \frac{1}{1 + 1 \times 1}$$

$$= 0.5$$

$$C = 0.5 \times R$$

$$C = 0.5 \times 5 \text{ V}$$

$$= 2.5 \text{ V}$$

EXAMPLE 2.7

Repeat Example 2.6 with the following values of forward gain:

a. $G = 2$
b. $G = 5$
c. $G = 10$
d. $G = 100$
e. $G = 1000$

Solution

$$\text{output } C = \text{TF} \times \text{input } R$$

$$C = \frac{G}{1 + GH} \times R$$

Output can be obtained by substituting the values of G, H, and R in this expression:

a. $G = 2$:

$$C = \frac{2}{1 + 2 \times 1} \times 5 \text{ V}$$

$$C = 3.333 \text{ V}$$

b. $G = 5$:

$$C = \frac{5}{1 + 5 \times 1} \times 5 \text{ V}$$

$$C = 4.166 \text{ V}$$

c. $G = 10$:

$$C = \frac{0}{1 + 10 \times 1} \times 5 \text{ V}$$

$$C = 4.545 \text{ V}$$

d. $G = 100$:

$$C = \frac{100}{1 + 100 \times 1} \times 5 \text{ V}$$

$$C = 4.995 \text{ V}$$

Observation: With an increase in forward gain, the output value approaches the input value but can never become equal to it.

EXAMPLE 2.8

Determine the transfer function for a positive-feedback closed-loop system with a forward gain of 1 and a feedback gain of 0.5.

Solution

Given $G = 1$ and $H = 0.5$. For a positive-feedback closed-loop control system, the transfer function is given by

$$\text{TF} = \frac{G}{1 - GH}$$

Substituting the values of G and H:

$$\text{TF} = \frac{1}{1 - 1 \times 0.5}$$

$$\text{TF} = 2.0$$

EXAMPLE 2.9

A positive-feedback closed-loop system was subjected to an input of 5 V. Determine the output voltage. The system has a forward gain of 1 and a feedback gain of 0.5.

Solution

Given $G = 1, H = 0.5$, and input $R = 5$ V.

$$\text{output } C = \text{transfer function} \times \text{input } R$$

$$C = \frac{G}{1 - GH} \times R$$

Substituting the values of G, and H, and R:

$$C = \frac{1}{1 - 1 \times 0.5} \times 5 \text{ V}$$

$$C = 10 \text{ V}$$

EXAMPLE 2.10

A positive-feedback closed-loop system was subjected to an input of 5 (units). Determine the output voltage for the following sets of gains:

a. $G = 1, H = 0.75$
b. $G = 1, H = 0.9$
c. $G = 1, H = 0.99$

d. $G = 1.9, H = 0.5$
e. $G = 1.99, H = 0.5$
f. $G = 1.999, H = 0.5$

Solution

$$\text{output } C = \text{TF} \times \text{input } R$$

$$C = \frac{G}{1 - GH} \times R$$

Output is obtained on substituting the values of G and H, and R in the preceding expression

a. $G = 1, H = 0.75, R = 5$ V

$$C = \frac{G}{1 - GH} \times R$$

$$C = \frac{1}{1 - 1 \times 0.75} \times 5 \text{ V}$$

$$C = 4 \times 5 \text{ V}$$

$$C = 20 \text{ V}$$

b. $G = 1, H = 0.9, R = 5$ V

$$C = \frac{1}{1 - 1 \times 0.9} \times 5 \text{ V}$$

$$C = 10 \times 5 \text{ V}$$

$$C = 50 \text{ V}$$

c. $G = 1, H = 0.99, R = 5$ V

$$C = \frac{1}{1 - 1 \times 0.99} \times 5 \text{ V}$$

$$C = 100 \times 5 \text{ V}$$

$$C = 500 \text{ V}$$

d. $G = 1.9, H = 0.5, R = 5$ V

$$C = \frac{1.9}{1 - 1.9 \times 0.5} \times 5 \text{ V}$$

$$C = 38 \times 5 \text{ V}$$

$$C = 190 \text{ V}$$

e. $G = 1.99$, $H = 0.5$, $R = 5$ V

$$C = \frac{1.99}{1 - 1.99 \times 0.5} \times 5\ \text{V}$$

$$C = 398 \times 5\ \text{V}$$

$$C = 1990\ \text{V}$$

f. $G = 1.999$, $H = 0.5$, $R = 5$ V

$$C = \frac{1.999}{1 - 1.999 \times 0.5} \times 5\ \text{V}$$

$$C = 3998 \times 5\ \text{V}$$

$$C = 19{,}990\ \text{V}$$

Observation: The output value increases as loop gain (GH) approaches 1. The output becomes very large and approaches infinity when the loop gain approaches a value of 1 ($GH = 1$).

EXERCISES

2.4. An input of value 1.5 is applied to a negative-feedback closed-loop system. Determine the TF if the output obtained is 0.75.

2.5. An input signal of 1.0 V is applied to a positive-feedback closed-loop system. Determine the TF if the output signal is 5.0 V.

2.6. Determine the value of error E given that $G = 1$, $H = 1$, and input $= 1$ (Figure 2.13).

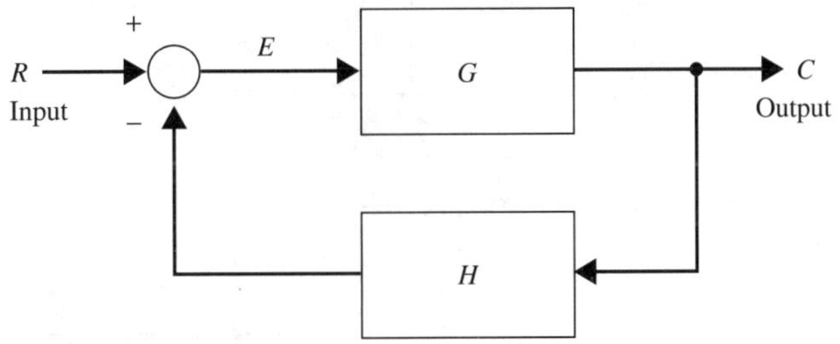

Figure 2.13 Diagram for Exercise 2.6.

UNITY FEEDBACK SYSTEMS

If feedback gain is unity ($H = 1$), the control system block diagram can be further simplified (Figure 2.14). In this case, the feedback signal reaching the summing point is the same as the controlled output. A number of control system analysis techniques are based upon the use of a unity feedback configuration.

$$\text{transfer function of unity feedback system} = \frac{G}{1 + GH}$$

When the feedback is unity ($H = 1$),

$$\text{TF} = \frac{G}{1 + G}$$

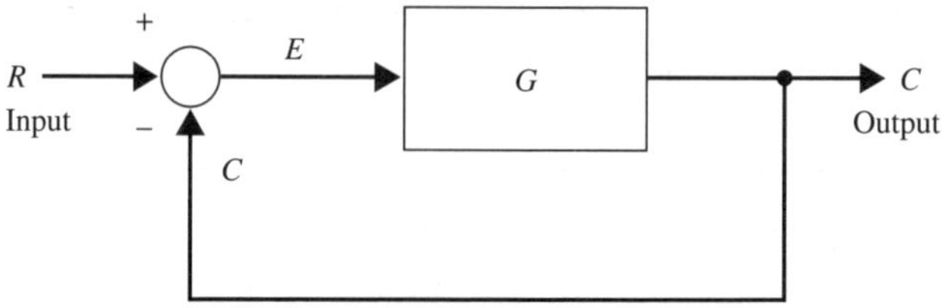

Figure 2.14 Unity gain feedback system.

Conversion to a Unity Feedback System

Any closed-loop system can be converted into an equivalent unity feedback system (Figure 2.15). The transfer function of a normal negative-feedback system with forward gain G and feedback gain H is given as

$$\text{TF} = \frac{G}{1 + GH}$$

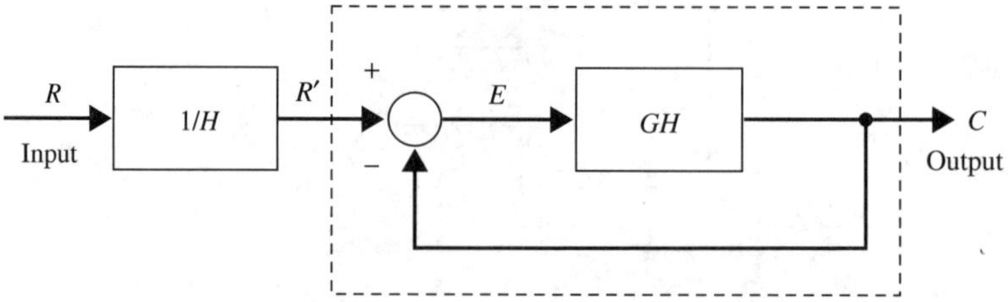

Figure 2.15 Equivalent unity gain feedback system.

Multiplying and dividing by feedback gain H gives

$$\mathrm{TF} = \frac{1}{H}\frac{GH}{1+GH}$$

$$\mathrm{TF} = \frac{1}{H}\frac{G'}{1+G'}, \quad \text{where} \quad G' = GH$$

BLOCK DIAGRAM SIMPLIFICATION

Even a simple closed-loop control system may contain a large number of blocks and may involve multiple feedback (or feedforward) paths. It is necessary to reduce the block diagram to a simplified form before an overall transfer function can be obtained. The following set of rules aids in simplification of block diagrams.

1. ***Cascading blocks:*** When blocks are connected in series, they can be replaced by a single block with its gain equal to the product of individual gains (Figure 2.16).

$$\text{overall gain } G = \text{gain } G_1 \times \text{gain } G_2 \times \text{gain } G_3 \times \cdots$$

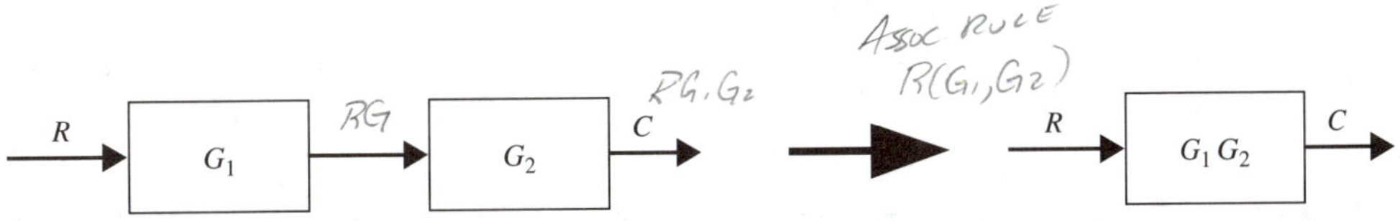

Figure 2.16 Cascading blocks.

2. ***Blocks in parallel*** (forward loop): See Figure 2.17.

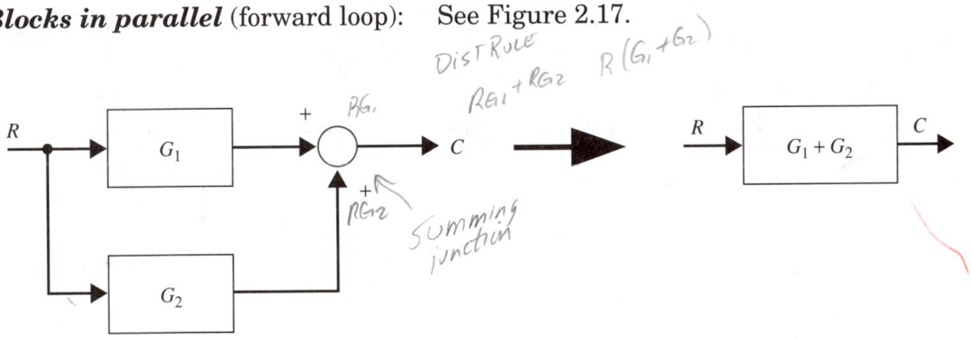

Figure 2.17 Cascading blocks.

3. ***Moving the summing point ahead of the block:*** See Figure 2.18.

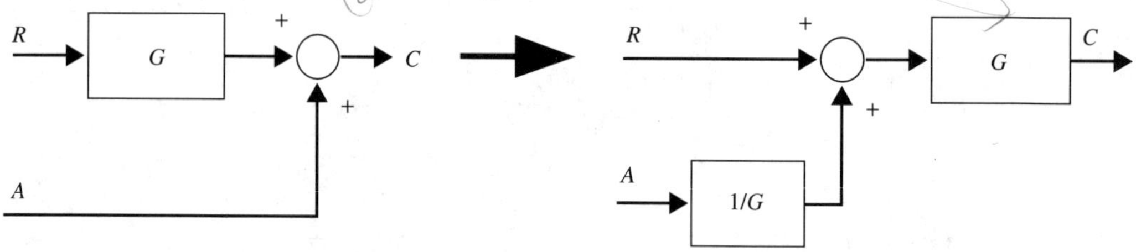

Figure 2.18 Moving a summing point ahead.

4. ***Moving the summing point beyond a block:*** See Figure 2.19.

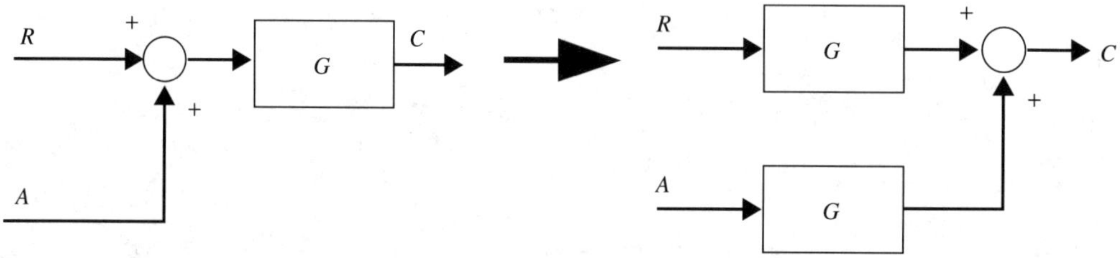

Figure 2.19 Moving the summing point ahead.

5. ***Moving the takeoff point ahead of a block:*** See Figure 2.20.

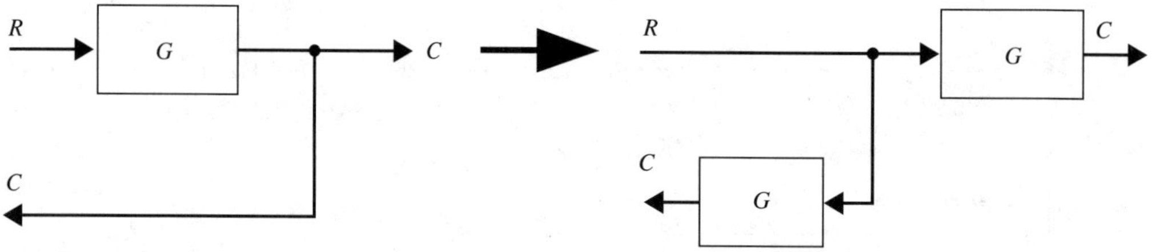

Figure 2.20 Moving the takeoff point ahead of the block.

6. ***Moving the takeoff point beyond a block:*** See Figure 2.21.

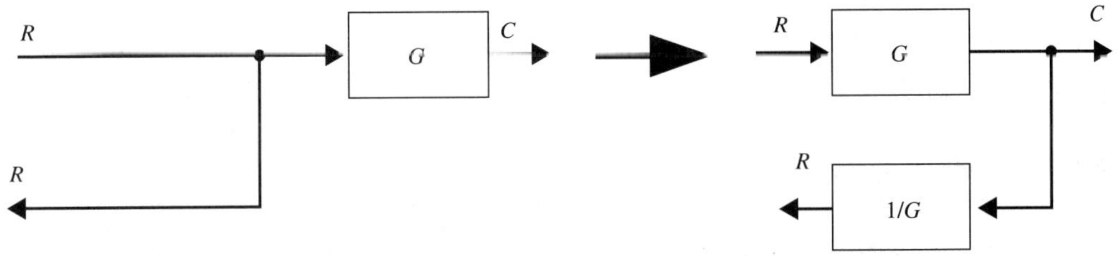

Figure 2.21 Moving the takeoff point beyond the block.

EXAMPLE 2.11

Simplify the block diagram in Figure 2.22 and determine the value of output C.

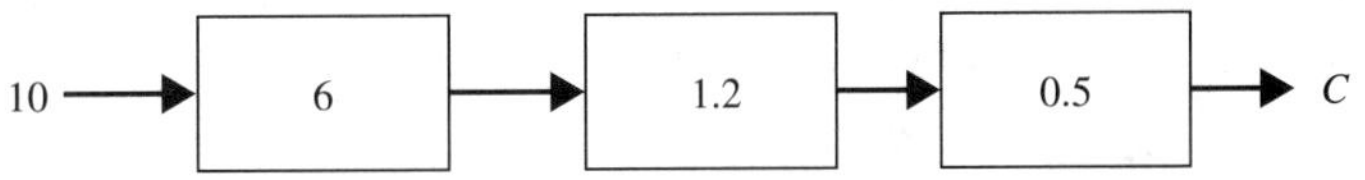

Figure 2.22 Block diagram for Example 2.11.

Solution

Because all three blocks are in series, rule 1 is applicable.

$$\text{overall gain } G = 6 \times 1.2 \times 0.5$$
$$= 3.6$$

Therefore,

$$\text{output } C = \text{overall gain (IF)} \times \text{input } R$$
$$= 3.6 \times 10$$
$$= 36$$

The result is shown in Figure 2.23.

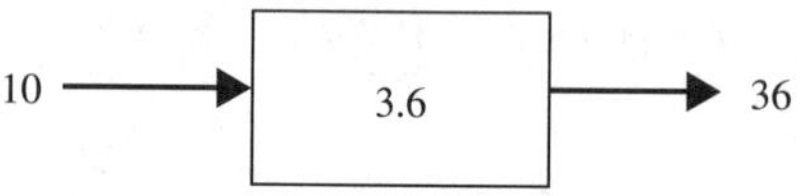

Figure 2.23 Reduced block for Example 2.11.

EXAMPLE 2.12

Determine the transfer function for the block diagram in Figure 2.24 and also determine the error value E given that $R = 2$, $G_1 = \frac{2}{3}$, and $G_2 = 12$.

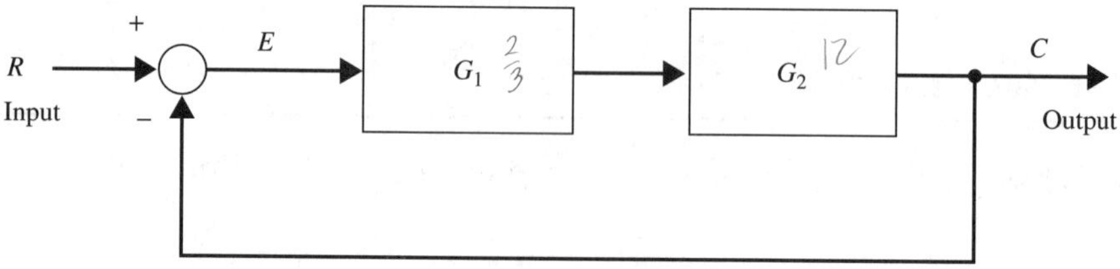

Figure 2.24 Block diagram for Example 2.13.

Solution

Blocks G_1 and G_2 are in cascade and thus can be combined into one block.

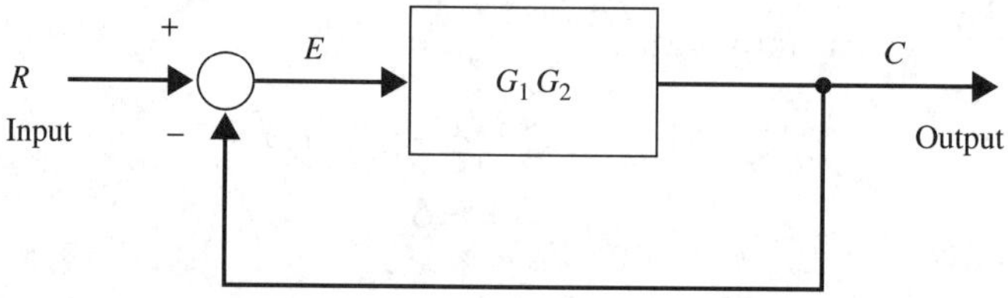

Figure 2.25 Reduced diagram for Example 2.13.

Figure 2.25 has been reduced to a unity feedback system, and its transfer function is given as

$$\text{TF} = \frac{G_1 G_2}{1 + G_1 G_2}$$

Substituting the values of G_1 and G_2,

$$\mathrm{TF} = \frac{\frac{2}{3} \cdot 12}{1 + \frac{2}{3} \cdot 12}.$$

$$\mathrm{TF} = 0.889$$

We know error E = reference R − feedback MV. Because it's a unity feedback system, the value of the feedback signal is the same as the output C.

$$\begin{aligned} \text{error } E &= R - C \\ &= R - [\mathrm{TF} \times R] \\ &= R(1 - \mathrm{TF}) \\ &= 2(1 - 0.889) \\ &= 0.222 \end{aligned}$$

EXAMPLE 2.13

Determine the transfer function of the block diagram in Figure 2.26.

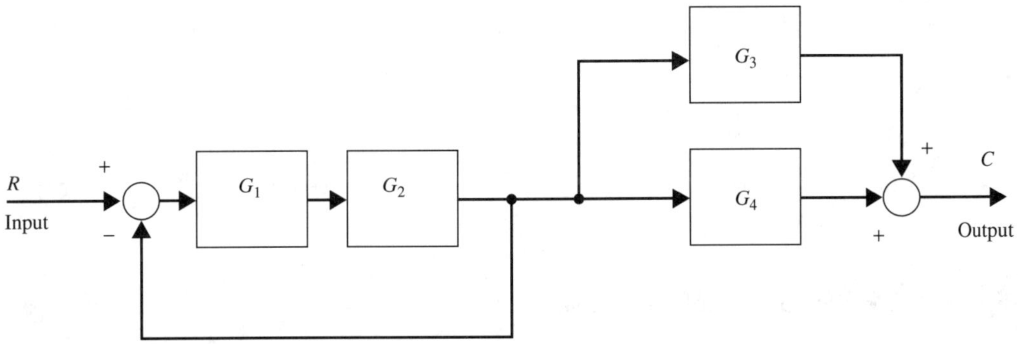

Figure 2.26 Diagram for Example 2.13.

Solution

Blocks G_1 and G_2 are in cascade and can be replaced by a single block, with a transfer function of $G_1 \times G_2$.

Blocks G_3 and G_4 also are in parallel and can be replaced by a single block, with a transfer function of $G_3 + G_4$ (Figure 2.27).

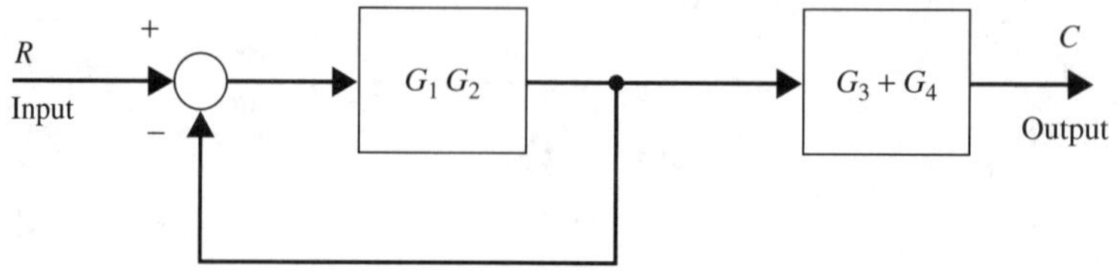

Figure 2.27 First stage of reduction in Example 2.13.

Unity feedback loop can now be eliminated (Figure 2.28).

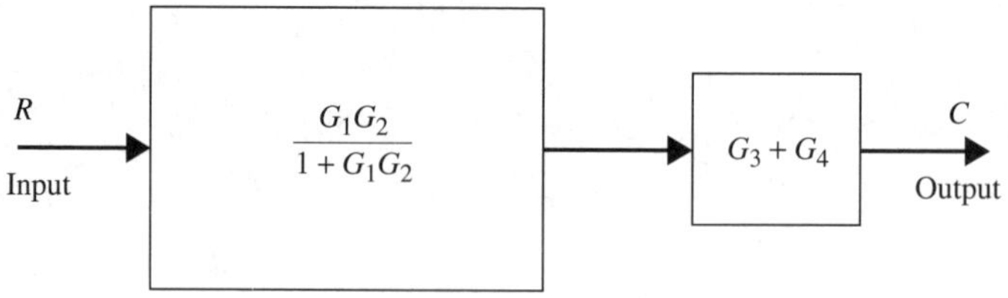

Figure 2.28 Second reduction for Example 2.13.

Again the two blocks in cascade can be combined (Figure 2.29). The overall transfer function then is

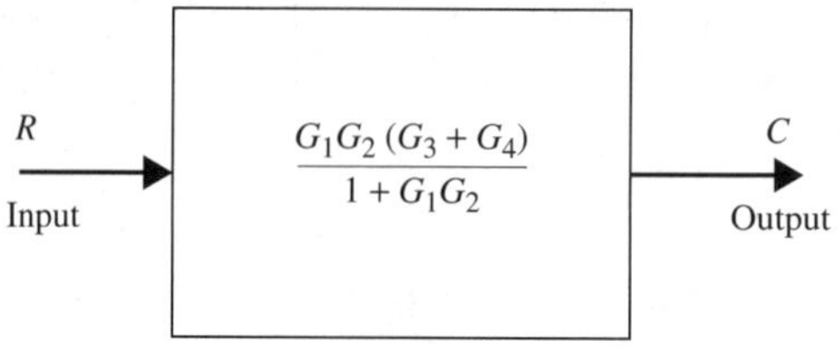

Figure 2.29 Final diagram for Example 2.13.

OP-AMP SIMULATION OF A BLOCK DIAGRAM

Operational amplifiers can be used for analog simulation of control systems. Once a circuit is built, its operation can reveal significant information about the operational behavior of the control system being simulated. A closed-loop control system can be completely described through a block diagram. A number of different types of blocks are used in a block diagram, such as a gain block, summing point, and takeoff point. Operational amplifiers can be used to implement each of these functions. Hence, the problem of control system simulation is reduced to representing each element of a control system block diagram by op-amps and connecting these through wires.

Summing Point

Figure 2.30 shows the op-amp implementation of a summing point. In a block diagram, a summing point accepts two input signals, SP (reference) and MV (feedback), and it generates an output signal E (error). The op-amp is configured as a difference amplifier. The output of amplifier is given as

$$V_{out} = \frac{R_4}{R_2} V_2 - \frac{R_3}{R_1} V_1$$

where V_2 = signal applied at the noninverting input
V_1 = signal applied at the inverting input

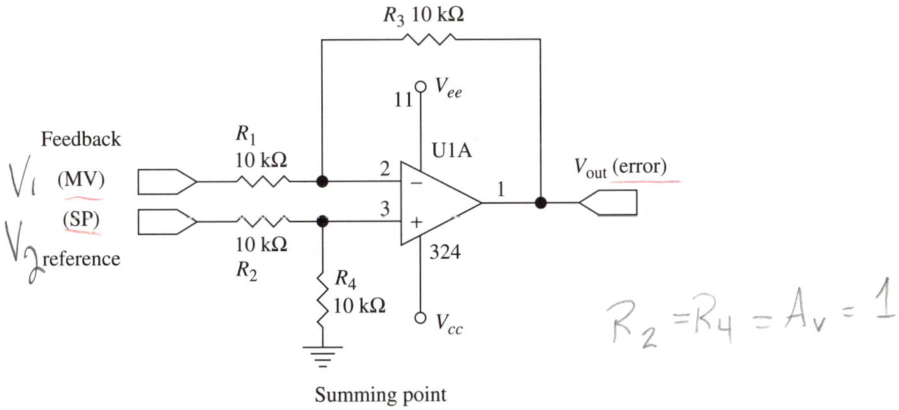

Figure 2.30 Summing point circuit.

Resistors R_4 and R_2 set the gain for the V_2 input. If these two resistors are equal in value, gain V_2 becomes equal to 1. Similarly, R_3 and R_1, which set the gain for the V_1 input, can be made equal for unity gain. Single resistors with 5% tolerance are common, which is satisfactory for the majority of work. For best performance (common mode rejection), the following should be considered:

- Select the resistors to be in the range of 10 kΩ to 100 kΩ.
- Select and match R_1 and R_2.
- Select and match R_3 and R_4.
- Make circuit voltages within the limits of op-amp output swing.

If all four resistors are chosen to be of the same value (10 kΩ in this case),

$$R_1 = R_2 = R_3 = R_4 = R$$

The output equation thus becomes

$$V_{out} = V_2 - V_1$$

If the signals are applied in such a manner that

$$V_2 = \text{SP (reference signal)}$$
$$V_1 = \text{MV (feedback signal)}$$

then output voltage is given by

$$V_{out} = \text{SP} - \text{MV}$$

which is equivalent to the error (E) signal in a block diagram (Figure 2.31).

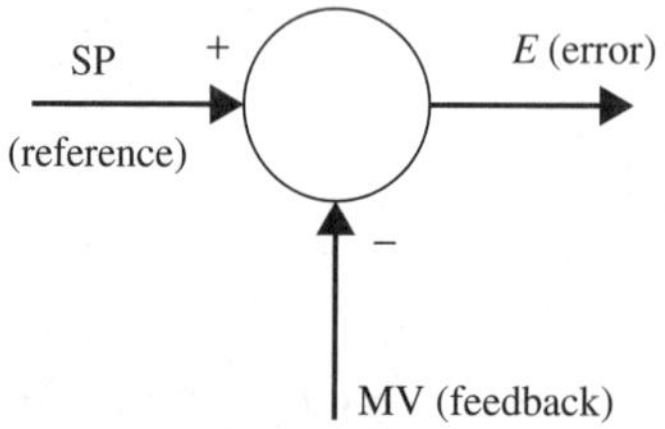

Figure 2.31 Equivalent summing point.

The preceding equation can now be written as

$$E(V_{out}) = \text{SP}(V_2) - \text{MV}(V_1)$$

Summing Point with Adjustable SP Input

In laboratory applications, a potentiometer is used to provide an adjustable set-point input to a closed-loop control system. A typical implementation is shown in Figure 2.32. The diagram is similar to one without an adjustable set-point control. All four resistors, R_1 to R_4, have their values increased to 33 kΩ. These resistors should generally be much larger than the resistors used for set-point adjustments (R_5, R_6, and R_7). The potentiometer R_7 allows the adjustment of set-point voltage. The set-point network (R_5, R_6, and R_7) is connected across a bipolar supply. This permits a bipolar SP voltage to be applied to the circuit. Resistors R_5 and R_6 allow the SP voltage range to be adjusted within the bipolar supply voltage range. Thus, a finer control can be exercised over SP adjustment. The SP voltage range can be set with the help of the following expressions:

$$\text{minimum SP voltage} = \left[\frac{R_6(V_{cc} - V_{ee})}{R_5 + R_6 + R_7} + V_{ee}\right]$$

$$\text{maximum SP voltage} = \left[V_{cc} - \frac{R_5(V_{cc} - V_{ee})}{R_5 + R_6 + R_7}\right]$$

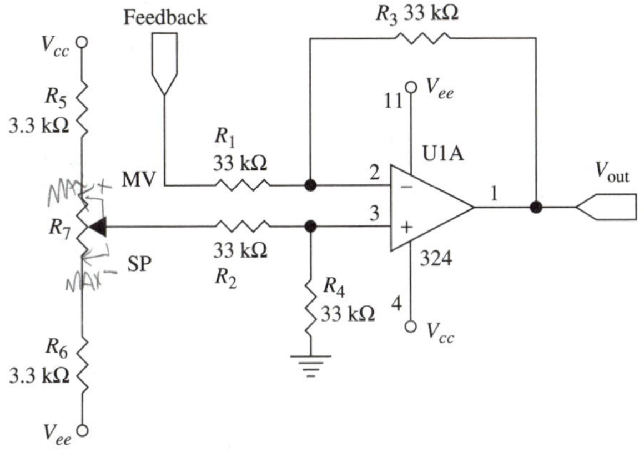

Figure 2.32 Summing point with adjustable set-point control.

EXAMPLE 2.14

Refer to Figure 2.32. Assuming a bipolar supply voltage of ±15 V, determine the value of potentiometer R_7, to provide an SP voltage range of ±1 V.

Solution

Given $R_5 = 3.3$ kΩ, $R_6 = 3.3$ kΩ, $V_{cc} = +15$ V, $V_{ee} = -15$ V, minimum SP voltage = −1 V, and maximum SP voltage = +1 V. Substituting the values in the expression for minimum SP voltage,

$$\text{minimum SP voltage} = \left[\frac{R_6(V_{cc} - V_{ee})}{R_5 + R_6 + R_7} + V_{ee}\right]$$

$$-1\text{ V} = \left[\frac{3.3\text{ k}\Omega\,(15\text{ V} - (-15\text{ V}))}{3.3\text{ k}\Omega + R_7 + 3.3\text{ k}\Omega} + (-15\text{ V})\right]$$

$$-1\text{ V} = \left[\frac{3.3\text{ k}\Omega \times 30\text{ V}}{6.6\text{ k}\Omega + R_7} - 15\text{ V}\right]$$

$$(1\text{ V})(6.6\text{ k}\Omega + R_7) = (3.3\text{ k}\Omega \times 30\text{ V}) - (15\text{ V})(6.6\text{ k}\Omega + R_7)$$

$$R_7 = 0.471\text{ k}\Omega$$

Check: Substituting this value of R_7 in the expression for maximum SP voltage gives

$$\text{maximum SP voltage} = \left[V_{cc} - \frac{R_5(V_{cc} - V_{ee})}{R_5 + R_6 + R_7}\right]$$

$$\text{maximum SP voltage} = \left[15\text{ V} - \frac{3.3\text{ k}\Omega\ (15\text{ V} - (-15\text{ V}))}{3.3\text{ k}\Omega + 0.471 + 3.3\text{ k}\Omega}\right]$$

$$= \left[15\text{ V} - \frac{3.3\text{ k}\Omega \times 30\text{ V}}{7.071\text{ k}\Omega}\right]$$

$$= 0.9992\text{ V} \quad \textbf{OK}$$

The closest standard value for a potentiometer is 500 Ω, which will be used. The actual SP range for this potentiometer will be ±1.056 V.

Gain Block

The input and output relationship of a control element can be described by an expression containing algebraic, differential, and integral terms. A block is used to correctly represent each of these types of relationships. The following types of blocks can be found in a block diagram representing a control system:

- Constant (inverting) gain block
- Constant (noninverting) gain block
- Derivative gain block
- Integral gain block

Development of circuits to represent blocks with integral and derivative gains is covered later. Here, only the blocks with constant inverting and noninverting gains are discussed.

Constant-Gain (Noninverting) Block

A constant-gain block is essentially an amplifier with a gain of G (Figure 2.33). The input/output relationship of a constant-gain block is given as

$$C\ (\text{output}) = G\ (\text{gain}) \times R\ (\text{input})$$

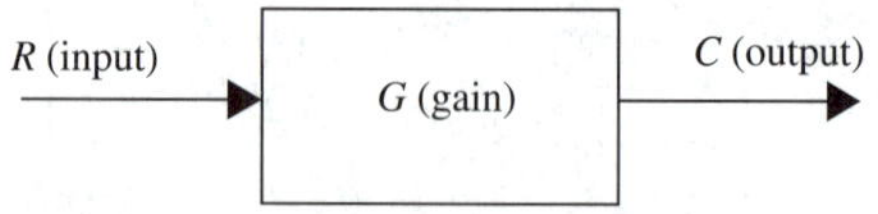

Figure 2.33 Constant-gain block.

Figure 2.34 shows op-amp representation of a constant-gain (noninverting) block. Resistors R_1 and R_2 set the amplifier gain, G:

$$G = 1 + \frac{R_2}{R_1}$$

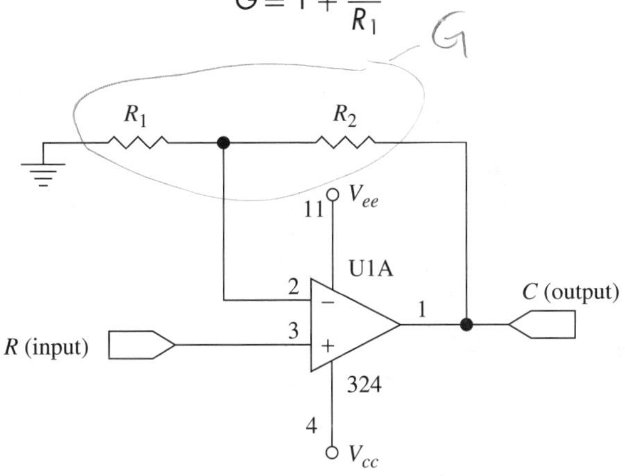

Figure 2.34 Constant-gain (noninverting) block.

For fixed-gain applications, resistors R_1 and R_2 can be selected from a wide range of standard resistor values. For more critical applications or those requiring gain adjustment, the feedback resistor R_2 can be replaced with a potentiometer.

EXAMPLE 2.15

Construct a circuit to represent the block in Figure 2.35. Identify the range of input signals over which the circuit will yield true representation.

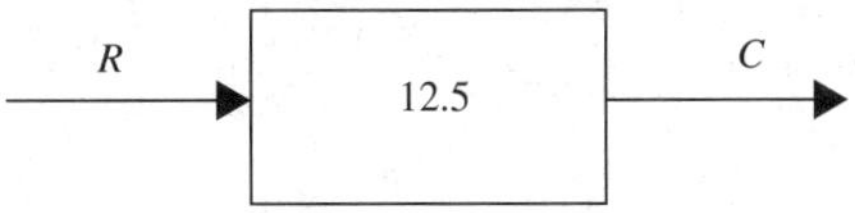

Figure 2.35 Diagram for Example 2.16.

Solution

The given block has a constant gain (noninverting) and can be represented by a noninverting amplifier with a gain, G, of 12.5. Amplifier gain G is

$$G = 1 + \frac{R_2 \text{ (feedback resistor)}}{R_1 \text{ (inverting terminal resistor)}}$$

$$12.5 = 1 + \frac{R_2}{R_1}$$

$$\frac{R_2}{R_1} = 11.5$$

Selecting a 2.0-kΩ value for R_1,

$$R_2 = 11.5 \times 2.0 \text{ k}\Omega$$
$$R_2 = 23 \text{ k}\Omega$$

A 23.2-kΩ 1% precision resistor can be used for R_2. Figure 2.36 shows the complete schematic diagram for implementing this gain block.

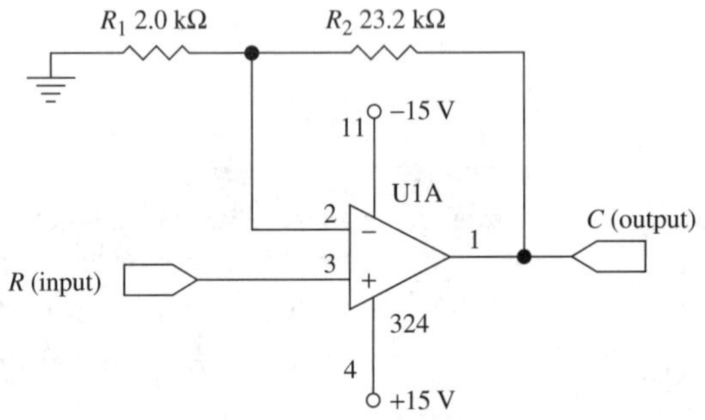

Figure 2.36 Op-amp circuit for Example 2.15.

Maximum Input Signal: A large supply voltage will result in a greater range for input signal. An LM324 op-amp is employed, which can accept a bipolar supply of ±15 V. A saturation loss of 1 to 1.5 V can be expected for LM324. This leaves a maximum output voltage swing of ±13.5 V. Because the amplifier gain is 12.5, the maximum range of input signal that can be safely handled by this op-amp is

$$\text{input signal swing} = \frac{\text{output saturation range}}{\text{amplifier gain}} = \frac{\pm 13.5 \text{ V}}{12.5} = \pm 1.08 \text{ V}$$

This amplifier is useful when input signal can be limited to within ±1.08 V.

Constant-Gain (Inverting) Block

Figure 2.37 shows an op-amp representation of a constant-gain (inverting) block. The input signal is now applied to the inverting terminal. Output signal is inverted and is 180° out of phase with the input signal. Resistors R_1 and R_2 set the amplifier gain, G:

$$G = -\left(\frac{R_2}{R_1}\right)$$

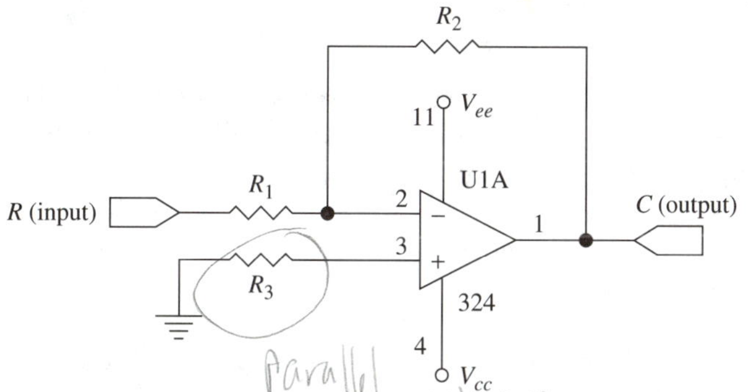

Figure 2.37 Constant-gain (inverting) block.

It is recommended that a resistor be used between the ground and noninverting terminal. This resistor helps to cancel the error due to op-amp input bias currents. Ideally R_3 is chosen to be equal to the parallel combination of R_1 and R_2. But, if feedback resistor R_2 is much larger than input resistor R_1, R_3 can be made equal to R_1.

$$R_3 = \left(\frac{R_1 + R_2}{R_1 R_2}\right)$$

Standard resistor values can be chosen for R_1 and R_2 for fixed gain or a potentiometer can be used for feedback resistor R_2 to allow gain adjustment.

Takeoff Point

A takeoff point allows signal information to be communicated to more than one point. In equivalent electrical systems, such communication can easily be done via simple interconnection between two points. Sometimes circuit components can be overloaded due to low input impedance of the destination. This can be prevented through the use of unity gain buffers, which reduce the loading on the source and maintain signal integrity.

Figure 2.38 shows a unity gain buffer used to isolate different sections of a circuit. An input signal is applied to the noninverting terminal. The inverting terminal is

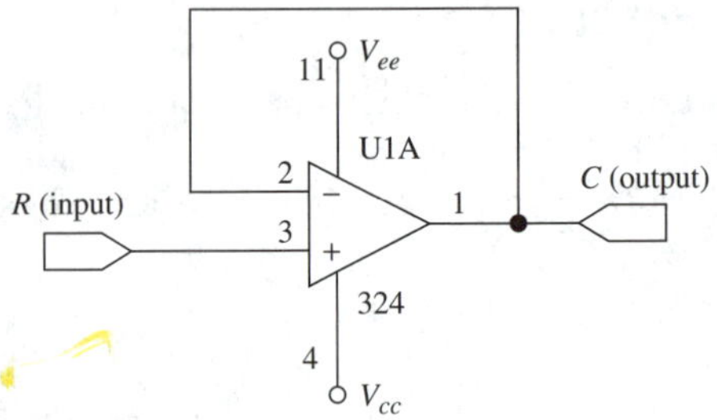

Figure 2.38 Unity gain buffer.

connected to the output terminal. The circuit can be considered a modified noninverting amplifier, with zero feedback resistance and an infinite resistance connected between the inverting terminal and ground.

$$\text{amplifier gain } G = 1 + \frac{0}{\infty}$$
$$G = 1$$

With a gain of 1, the output signal is of the same magnitude and polarity as the input signal. The input impedance of a noninverting amplifier (and, thus, unity gain buffer) is very high. This reduces the amount of power drawn from the source circuit. It should, however, not be applied without proper consideration, as an actual op-amp does not have an exact gain of 1, and at the same time, additional offset errors, although small, may be introduced in the circuit.

Summing Point with Multiple Inputs

The difference amplifier used earlier for summing-point implementation is not suitable for obtaining the algebraic sum of multiple signals (Figure 2.39). For such a case, a summing amplifier is needed.

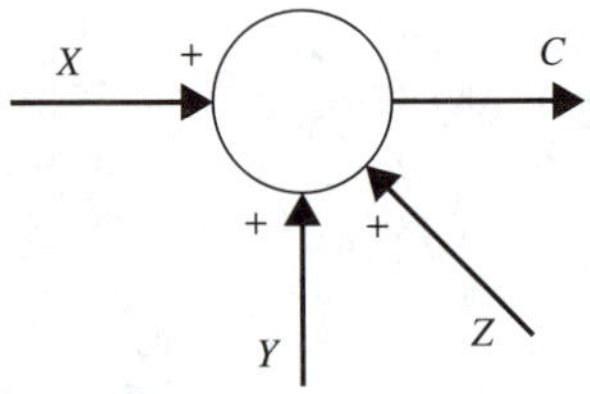

Figure 2.39 Summing point: multiple inputs.

Figure 2.40 shows an op-amp implementation of a multiple-input summing point. It is basically an inverting amplifier. Each input signal is connected to inverting terminal

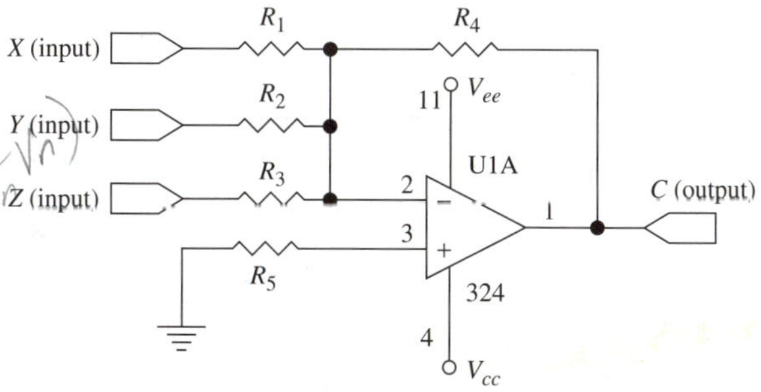

Figure 2.40 Multiple-input summing point configuration.

through an input resistor. Multiple inputs can be applied to the op-amp. The ratio of feedback resistor to the input resistor sets the gain for that input signal.

$$\text{output } C = -\left(\frac{R_4}{R_1}X + \frac{R_4}{R_2}Y + \frac{R_4}{R_3}Z\right)$$

If $R_1 = R_2 = R_3 = R_4$, the expression simplifies to

$$\text{output } C = -(X + Y + Z)$$

The inverting sign can be removed by feeding the output signal to another inverting amplifier with a unity gain. Figure 2.41 shows a complete implementation.

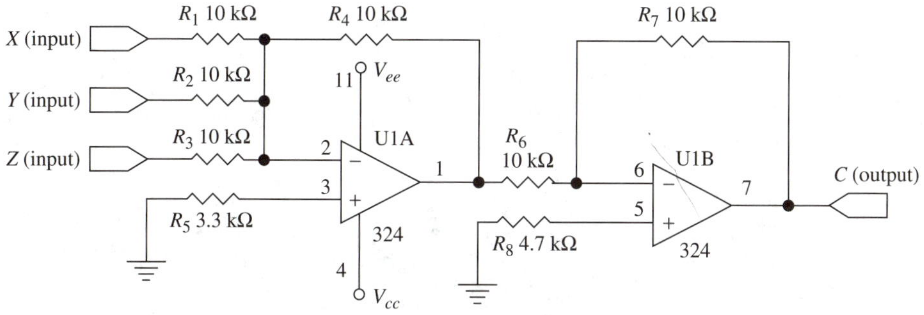

Figure 2.41 Complete implementation: multiple-input summing point.

EXAMPLE 2.16

Construct a circuit to implement Figure 2.42.

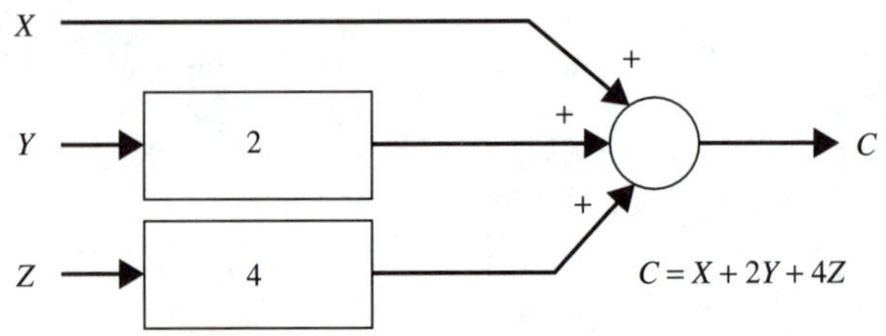

Figure 2.42 Diagram for Example 2.16.

Solution

The block diagram (Figure 2.42) represents a summing point with multiple input. Input A is added without any amplification, whereas input B has been subjected to a gain of 2, and input C is amplified by 4.

There are several different ways in which the circuit can be implemented. Perhaps the simplest approach is to use the configuration of Figure 2.41, with one modification. The second inverting amplifier will reverse the gains back to positive values. The input resistors R_1, R_2, and R_3 will be so chosen that the gain requirement of all three input signals are satisfied. Thus, the gain values will be −1, −2, and −4 for X, Y, and Z signals, respectively.

Using a feedback resistor of 10 kΩ, values of input resistors can now be calculated,

R_1:

$$\text{gain} = -\frac{R_4}{R_1}$$
$$-1 = -\frac{10\,\text{k}\Omega}{R_1}$$
$$R_1 = 10\ \text{k}\Omega$$

R_2:

$$\text{gain} = -\frac{R_4}{R_2}$$
$$-2 = -\frac{10\,\text{k}\Omega}{R_2}$$
$$R_2 = 5\ \text{k}\Omega$$

R_3:

$$\text{gain} = -\frac{R_4}{R_3}$$
$$-4 = -\frac{10\,\text{k}\Omega}{R_3}$$
$$R_3 = 2.5\ \text{k}\Omega \quad \text{(use standard 1\% 2.49 k}\Omega\text{)}$$

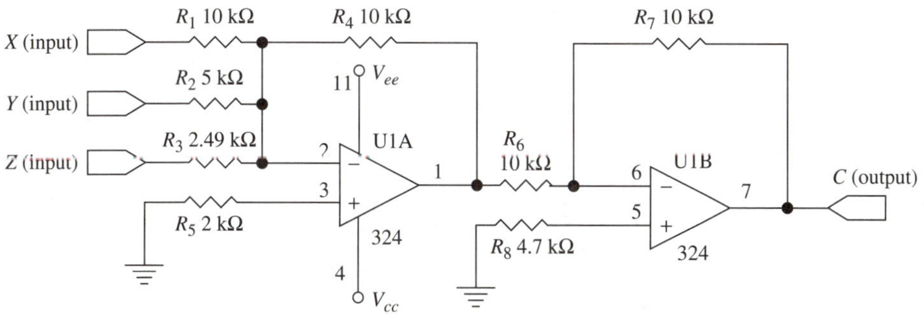

Figure 2.43 Diagram for Example 2.16.

Figure 2.43 implements the preceding block diagram.

EXAMPLE 2.17

Draw a schematic diagram to represent Figure 2.44 unity gain closed-loop control system with a forward gain of 10.

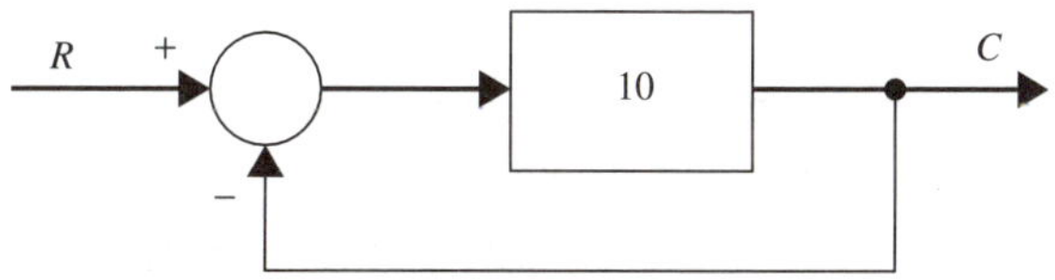

Figure 2.44 Diagram for Example 2.17.

Solution

The set point is not required to be adjustable in Figure 2.44. It contains the following two main elements:

- Summing block
- Gain (noninverting) block

The schematic diagram in Figure 2.45 shows the complete implementation.

$$\text{gain of constant-gain block (amplifier U1B)} = 1 + \frac{91\ \text{k}\Omega}{10\ \text{k}\Omega}$$
$$= 10.1$$

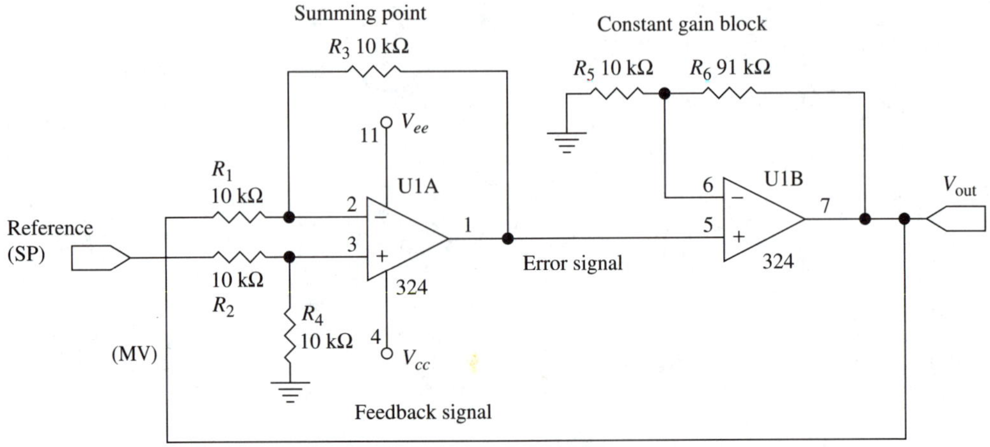

Figure 2.45 Circuit for Example 2.27.

For more precision, use a 100-kΩ potentiometer for R_2, and set it to the precise gain value.

EXAMPLE 2.18

Develop a circuit to simulate the operation of a closed-loop control system represented by the closed-loop block diagram in Figure 2.46. The circuit has to be tested for an input voltage ranging from 0 to 2 V dc. Given $G_1 = 2.5$, $G_2 = 2$, and $H = 1.25$.

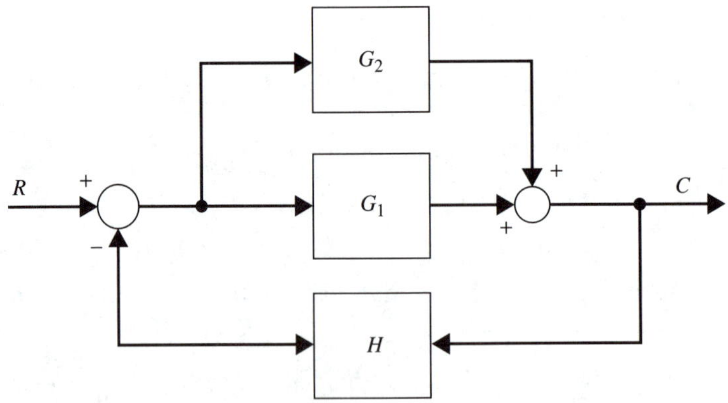

Figure 2.46 Block diagram for Example 2.18.

Solution

The circuit requirement can be stated as follows:

- Summing point with SP input
- Gain (noninverting) block G_1
- Gain (noninverting) block G_2
- Gain (noninverting) block H
- A second two-input summing point

The operations of the G_1 and G_2 blocks can be combined with that of signal summing.

The op-amp LM324 with a bipolar power supply of ±15 V will be used. This IC is available in a low-cost DIP package, containing four op-amps.

Summing point with SP input: The applied set-point voltage ranges from 0 to 2 V and does not swing to negative, so the input potentiometer is connected between positive supply (+15 V) and ground.

$$\text{maximum input voltage} = 2.0\ \text{V}$$
$$= \frac{R_6}{R_5 + R_6} \times V_{cc}$$

Therefore,

$$2.0\ \text{V} = \frac{R_6}{3.3\ \text{k}\Omega + R_6} \times 15\ \text{V}$$
$$R_6 = 0.5\ \text{k}\Omega$$

Choose a (standard) 500-Ω potentiometer. See Figure 2.47.

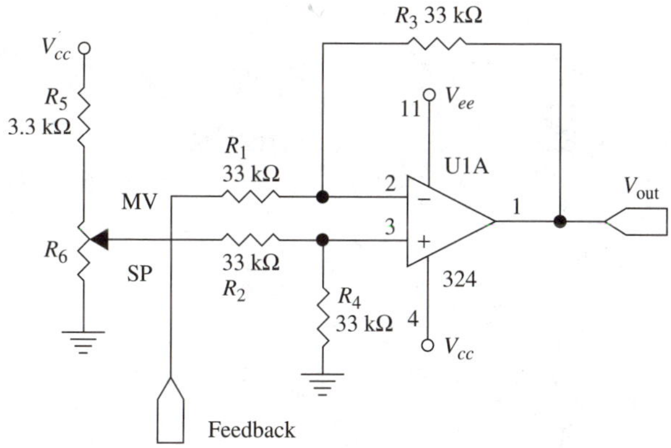

Figure 2.47 Summing point circuit for Example 2.18.

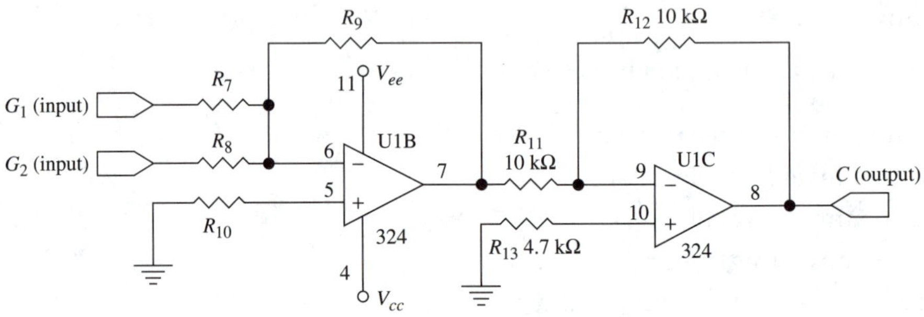

Figure 2.48 G_1, G_2 blocks and second summing point for Example 2.18.

G_1, G_2, and second summing point: See Figure 2.48. Resistors R_7 and R_9 set the gain for the G_1 block, and resistors R_8 and R_9 set the gain for the G_2 block. The voltage at terminal 7 is the algebraic sum of the voltages G_1 and G_2 applied to the inverting terminal. Inverting amplifier U1C with a gain of 1 further inverts the signal polarity. The resistor values can now be calculated.

Let the feedback resistor R_9 be 10 kΩ.

$$\text{gain } G_1 = \frac{R_9}{R_7}$$

$$2.5 = \frac{10 \text{ k}\Omega}{R_7}$$

$$R_7 = 4 \text{ k}\Omega \quad (\text{use } 3.9 \text{ k}\Omega)$$

Similarly, resistor R_8 can be determined.

$$\text{gain } G_2 = \frac{R_9}{R_8}$$

$$2.0 = \frac{10 \text{ k}\Omega}{R_8}$$

$$R_8 = 5 \text{ k}\Omega$$

Gain block *H*: See Figure 2.49. An inverting amplifier serves the function of a feedback block with a gain of 2. Choosing the feedback resistor R_{15} to be 10 kΩ, the

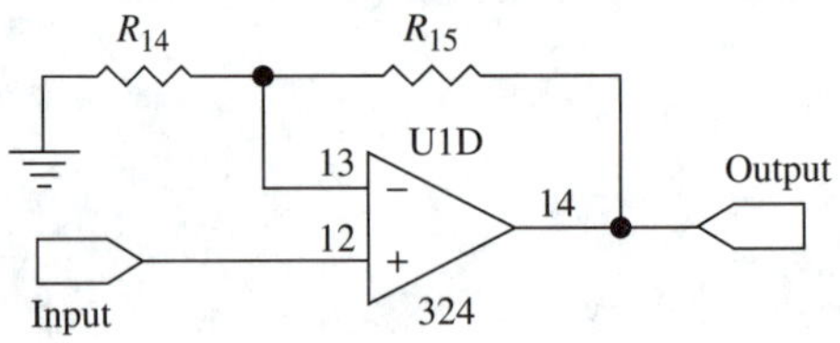

Figure 2.49 Circuit for block H for Example 2.18.

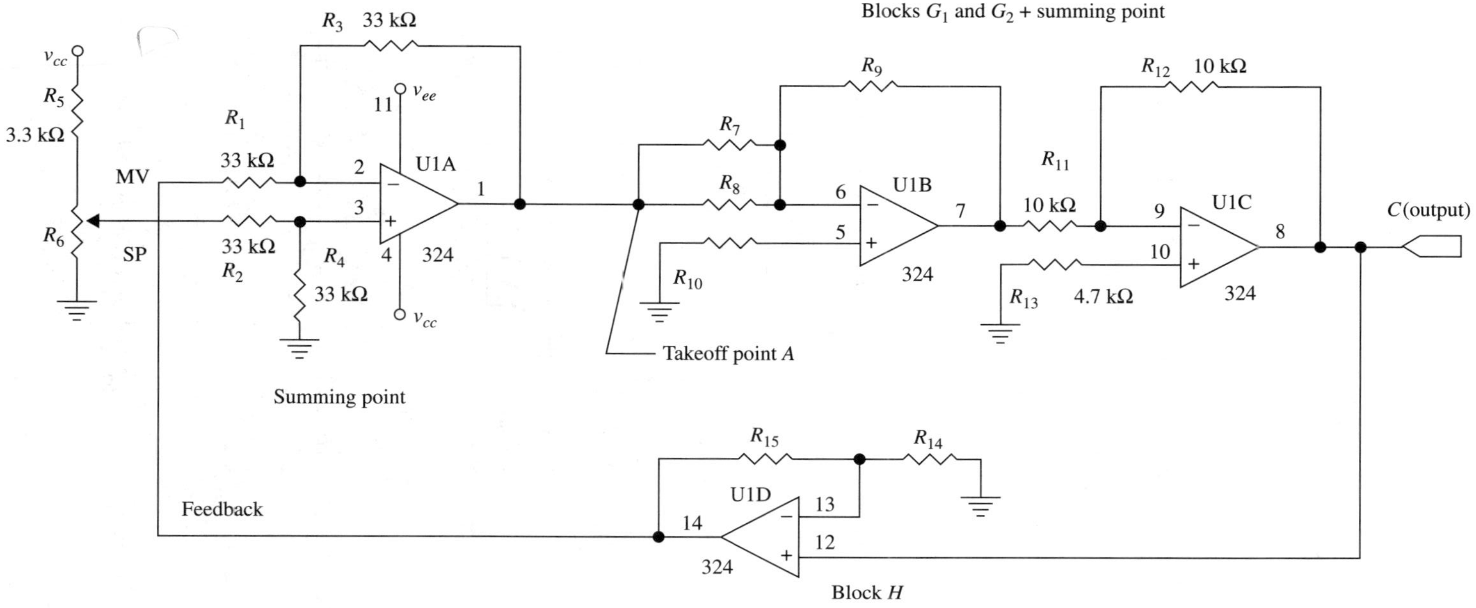

Figure 2.50 Complete schematic for Example 2.18.

value of resistor R_{14} can be determined.

$$\text{gain} = 1 + \frac{R_{15}}{R_{14}}$$

$$2.25 = 1 + \frac{10\ \text{k}\Omega}{R_{14}}$$

$$R_{14} = 8\ \text{k}\Omega \quad (\text{use } 7.9\ \text{k}\Omega)$$

With all the functions implemented, a complete schematic (Figure 2.50) can now be drawn.

PROBLEMS

1. Write the equation(s) describing the following block diagrams.

   a.

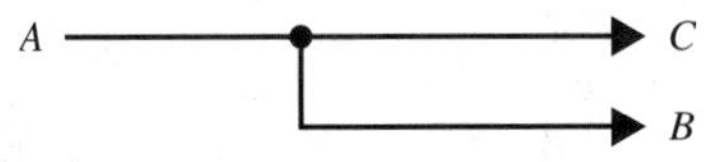

   b.

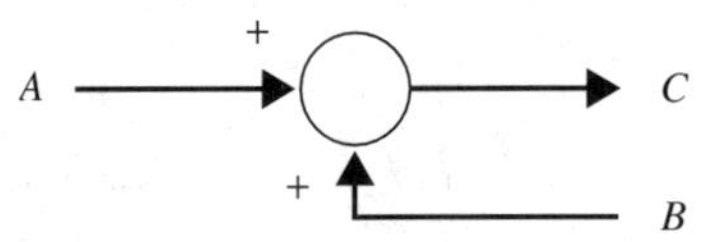

   c.

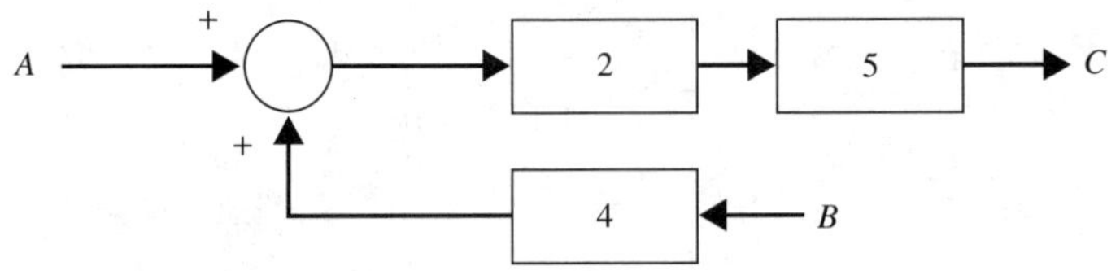

2. Draw a block diagram representing each expression.
   a. $Z = 2X + Y$
   b. $C = 0.1A - 3$
   c. $C = 10e + 2.5$ V

3. Prove that for a negative-feedback closed-loop control system, the system transfer function is given by

$$\text{CLTF} = \frac{G}{1 + GH}$$

where

G = forward gain
H = feedback gain

4. Determine the CLTF for a negative-feedback closed-loop system, with forward gain = 1 and feedback gain = 0.5.
5. Derive an expression for error in a closed-loop system in terms of input R, forward gain G, and feedback gain H.
6. Modify the expression in Problem 3 for error for a unity feedback system.
7. Explain how error is affected with a change in each of the following.
   a. Forward gain G
   b. Feedback gain H
   c. Input R
8. What effect does each of the following have on transfer function?
   a. Input R
   b. Forward gain G
   c. Feedback gain H
9. A positive-feedback closed-loop system is subjected to an input voltage of 1 V dc. Determine the output voltage. Assume the following:

   forward path gain = 1

   feedback path gain = 0.5

10. Identify the correct statement.
    a. The TF of a control system is dependent upon the input signal.
    b. The TF of a control system is dependent upon the output signal.
    c. The TF increases with an increase in input signal.
    d. The TF is independent of the input signal.
    e. The TF is totally dependent upon the input signal.
11. A negative-feedback closed-loop system was subjected to an input of 5 V. Determine the output voltage (volts). System has a forward gain of 1 and feedback gain of 1.
12. Identify the correct statement related to a negative-feedback closed-loop control system.
    a. With an increase in forward gain, the output value approaches the input value.

b. With an increase in forward gain, the output value becomes smaller.
c. Output value is always greater than input value.
d. Output value is independent of forward gain.

13. An input of value 1.5 is applied to a negative-feedback closed-loop system. Determine the TF if the output obtained is 0.75.

14. What is the value of the input *R* if the output is 1 V?

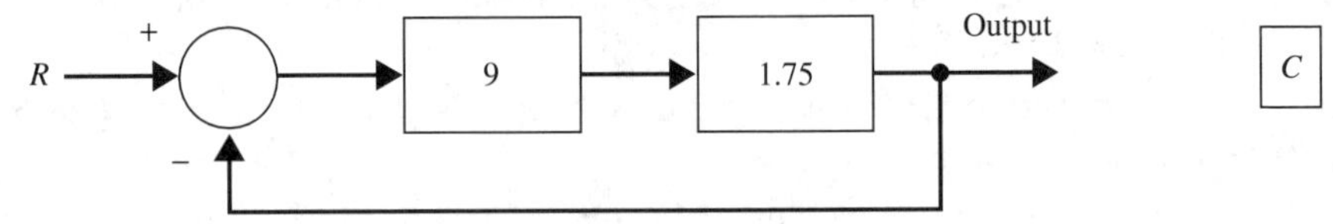

15. Draw a circuit diagram to simulate the block diagram in Problem 14.

16. Apply a variable voltage to the input of the circuit developed in Problem 15, and determine the value of input signal when an output of 1 V is obtained.

17. Develop a circuit diagram to simulate a negative-feedback closed-loop system with unity feedback and a forward gain of 10. Assuming a power supply of ±15 V, what is the maximum input signal that can be applied to the circuit, with the circuit functioning as required?

18. Develop a circuit to simulate a positive-feedback closed-loop system, with forward gain of 2.5 and feedback gain of 0.4.

19. Develop a scheme to test the operation of the circuit in Problem 18.

20. Simplify the following block diagram and determine the overall transfer function.

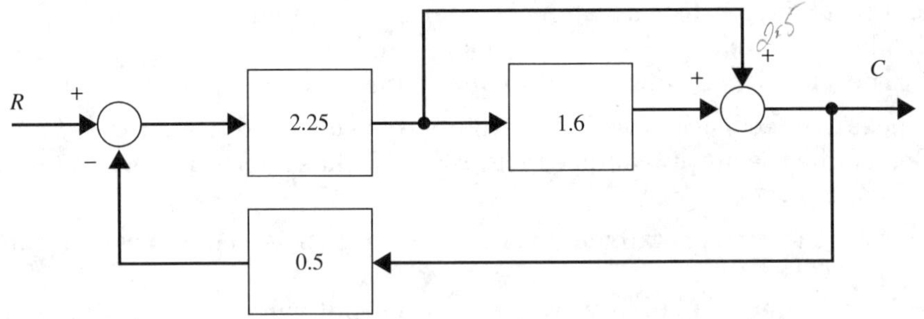

21. Determine the transfer function of the following system.

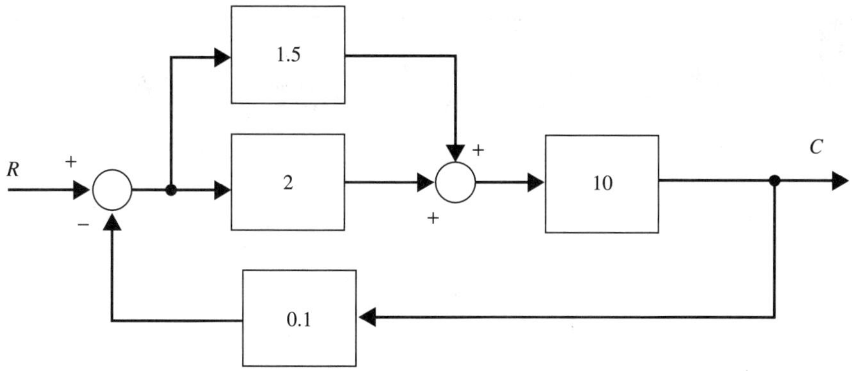

22. Simplify the following diagram.

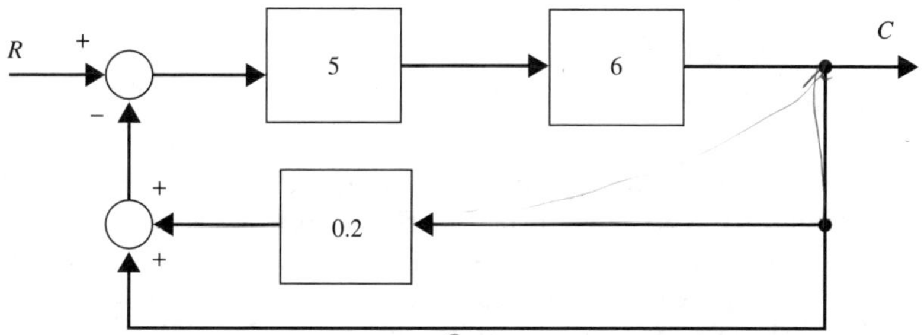

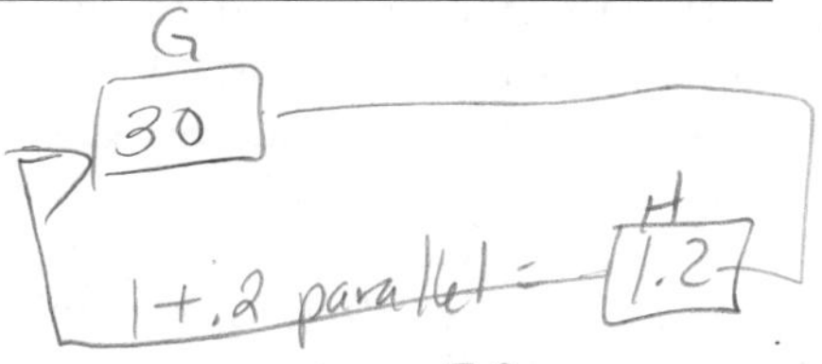

$$TF = \frac{30}{1+(30\times 1.2)}$$

3

LAPLACE TRANSFORMS

INTRODUCTION

A linear real-time control system can be replaced with a mathematical model for the purpose of analysis. The model is in the form of linear differential equations. Solutions of these differential equations completely describe the control system characteristics, including the transient response. Several techniques, available in calculus for obtaining the solutions of these differential equations, some of which are quite demanding. *Laplace transformation* is a method that allows the solutions of linear differential equations to be obtained without much complexity.

In this section, the concept of Laplace transforms is introduced, and some of the transform properties are discussed. A rigorous treatment of transforms is not intended and is beyond the scope of this text. The emphasis is on using a transform table to perform forward and inverse transformations for determining the response of control systems.

TRANSFORMATIONS

The Laplace transform is a method of operational calculus that takes a function of time (*time domain*) and converts it into a function of complex variable s (*frequency domain*, or *s domain*). In some ways, using a Laplace transform is analogous to logarithmic transformation. Before the advent of calculators and computers, logarithms were used for performing multiplication, division, and exponential calculations. These two processes are fairly comparable.

Logarithmic Process

- Logarithm tables are used to convert numerical terms.
- Arithmetic operations are carried out to reduce the expression to a single numerical term.
- Logarithm tables are used to find antilogarithms, yielding the desired result.

Laplace Transformation Process

- Laplace transform tables are used to transform a differential equation.
- The transformed equation is simplified to isolate the desired variable.
- Inverse transformation is applied to obtain the desired time equation.

Laplace transforms are useful in control system analysis. The time response of a control system can be obtained by first applying a Laplace transform and then taking its inverse transform. Because the forward transformation leads into a frequency domain, the frequency response of a control system can be obtained directly from the transformed expression. The transfer function of a control system is defined in s domain and provides valuable information about stability and performance of a closed-loop system.

LAPLACE TRANSFORM INTEGRAL

The Laplace transform of a function of time, $f(t)$, is defined as the integral

$$L[f(t)] = \int_0^{\infty} f(t)e^{-st}\,dt$$

This function is defined for every s, which results in convergence of the integral. The variable s is a complex variable ($s = \sigma + j\omega$).

Forward Laplace Transformation

The process of converting a time domain function into s domain is known as **forward Laplace transformation,** or simply *forward transformation*.

Inverse Laplace Transformation

The process of converting an s-domain function back into a time-domain function is called the **inverse Laplace transformation,** or simply *inverse transformation*.

Transform Notation

- The forward Laplace transformation process is indicated by the letter **L**; for example, $\mathbf{L}[f(t)] = F(s)$.
- The inverse Laplace transformation process is indicated by $\mathbf{L}^{-1}$; for example, $f(t) = \mathbf{L}^{-1}[F(s)]$.
- Lowercase letters with or without a t in parentheses are used to indicate functions of time; for example, $f(t)$, $x(t)$, $y(t)$, f, x, and y.
- Uppercase letters with an s in parentheses are used to indicate transformed functions; for example, $F(s)$, $X(s)$, and $Y(s)$.
- Uppercase letters without a t or an s in parentheses are used to indicate constants; for example, F, X, and Y.

EXAMPLE 3.1

The following are some examples of constant, time-domain, and s-domain terms.

a. Y constant
b. $x(t)$ time-domain function
c. v time-domain function
d. $X(s)$ s-domain function
e. E constant
f. $Z(s)$ s-domain function

Rules of Transformation

A number of rules have been developed to aid in performing forward and inverse Laplace transformation.

Rule 1. Multiplication (or Division) by a Constant

When a function is multiplied by a constant, the Laplace transformation is the product of the original transform and the constant.

Let K be a constant and let $\mathbf{L}[f(t)] = F(s)$, then

$$\mathbf{L}[Kf(t)] = KF(s)$$

Similarly, if C is a constant and $\mathbf{L}[f(t)] = F(s)$, then

$$\mathbf{L}\left[\frac{f(t)}{C}\right] = \frac{F(s)}{C}$$

Rule 2. Sum (or Difference) of Two Functions

The transform of a sum of two time functions is equal to the sum of their individual transforms.

Let $\mathbf{L}[f_1(t)] = F_1(s)$, and $\mathbf{L}[f_2(t)] = F_2(s)$, then

$$\begin{aligned}\mathbf{L}[f_1(t) + f_2(t)] &= \mathbf{L}[f_1(t)] + \mathbf{L}[f_2(t)] \\ &= F_1(s) + F_2(s)\end{aligned}$$

Rule 3. Derivative of a Function

First Derivative The transform of the first derivative of a time function is given as

$$\mathbf{L}\left[\frac{df(t)}{dt}\right] = sF(s) - f(0)$$

where $F(s) = \mathbf{L}[f(t)]$ and the term $f(0)$ is the value of function $f(t)$ at time $t = 0$ (also known as the initial value).

Second Derivative The transform of the second derivative of a time function is given as

$$\mathbf{L}\left[\frac{d^2 f(t)}{dt^2}\right] = s^2 F(s) - sf(0) - \frac{df(0)}{dt}$$

where $F(s) = \mathbf{L}[f(t)]$, $f(0)$ is the value of function $f(t)$ at time $t = 0$ (initial value), and $df(0)/dt$ is the value of the first derivative of function $f(t)$ at time $t = 0$. For a function with zero initial values, the transformation simplifies to

$$\mathbf{L}\left[\frac{df(t)}{dt}\right] = sF(s) \qquad \text{first derivative}$$

$$\mathbf{L}\left[\frac{d^2 f(t)}{dt^2}\right] = s^2 F(s) \qquad \text{second derivative}$$

$$\mathbf{L}\left[\frac{d^n f(t)}{dt^n}\right] = s^n F(s) \qquad n\text{th derivative}$$

EXAMPLE 3.2

The current $i(t)$ flowing into a 1-μF capacitor is related to capacitor voltage v through the following equation.

$$i(t) = \frac{dv}{dt}$$

Determine the Laplace transform of the current. Initially, the capacitor has no voltage across it ($v(0) = 0$, zero initial condition).

Solution

Take the Laplace transform of both sides of the equation.

$$\mathbf{L}[i(t)] = \mathbf{L}\left[\frac{dv}{dt}\right]$$

$$I(s) = sV(s) - v(0)$$

Because the initial value of capacitor voltage $v(0)$ is zero, the final expression is

$$I(s) = sV(s)$$

EXAMPLE 3.3

Repeat Example 3.2, assuming that the capacitor was initially (at time $t = 0$) charged, with the voltage across capacitor being 1.5 V.

Solution

From the previous example, the transformed equation is

$$I(s) = sV(s) - v(0)$$

Substituting the value of $v(0)$ (initial condition),

$$I(s) = sV(s) - 1.5$$

Rule 4. Integral of a Function

The transform of the first integral of a time function is given as

$$\mathbf{L}\left[\int f(t)\,dt\right] = \frac{F(s)}{s} - \frac{\int f(0)\,dt}{s}$$

where $F(s) = \mathbf{L}[f(t)]$, $f(0)$ is the value of function $f(t)$ at time $t = 0$ (initial value), and $\int f(0)\,dt$ is the value of integral of function at time $t = 0$ (initial value). For functions with zero initial values, the transformation simplifies to

$$\mathbf{L}\left[\int f(t)\,dt\right] = \frac{F(s)}{s}$$

EXAMPLE 3.4

Take the Laplace transform of the following. Assume zero initial conditions.

a. $\int i(t)\,dt$
b. $10\int i(t)\,dt$

Solution

a. Here $i(t)$ is the time function. The Laplace transform of $i(t)$ is

$$\mathbf{L}[i(t)] = I(s)$$

Thus,

$$\mathbf{L}\left[\int i(t)\,dt\right] = \frac{I(s)}{s}$$

b. Because 10 is a constant, it is transparent to the transformation process (rule 2).

$$\mathbf{L}\left[10\int i(t)\,dt\right] = 10\mathbf{L}\left[\int i(t)\,dt\right] = 10\frac{I(s)}{s}$$

Rule 5. Initial Value Theorem

The initial value ($t \to 0$) of a time function $f(t)$ whose transform is $F(s)$ is given by the following limit:

$$f(0^+) = \lim_{t\to 0} f(t) = \lim_{s\to\infty} sF(s)$$

This theorem is useful in determining the initial value of the function $f(t)$ from $F(s)$ without performing the inverse Laplace transformation.

EXAMPLE 3.5

Current through a series RL circuit, when subjected to an applied voltage, is given by the following equation:

$$I(s) = \frac{E}{s(sL + R)}$$

where E = battery voltage (10 V)
R = series resistance (100 Ω)
L = series inductance (10 mH)

Determine the (instantaneous) value of current immediately after the application of voltage. Assume that before the voltage was applied, no current was flowing in the circuit and that there was no magnetic field present across the coil.

Solution

Because the Laplace transform of the current $I(s)$ is given, the initial value of the current can be obtained simply by applying the initial value theorem.

$$
\begin{aligned}
\text{initial current value } (t = 0+) &= \lim_{s\to\infty} [sI(s)] \\
&= \lim_{s\to\infty} \left[s \frac{E}{s(sL + R)} \right] \\
&= \lim_{s\to\infty} \left[\frac{E}{(sL + R)} \right] \\
&= 0
\end{aligned}
$$

This is the expected result, because inductance offers almost an infinite resistance to current buildup from zero (initial state) value.

Rule 6. Final Value Theorem

The final value ($t \to \infty$) of a time function $f(t)$ with transform $F(s)$ is given by the following limit:

$$
f(\infty) = \lim_{t\to\infty} f(t) = \lim_{s\to 0} sF(s)
$$

This theorem is useful in determining the final value (steady state) of the function $f(t)$ from $F(s)$ without performing the inverse Laplace transformation.

EXAMPLE 3.6

Determine the final (steady-state) value of current in Example 3.5.

Solution

The final value of current in the circuit (time $t \to \infty$) can be obtained by applying the final value theorem.

$$\text{final current value} = \lim_{s \to 0}\left[s\frac{E}{s(sL + R)}\right]$$

$$= \lim_{s \to 0}\left[\frac{E}{(sL + R)}\right]$$

$$= \frac{E}{R}$$

$$= \frac{10\ \text{V}}{100\ \Omega}$$

$$= 0.1\ \text{A}$$

FORWARD TRANSFORMATION PROCEDURE

The following procedure can be used for carrying out the forward transformation.

Transform of an Equation

The transform of an equation is obtained by taking the transform of the expressions on each side of the equation.

Transform of an Expression

1. Identify the terms, which are connected by sum (+) or difference (−) signs.
2. Categorize the terms as follows and take the transform of each term.
   - *Constant Term* Term containing only constants and no variables of time.

EXAMPLE 3.7

Take the Laplace transform of 12, 256, 0.25, and $\frac{1}{40}$.

Solution

The transformed term is represented by the constant divided by s. The corresponding terms are

$$\frac{12}{s}, \quad \frac{256}{s}, \quad \frac{0.25}{s}, \quad \frac{1}{40s}$$

Undefined Variable Term without Any Coefficient

Term containing an undefined variable of time with no coefficient.

EXAMPLE 3.8

Take the Laplace transform of e, x, $x(t)$, y, and $i(t)$.

Solution

The transform is represented by the capital letter representing the variable, followed by s enclosed in parentheses: $E(s)$, $X(s)$, $X(s)$, $Y(s)$, $I(s)$.

Undefined Variable Term with Constant Coefficient

Term containing an undefined variable of time multiplied or divided by a constant.

EXAMPLE 3.9

Take the Laplace transform of $10e$, $0.5x$, $\frac{x(t)}{25}$, and $\frac{2y(t)}{3}$.

Solution

The transform is represented by the constant term (as is) followed by the capital letter representing the variable, with s enclosed in parentheses. The transformed terms are $10E(s)$, $0.5X(s)$, $\frac{X(s)}{25}$, and $\frac{2Y(s)}{3}$.

Defined Variable Term without Any Constant Coefficient

A term that is defined as a variable of time and is not multiplied or divided by a constant.

EXAMPLE 3.10

Take the Laplace transform of $\sin 10t$, t, e^{-20t}, and $\sin \pi t$.

Solution

The Transform is determined using the transform table. The corresponding entries from the table of transform pairs are

$$\frac{100}{s^2+100}, \quad \frac{1}{s^2}, \quad \frac{1}{s+20}, \quad \frac{\pi}{s^2+\pi^2}$$

Defined Variable Term with Constant Coefficient
A term defined as a variable of time and multiplied and/or divided by a constant.

EXAMPLE 3.11

Take the Laplace transform of $5 \sin 10t$, $100t$, $2.5e^{-20t}$, and $6 \sin \pi t$.

Solution

The transform of each term is the product of the constant term and the corresponding entry from the transform table.

$$5 \times \frac{100}{s^2+100}, \quad 100 \times \frac{1}{s^2}, \quad 2.5 \times \frac{1}{s+20}, \quad 6 \times \frac{\pi}{s^2+\pi^2}$$

or

$$\frac{500}{s^2+100}, \quad \frac{100}{s^2}, \quad \frac{2.5}{s+20}, \quad \frac{6\pi}{s^2+\pi^2}$$

EXAMPLE 3.12

Find the Laplace transform of each of the following.

a. $i(t)$
b. $v(t)$
c. a
d. x

Solution

a. $\mathbf{L}[i(t)] = I(s)$
The Laplace transform of a function of time is indicated simply by changing the letter to uppercase and replacing t in parentheses with s.
b. Similarly, $\mathbf{L}[v(t)] = V(s)$.
c. $\mathbf{L}[a] = A(s)$
By convention, a is a function of time and can also be expressed as $a(t)$.
d. $\mathbf{L}[z] = Z(s)$

EXAMPLE 3.13

a. Determine the Laplace transform of t.
b. Determine the Laplace transform of t^3.

Solution

a. From Table 3.1, the Laplace transform table (entry 4),

$$L\,[t] = \frac{1}{s^2}$$

b. From the transform table (entry 7),

$$L\,[t^3] = \frac{6}{s^4}.$$

EXAMPLE 3.14

a. Determine the Laplace transform of $\sin 10t$.
b. Determine the Laplace transform of $\sin t$.
c. Determine the Laplace transform of $\cos 0.1t$.

Solution

a. Entry 13 of the transform table gives

$$\mathbf{L}[\sin wt] = \frac{w}{s^2 + w^2}$$

Table 3.1 Laplace Transform Pairs

	Time-Domain Function: $f(t)$	Laplace-Domain Function: $F(s)$
1	Unit impulse $\delta(t)$	1
2	Unit step function $u(t)$	$\frac{1}{s}$
3	Constant K	$\frac{K}{s}$
4	t	$\frac{1}{s^2}$
5	t^2	$\frac{2!}{s^3}$
6	$\frac{t^2}{2!}$	$\frac{1}{s^3}$
7	t^n	$\frac{n!}{s^{n+1}}$
8	$\frac{t^{n-1}}{n!}$	$\frac{1}{s^n}$
9	e^{-at}	$\frac{1}{s+a}$
10	te^{-at}	$\frac{1}{(s+a)^2}$
11	$\frac{t^2 e^{-at}}{2!}$	$\frac{1}{(s+a)^3}$
12	$\frac{t^{n-1} e^{-at}}{n-1!}$	$\frac{1}{(s+a)^n}$
13	$\sin \omega t$	$\frac{\omega}{s^2+\omega^2}$
14	$\cos \omega t$	$\frac{S}{s^2+\omega^2}$
15	$\frac{1}{a}(1-e^{-at})$	$\frac{1}{s(s+a)}$
16	$\frac{1}{a^2}(at-1+e^{-at})$	$\frac{1}{s^2(s+a)}$
17	$\frac{1}{b-a}(e^{-at}-e^{-bt})$	$\frac{1}{(s+a)(s+b)}$
18	$e^{-at}\sin \omega t$	$\frac{\omega}{(s+a)^2+\omega^2}$
19	$e^{-at}\cos \omega t$	$\frac{s+a}{(s+a)^2+\omega^2}$

Here $w = 10$, so substituting the value of w in this expression gives

$$\mathbf{L}[\sin 10t] = \frac{10}{s^2 + 10^2}$$

$$\mathbf{L}[\sin 10t] = \frac{10}{s^2 + 100}$$

b. Entry 13 of the transform table is applicable with $w = 1$. Substituting this value of w,

$$\mathbf{L}[\sin t] = \frac{1}{s^2 + 1^2}$$

$$\mathbf{L}[\sin t] = \frac{1}{s^2 + 1}$$

c. Entry 14 of the transform table gives

$$\mathbf{L}[\cos wt] = \frac{s}{s^2 + w^2}$$

Here $w = 0.1$, so substituting the value of w in this expression gives

$$\mathbf{L}[\cos 0.1t] = \frac{s}{s^2 + 0.1^2}$$

$$\mathbf{L}[\cos 0.1t] = \frac{s}{s^2 + 0.01}$$

EXAMPLE 3.15

Determine the Laplace transform of e^{-2t}.

Solution

From entry 9 of the transform table:

$$\mathbf{L}[e^{-at}] = \frac{1}{s + a}$$

Here $a = 2$, so substituting the value of a in this expression gives

$$\mathbf{L}[e^{-2t}] = \frac{1}{s + 2}$$

EXERCISES

3.1 Take the Laplace transform of $e^{-t/10}$.

3.2 Take the Laplace transform of te^{-t}.

3.3 Take the Laplace transform of $(1 - e^{-10t})$.

INVERSE TRANSFORMATION PROCEDURE

Inverse transformation is performed to convert the s-domain terms back to real time (time domain). The procedure for inverse transformation is very similar to that of forward transformation.

Inverse Transform of an Equation

Inverse transformation of an equation is done by taking the inverse transform of the expressions on each side of the equation.

Inverse Transform of an Expression

The inverse transformation process can be as simple as the forward transformation process. Ideally, an entry can be found in transform table that matches the problem at hand. It then becomes a simple matter of substitution of the terms. Unfortunately, it is rare to find an exact match, and some manipulation of terms is needed. In general, the following approach is recommended.

1. Identify the terms, which are connected by sum (+) or difference (−) signs.
2. Categorize the terms as follows and take the inverse transform of each term
   - *Constant Term* A term containing only constants and no s term (variables of time). A purely constant term represents an impulse function in time.

EXAMPLE 3.16

Determine the inverse Laplace transform of 1, 2.5, and 10.

Solution

The corresponding time-domain term (inverse transformation) is a product of a constant with the *delta* function. (*See Chapter 6 for the delta/impulse function.*) The inverse transformed terms are $\delta(t)$, $2.5\delta(t)$, and $10\delta(t)$.

In the following discussions, multiplication or division by a constant is not considered. A constant coefficient does not change the forward or inverse transformation process and can be considered transparent to both. The only time a constant term is treated differently is when it appears on its own and not as a coefficient to another term.

Term Containing Division by s

It relates to a constant term in the time domain.

EXAMPLE 3.17

Determine the inverse Laplace transform of $\frac{1}{s}$, $\frac{256}{s}$, $\frac{0.25}{s}$, and $\frac{1}{40s}$.

Solution

The corresponding time-domain term in each case is expressed by simply removing the s term from the expression: 1, 256, 0.25, and $\frac{1}{40}$.

Undefined Variable Term

A term containing a capital letter representing an undefined variable followed by s enclosed in parentheses.

EXAMPLE 3.18

Determine the inverse Laplace transform of $E(s)$, $X(s)$, $\frac{X(s)}{25}$, and $\frac{2Y(s)}{3}$.

Solution

The time-domain term is expressed as the lowercase letter representing the variable. To explicitly indicate that the inverse quantity is in the time domain, it can be expressed as the lowercase variable followed by t enclosed in parentheses.

$$e \text{ or } e(t), \quad x \text{ or } x(t), \quad \frac{x(t)}{25}, \quad \frac{2y(t)}{3}$$

Other Terms Containing s

1. Find an entry in the transform table similar to the s-domain term under consideration. The entry should have identical s terms in both the numerator and denominator but can differ in the numerical constant term.

2. If no match can be found, use partial fractions to factor the s-domain term into simpler terms.
3. Manipulate the s term so that the coefficient of the s term is the same as in a suitable entry in transform pair table. This can be done by dividing and/or multiplying both numerator and denominator by a suitable constant.
4. If necessary, take the constant terms outside the transformation (rule 1).
5. Using one-to-one correspondence with the table entry, convert the s-domain term into time domain.

EXAMPLE 3.19

Find the inverse Laplace transform of the following.

a. $M(s)$
b. $P(s)$
c. $X(s)$

Solution

a. $\mathbf{L}^{-1}[M(s)] = m(t)$
   Simply changing the uppercase letter to lowercase and replacing s in parentheses with t indicate the inverse Laplace transform of a function.
b. Similarly, $\mathbf{L}^{-1}[P(s)] = p(t)$.
c. $\mathbf{L}^{-1}[X(s)] = x(t)$

EXAMPLE 3.20

Determine the inverse Laplace transform of $\frac{s}{s^2+100}$.

Solution

On checking the entries in Table 3.1, entry 14 is found to be a close match:

$$\mathbf{L}[\cos \omega t] = \frac{s}{s^2+\omega^2}$$

Term-for-term comparison gives $\omega^2 = 100$. Therefore,

$$\begin{aligned} \omega &= 100^{0.5} \\ &= 10 \end{aligned}$$

Thus,

$$\mathbf{L}^{-1}\left[\frac{s}{s^2+100}\right] = \cos 10t$$

EXAMPLE 3.21

Determine the inverse Laplace transform of $\frac{1}{s+5}$.

Solution

Entry 9 of the transform table gives

$$\mathbf{L}[e^{-at}] = \frac{1}{s+a}$$

Here $a = 5$, so substituting the value of a in the above expression gives

$$\mathbf{L}^{-1}\left[\frac{1}{s+5}\right] = e^{-5t}$$

EXERCISES

3.4 Take the inverse Laplace transform of $\frac{1}{s+0.16}$.

3.5 Take the inverse Laplace transform of $\frac{1.25}{s}$.

3.6 Take the inverse Laplace transform of $\frac{24}{s^4}$.

EXAMPLE 3.22

Determine the Laplace transform of each of the following functions.

a. 10
b. $1.2y(t)$
c. $5t$
d. $10 \sin t$

e. $10 \cos 10t$

f. $\dfrac{\sin 10t}{10}$

g. $5 \cos \dfrac{3t}{10}$

h. $1.5\delta(t)$

Solution

a. From entry 3 in Table 3.1,

$$F(s) = \mathbf{L}[10] = \frac{10}{s}$$

b. Here, the time function $y(t)$ is being multiplied by a constant 1.2; so from rule 1,

$$\begin{aligned} F(s) &= \mathbf{L}[1.2y(t)] \\ &= 1.2\mathbf{L}[y(t)] \\ &= 1.2Y(s) \end{aligned}$$

c.

$$\begin{aligned} F(s) &= \mathbf{L}[5t] \\ &= 5\mathbf{L}[t] \\ &= \frac{5}{s^2} \quad \text{(entry 4)} \end{aligned}$$

d.

$$\begin{aligned} F(s) &= \mathbf{L}[10 \sin t] \\ &= 10\mathbf{L}[\sin t] \end{aligned}$$

Entry 13 is applicable with $\omega = 1$:

$$\begin{aligned} F(s) &= 10\frac{1}{s^2 + 1^2} \\ &= 10\frac{1}{s^2 + 1} \end{aligned}$$

e.

$$F(s) = \mathbf{L}[10\cos 10t] = 10\mathbf{L}[\cos 10t]$$

Entry 14 is applicable with $\omega = 10$:

$$F(s) = 10\frac{s}{s^2 + 10^2} = \frac{10s}{s^2 + 100}$$

f.

$$F(s) = L\left[\frac{\sin 10t}{10}\right] = \frac{1}{10}L[\sin 10t]$$

Entry 13 is applicable with $\omega = 10$:

$$F(s) = \frac{1}{10}\frac{10}{s^2 + 10^2} = \frac{1}{10}\frac{10}{s^2 + 100} = \frac{1}{s^2 + 100}$$

g.

$$F(s) = \mathbf{L}\left[5\cos\frac{3t}{10}\right] = 5\mathbf{L}\left[\cos\frac{3t}{10}\right] = 5\mathbf{L}[\cos 0.3t]$$

Entry 14 is applicable with $\omega = 0.3$:

$$F(s) = 5\frac{s}{s^2 + 0.3^2} = 5\frac{s}{s^2 + 0.09} = \frac{5s}{s^2 + 0.09}$$

h.

$$
\begin{aligned}
F(s) &= \mathbf{L}[1.5\delta(t)] \\
&= 1.5\mathbf{L}[\delta(t)] \\
&= 1.5 \times 1 \quad \text{(entry 1)} \\
&= 1.5
\end{aligned}
$$

EXAMPLE 3.23

Take the inverse Laplace transform of each of the following:

a. $\dfrac{1}{s}$

b. $\dfrac{2.5}{s}$

c. $\dfrac{10}{s^2}$

d. $\dfrac{s}{s^2 + 5^2}$

e. $\dfrac{s}{s^2 + 25}$

f. $\dfrac{100s}{s^2 + 100}$

g. $\dfrac{100s}{s^2 + 2}$

h. $\dfrac{10}{(s+2)^2 + 100}$

i. $\dfrac{10}{s^2 + 2s + 101}$

j. $\dfrac{1}{10(s^2 + 4)}$

Solution

a.

$$f(t) = \mathbf{L}^{-1}\left[\frac{1}{s}\right]$$

Scanning the table, entry 2 is found to be similar to $F(s)$:

$$f(t) = 1$$

b.

$$f(t) = \mathbf{L}^{-1}\left[\frac{2.5}{s}\right]$$

The table reveals an exact match, with entry 3. Here $k = 2.5$, thus

$$f(t) = 2.5$$

c. An exact match is not available, but taking the constant term 10 outside the inverse transformation allows entry 4 to be used.

$$f(t) = \mathbf{L}^{-1}\left[\frac{10}{s^2}\right]$$

$$f(t) = 10\mathbf{L}^{-1}\left[\frac{1}{s^2}\right]$$

$$f(t) = 10t$$

d. Entry 14 seems to be a match. Both the numerator and denominator have identical s terms. This leaves ω^2 to be equated to 5^2:

$$\omega^2 = 5^2$$

$$\omega = 5$$

Substituting the value of ω into the $f(t)$ entry in the table gives

$$f(t) = \cos 5t$$

e. There is no exact match in the table, but entry 14 is identical in s terms, so equating the constant terms gives

$$\omega^2 = 25$$

$$\omega = \sqrt{25}$$

$$\omega = 5$$

$$f(t) = \mathbf{L}^{-1}\left[\frac{s}{s^2 + 25}\right]$$

$$f(t) = \mathbf{L}^{-1}\left[\frac{s}{s^2 + 5^2}\right]$$

This, in fact, is the same as the last problem, so

$$f(t) = \cos 5t$$

f. There is no exact match in the table, but entry 14 is identical in s terms.

- Numerator: The constant (100) can be moved outside the inverse transformation.
- Denominator: ω^2 is equal to 100.

$$f(t) = \mathbf{L}^{-1}\left[\frac{100s}{s^2 + 100}\right]$$

$$f(t) = 100\mathbf{L}^{-1}\left[\frac{s}{s^2 + (\sqrt{100})^2}\right]$$

$$f(t) = 100\mathbf{L}^{-1}\left[\frac{s}{s^2 + 10^2}\right]$$

$$f(t) = 100 \cos 10t$$

g. This is similar to Part f.

$$f(t) = \mathbf{L}^{-1}\left[\frac{100s}{s^2 + 2}\right]$$

$$f(t) = 100\mathbf{L}^{-1}\left[\frac{s}{s^2 + (\sqrt{2})^2}\right]$$

$$f(t) = 100 \cos \sqrt{2}t$$

$$f(t) = 100 \cos 1.414t$$

h.

$$f(t) = \mathbf{L}^{-1}\left[\frac{10}{(s+2)^2 + 100}\right]$$

Entry 18 has a similar format and is a match for s terms. Rewriting 100 gives

$$f(t) = \mathbf{L}^{-1}\left[\frac{10}{(s+2)^2 + 10^2}\right]$$

Now $F(s)$ looks like a perfect match, with $a = 2$, and $\omega = 10$. Substituting the values of a and ω in $f(t)$ gives

$$f(t) = e^{-2t} \sin 10t$$

i.

$$f(t) = \mathbf{L}^{-1}\left[\frac{10}{s^2 + 2s + 104}\right]$$

There is no entry that matches the expression at hand, so rearranging the terms gives

$$f(t) = \mathbf{L}^{-1}\left[\frac{10}{(s^2 + 2s + 4) + 100}\right]$$

$$f(t) = \mathbf{L}^{-1}\left[\frac{10}{(s + 2)^2 + 10^2}\right]$$

This is exactly the same expression as in the previous problem, so

$$f(t) = e^{-2t} \sin 10t$$

Thus, at first an expression may seem very different and perhaps difficult to solve. But with some simple rearrangement of terms, practically any expression can be formatted to use transform tables. Comprehensive Laplace transform tables are available for complex expressions. At the same time, several mathematical techniques—such as partial fraction expansion and complex integration—are available for demanding problems.

j.

$$f(t) = \mathbf{L}^{-1}\left[\frac{1}{10(s^2 + 4)}\right]$$

Entry 13 has similar s terms. Rearranging gives

$$f(t) = \frac{1}{10}\mathbf{L}^{-1}\left[\frac{1}{(s^2 + 4)}\right]$$

$$f(t) = \frac{1}{10}\mathbf{L}^{-1}\left[\frac{1}{(s^2 + 2^2)}\right]$$

Multiplying and dividing by 2 (because $\omega = 2$) gives

$$f(t) = \frac{1}{10} \cdot \frac{1}{2} \mathbf{L}^{-1}\left[\frac{2}{(s^2 + 2^2)}\right]$$

$$f(t) = \frac{1}{20} \mathbf{L}^{-1}\left[\frac{2}{(s^2 + 2^2)}\right]$$

Entry 13 can now be applied:

$$f(t) = 0.05 \sin 2t$$

EXAMPLE 3.24

Take the Laplace transform of the following.

a. $\dfrac{dx(t)}{dt} + 2x(t)$

b. $10\dfrac{dx}{dt} + 3.5$

c. $\dfrac{d^2x}{dt^2} + 5\dfrac{dx(t)}{dl} + 2x(t)$

d. $2\displaystyle\int x(t)\,dt$

Solution

a.

$$F(s) = \mathbf{L}\left[\frac{dx(t)}{dt} + 2x(t)\right]$$

$$= \mathbf{L}\left[\frac{dx(t)}{dt}\right] + \mathbf{L}[2x(t)]$$

$$= \mathbf{L}\left[\frac{dx(t)}{dt}\right] + 2\mathbf{L}[x(t)]$$

$$= sX(s) + 2X(s)$$

$$= X(s)(s + 2)$$

b.

$$
\begin{aligned}
F(s) &= \mathbf{L}\left[10\frac{dx}{dt} + 3.5\right] \\
&= \mathbf{L}\left[10\frac{dx}{dt}\right] + \mathbf{L}[3.5] \\
&= 10\mathbf{L}\left[\frac{dx}{dt}\right] + 3.5\mathbf{L}[1] \\
&= 10sX(s) + 3.5\frac{1}{s} \\
&= \frac{10}{s}(s^2X(s) + 3.5)
\end{aligned}
$$

c.

$$
\begin{aligned}
F(s) &= \mathbf{L}\left[\frac{d^2x}{dt^2} + 5\frac{dx(t)}{dt} + 2x(t)\right] \\
&= \mathbf{L}\left[\frac{d^2x}{dt^2}\right] + \mathbf{L}\left[5\frac{dx(t)}{dt}\right] + \mathbf{L}[2x(t)] \\
&= s^2X(s) + 5sX(s) + 2X(s) \\
&= X(s)(s^2 + 5s + 2)
\end{aligned}
$$

d.

$$
\begin{aligned}
F(s) &= \mathbf{L}\left[2\int x(t)\,dt\right] \\
&= 2\mathbf{L}\left[\int x(t)\,dt\right] \\
&= 2\frac{X(s)}{s}
\end{aligned}
$$

EXAMPLE 3.25

Find the inverse Laplace transform of the following functions.

a. 2

b. $sI(s)$

c. $s^2X(s) + Y(s)$

d. $4\frac{X(s)}{s}$

Solution

a.

$$\begin{aligned} f(t) &= \mathbf{L}^{-1}[2] \\ &= 2\mathbf{L}^{-1}[1] \\ &= 2\delta(t) \qquad (\delta(t) \text{ is the impulse function}) \end{aligned}$$

b.

$$\begin{aligned} f(t) &= \mathbf{L}^{-1}[sI(s)] \\ &= \frac{di(t)}{dt} \end{aligned}$$

c.

$$\begin{aligned} f(t) &= \mathbf{L}^{-1}[s^2X(s)] + \mathbf{L}^{-1}[Y(s)] \\ &= \mathbf{L}^{-1}[s^2X(s)] + \mathbf{L}^{-1}[Y(s)] \\ &= \frac{d^2x(t)}{dt^2} + y(t) \end{aligned}$$

d.

$$\begin{aligned} f(t) &= \mathbf{L}^{-1}\left[4\frac{X(s)}{s}\right] \\ &= 4\int x(t)\,dt \end{aligned}$$

PARTIAL-FRACTION EXPANSION

The primary reason for using Laplace transformation is the ease with which solution to a differential equation can be found. A differential equation is converted into an algebraic equation. After some simple manipulation, the desired variable is isolated, and an inverse transformation yields a real-time equation for the desired variable. The forward-transformation process is relatively simple, and if a matching entry can be located in transform-pair tables, reverse transformation is also easy. Unfortunately, it is not always possible to find a suitable entry. For all such cases, the algebraic technique of partial-fraction expansion can be applied to factor the s-domain function into simpler terms. Having achieved that, inverse transformation can be carried out.

For all physically realizable systems, the Laplace transform of the desired variable can be reduced to a rational function of the following form.

$$F(s) = \frac{a_m s^m + a_{m-1} s^{m-1} + \cdots + a_1 s^1 + a_0 s^0}{b_n s^n + b_{n-1} s^{n-1} + \cdots + b_1 s^1 + b_0 s^0}$$

where a's and b's are real, and $n \geq m$. A polynomial can also be factored and expressed as the product of its roots. Thus the preceding rational function can be expressed as

$$F(s) = \frac{a_m s^m + a_{m-1} s^{m-1} + \cdots + a_1 s^1 + a_0 s^0}{(s - p_n)(s - p_{n-1}) \cdots (s - p_1)(s - p_0)}$$

The p's are the roots of the denominator polynomial. The roots are actually the solutions (also known as the poles) of the equation (known as the characteristic equation of function $F(s)$), formed by equating the denominator to zero. The characteristic equation of the function $F(s)$ can thus be written as

$$(s - p_n)(s - p_{n-1}) \cdots (s - p_1)(s - p_0) = 0$$

Now there are two distinct possibilities, depending upon the roots of the characteristic polynomial.

Distinct Roots

All roots are distinct; that is, every root is numerically different from other roots.

EXAMPLE 3.26

In each of the following cases, the roots of denominator are different.

a. $F(s) = \dfrac{8.5}{(s + 10)(s + 12.25)}$ (two roots: -10, -12.25)

b. $F(s) = \dfrac{s + 3}{s(s + 0.5)(s + 56)}$ (three roots: 0, -0.5, -56)

c. $F(s) = \dfrac{(s + 1)(s + 7)(s + 8)}{s(s - 1)(s + 25)}$ (three roots: 0, 1, -25)

Using partial-fraction expansion techniques, the function can be further expressed as

$$F(s) = \frac{C_0}{s - p_0} + \frac{C_1}{s - p_1} + \cdots + \frac{C_{n-1}}{s - p_{n-1}} + \frac{C_n}{s - p_n}$$

where C's are numerical coefficients, and p's are roots of denominator polynomial. Both p's and C's may be real and/or complex.

Coefficient Evaluation Each coefficient is evaluated separately and independently of the other coefficients.

- Obtain a partial product by multiplying the function by the denominator of that term.
- Take the limit of the resultant product as $s \to p_i$.
- Set coefficient C_i equal to this value.

Coefficient C_0:

$$C_0 = [F(s) \cdot (s + p_0)]\big|_{s=-p_0}$$

Coefficient C_1:

$$C_1 = [F(s) \cdot (s + p_1)]\big|_{s=-p_1}$$

$$\vdots$$

Coefficient C_{n-1}:

$$C_{n-1} = [F(s) \cdot (s + p_{n-1})]\big|_{s=-p_{n-1}}$$

Coefficient C_n:

$$C_n = [F(s) \cdot (s + p_n)]\big|_{s=-p_n}$$

Once the function $F(s)$ has been factored to this form, inverse Laplace transformation can be carried out. The time-domain function, $f(t)$, is always of the form

$$f(t) = C_0 e^{-p_0 t} + C_1 e^{-p_1 t} + \cdots + C_{n-1} e^{-p_{n-1} t} + C_n e^{-p_n t}$$

Repeated Roots

In this case, most of the roots are distinct, but there is at least one root that is repeated one or more times.

EXAMPLE 3.27

In each of the following cases, there is at least one repeated root.

a. $F(s) = \dfrac{s+2}{s(s+1)^2}$ (three roots total, with two repeated roots: 0, −1, and −1)

b. $F(s) = \dfrac{(s+1)(s+10)(s+15)}{s^3(s+2)(s+3)}$ (five roots total, with three repeated roots: 0, 0, 0, −2, −3)

Partial-fraction expansion of such a function takes a slightly different form than that of the function with distinct roots. The function $F(s)$ is repeated here, shown with a single repeated root.

$$F(s) = \frac{a_m s^m + a_{m-1} s^{m-1} + \cdots + a_1 s^1 + a_0 s^0}{(s - p_{n-r})(s - p_{n-r-1}) \cdots (s - p_1)(s - p_0)^r}$$

There are still a total of n roots, but root p_0 is repeated r times. The characteristic equation of the function $F(s)$ is

$$(s - p_{n-r})(s - p_{n-r-1}) \cdots (s - p_1)(s - p_0)^r = 0$$

Using partial fractions, the function $F(s)$ can be expanded into

$$F(s) = \frac{C_0}{(s - p_0)} + \frac{C_1}{(s - p_0)^2} + \cdots + \frac{C_{r-1}}{(s - p_0)^r}$$
$$+ \frac{C_r}{(s - p_1)} + \cdots + \frac{C_{n-1}}{(s - p_{n-r-1})} + \frac{C_n}{(s - p_{n-r})}$$

The C's are numerical coefficients. Both p's and C's may be real or complex.

Coefficient Evaluation When the denominator polynomial contains repeated roots, the coefficient evaluation is carried out in two steps.

1. All coefficients with distinct roots are evaluated first, using an identical procedure.
   - Obtain a partial product by multiplying the function by the denominator of that term.
   - Take the limit of the resultant product as $s \to p_i$.
   - Set coefficient C_i equal to this value.
2. Coefficients with repeated roots are evaluated next.

First coefficient of repeated root

- Obtain a partial product by multiplying the function by the denominator of that term.
- Cancel the common terms.
- Take the derivative of the resultant product with respect to s.
- Apply the limit of the resultant product as $s \to p_i$.
- Set coefficient C_i equal to this value.

Second coefficient of repeated root

- Obtain a partial product by multiplying the function by the second power of the term in the denominator.
- Cancel the common terms.
- Take the derivative of the resultant product with respect to s.
- Apply the limit of the resultant product as $s \to p_i$.
- Set coefficient C_i equal to this value.

Repeat until all the coefficients for that repeated root are resolved.

Coefficient C_0:

$$C_0 = \left.\frac{d[F(s) \cdot (s + p_0)]}{ds}\right|_{s=-p_0}$$

Coefficient C_1:

$$C_1 = \left.\frac{d[F(s) \cdot (s + p_0)^2]}{ds}\right|_{s=-p_0}$$

Coefficient C_2:

$$C_2 = \left.\frac{d[F(s) \cdot (s + p_0)^3]}{ds}\right|_{s=-p_0}$$

$$\vdots$$

Coefficient C_{r-2}:

$$C_{r-2} = \left.\frac{d[F(s) \cdot (s + p_0)^{r-1}]}{ds}\right|_{s=-p_0}$$

Coefficient C_{r-1}:

$$C_{r-1} = [F(s) \cdot (s + p_0)^r]|_{s=-p_0}$$ (derivative not required for evaluating this coefficient)

Once all the coefficients have been determined, inverse Laplace transformation can be carried out and time function, $f(t)$, determined. The resulting function $f(t)$ is always of the form

$$f(t) = C_0 e^{-p_0 t} + C_1 t e^{-p_0 t} + C_2 t^2 e^{-p_0 t} + \cdots + C_{r-1} t^r e^{-p_0 t}$$
$$+ C_r e^{-p_1 t} + \cdots + C_{n-1} e^{-p_{n-r-1} t} + C_n e^{-p_{n-r} t}$$

EXAMPLE 3.28

Take the inverse transform of the following using partial-fraction expansion.

$$F(s) = \frac{1}{(s+2)(s+3)}$$

Solution

Obviously, the function $F(s)$ is a simple one and its inverse transform can be easily determined using the transform table. At the same time, it serves to illustrate the partial-fraction expansion procedure. The denominator of the function is already in factored form, so the general expanded form is

$$F(s) = \frac{C_0}{s+2} + \frac{C_1}{s+3}$$

Evaluation of coefficient C_0:

$$C_0 = F(s) \cdot (s+2)|_{s=-2}$$

$$C_0 = \left[\frac{1}{(s+2)(s+3)}\right] \cdot (s+2)\Big|_{s=-2}$$

$$C_0 = \left[\frac{1}{(s+3)}\right]\Big|_{s=-2}$$

$$C_0 = \frac{1}{(-2+3)}$$

$$C_0 = 1$$

Evaluation of coefficient C_1:

$$C_1 = F(s) \cdot (s+3)\big|_{s=-3}$$

$$C_1 = \left[\frac{1}{(s+2)(s+3)}\right] \cdot (s+3)\bigg|_{s=-3}$$

$$C_1 = \left[\frac{1}{(s+2)}\right]\bigg|_{s=-3}$$

$$C_1 = \frac{1}{(-3+2)}$$

$$C_1 = -1$$

Thus, the expanded form is

$$F(s) = \frac{1}{s+2} - \frac{1}{s+3}$$

Check: Taking the common denominator and combining the two terms gives

$$F(s) = \frac{(s+3)-(s+2)}{(s+2)(s+3)} = \frac{1}{(s+2)(s+3)} \quad \textbf{OK}$$

The inverse Laplace transform thus is

$$f(t) = e^{-2t} - e^{-3t}$$

EXAMPLE 3.29

Take the inverse transform of the following using partial-fraction expansion.

$$F(s) = \frac{s+3}{s(s+1)(s+2)}$$

Solution

There are three roots of the denominator, $s = 1$, $s = -1$, and $s = -2$. Thus, the general expanded form will be

$$F(s) = \frac{C_0}{s} + \frac{C_1}{s+1} + \frac{C_2}{s+2}$$

Coefficient C_0:

$$C_0 = F(s) \cdot s\big|_{s=0}$$

$$C_0 = \left[\frac{(s+3)}{(s+1)(s+2)}\right]\Bigg|_{s=0}$$

$$C_0 = 1.5$$

Coefficient C_1:

$$C_1 = F(s) \cdot (s+1)\big|_{s=-1}$$

$$C_1 = \left[\frac{(s+3)}{s(s+2)}\right]\Bigg|_{s=-1}$$

$$C_1 = -2$$

Coefficient C_2:

$$C_2 = F(s) \cdot (s+2)\big|_{s=-2}$$

$$C_2 = \left[\frac{(s+3)}{s(s+1)}\right]\Bigg|_{s=-2}$$

$$C_2 = 0.5$$

The expanded form is

$$F(s) = \frac{1.5}{s} - \frac{2}{s+1} + \frac{0.5}{s+2}$$

The real-time function, $f(t)$, is thus

$$f(t) = 1.5e^{-0t} - 2e^{-1t} + 0.5e^{-2t}$$

$$f(t) = 1.5 - 2e^{-t} + 0.5e^{-2t}$$

EXAMPLE 3.30

Determine an expression for the time function $f(t)$ from the s-domain function

$$F(s) = \frac{s+3}{s^2(s+1)}$$

Solution

There are three roots of the denominator, $s = 0$, $s = 0$, and $s = -1$. Two of the roots are repeated. The general expanded form will be

$$F(s) = \frac{C_0}{s} + \frac{C_1}{s^2} + \frac{C_2}{s+2}$$

Coefficients C_1 and C_2 can be evaluated using the procedure for nonrepeated roots. Evaluation of C_0 demands a modified procedure with the derivative of a partial product. Coefficient C_1:

$$C_1 = F(s) \cdot s^2\big|_{s=0}$$

$$C_1 = \left[\frac{(s+3)}{s^2(s+1)} \cdot s^2\right]\Bigg|_{s=0}$$

$$C_1 = \frac{(s+3)}{(s+1)}\Bigg|_{s=0}$$

$$C_1 = 3$$

Coefficient C_2:

$$C_2 = F(s) \cdot (s+1)\big|_{s=-1}$$

$$C_2 = \left[\frac{(s+3)}{s^2}\right]\Bigg|_{s=-1}$$

$$C_2 = 2$$

Coefficient C_0: The coefficient is evaluated using the expression

$$C_0 = \frac{d[F(s) \cdot (s+p_0)]}{ds}\Bigg|_{s=-p_0}$$

$$C_0 = \frac{d\left[\left(\frac{s+3}{s^2(s+1)}\right) \cdot s\right]}{ds}\Bigg|_{s=0}$$

$$C_0 = \left[\frac{d\left(\frac{s+3}{s(s+1)}\right)}{ds}\right]\Bigg|_{s=0}$$

Taking the derivative using the quotient rule gives

$$C_0 = \left[\frac{1 \cdot (s+1) - (s+3) \cdot 1}{(s+1)^2}\right]\Bigg|_{s=0}$$

$$C_0 = \left[\frac{-2}{(s+1)^2}\right]\Bigg|_{s=0}$$

$$C_0 = -2$$

The expanded form is

$$F(s) = \frac{-2}{s} + \frac{3}{s^2} + \frac{2}{s+1}$$

The real-time function $f(t)$ is, then,

$$f(t) = -2e^{-0t} + 3te^{-0t} + 2e^{-1t}$$

$$f(t) = -2 + 3t + 2e^{-1t}$$

APPLICATIONS OF LAPLACE TRANSFORMS: DIFFERENTIAL EQUATIONS

Linear, constant-coefficients differential equations are used to describe the dynamic behavior of linear control systems. These equations can be easily solved using the Laplace transforms. The procedure can be broken down in the following steps.

- Obtain the differential equation.
- Transform both sides of the differential equation using a transform table.
- Simplify the equation, isolating the variable of interest.
- Perform inverse transformation to obtain the desired variable as a function of time.

EXAMPLE 3.31

Current flow i in a series RL circuit is given by the following differential equation. Determine the time equation describing the dynamic variation in the current.

$$L\frac{di}{dt} + Ri = E$$

where

$$L = \text{coil inductance} = 1 \text{ mH}$$
$$R = \text{resistance} = 1000\ \Omega$$
$$E = \text{applied voltage} = 10 \text{ V}$$

Solution

The time variable $i(t)$ describes the current flow in the circuit. Unfortunately, it is difficult to isolate $i(t)$ from the differential equation, because derivative terms are involved. Laplace transformation thus will be used to extract $i(t)$ from the differential equation.

If the values of L, R, and E are not substituted, then a general solution will be obtained, which may be more useful.

Taking the Laplace transformation of both sides gives

$$\mathbf{L}\left[L\frac{di}{dt} + Ri\right] = \mathbf{L}[E]$$

$$\mathbf{L}\left[L\frac{di}{dt}\right] + \mathbf{L}[Ri] = \mathbf{L}[E]$$

$$L\,\mathbf{L}\left[\frac{di}{dt}\right] + R\mathbf{L}[i] = \mathbf{L}[E]$$

$$L\,sI(s) + RI(s) = \frac{E}{s}$$

Simplifying to isolate the desired current variable $I(s)$, we have

$$I(s)[sL + R] = \frac{E}{s}$$

$$I(s) = \frac{E}{s(sL + R)}$$

Taking the inverse Laplace transform,

$$i(t) = \mathbf{L}^{-1}[I(s)] = \mathbf{L}^{-1}\left[\frac{E}{s(sL + R)}\right]$$

The function $I(s)$ can be transformed back by entry 15 of Table 3.1. Rearranging terms gives

$$i(t) = \mathbf{L}^{-1}\left[\frac{E}{L}\frac{1}{s(s + R/L)}\right]$$

$$i(t) = \frac{E}{L}\mathbf{L}^{-1}\left[\frac{1}{s(s + R/L)}\right]$$

E/L is a constant, so we take it out of the transformation.

$$i(t) = \frac{E}{L}\left\{\frac{L}{R}\left(1 - e^{-(R/L)t}\right)\right\}$$

$$i(t) = \frac{E}{R}\left(1 - e^{-(R/L)t}\right)$$

This equation shows that the current rises in the circuit at an exponential rate, with L/R as the time constant. Substituting the values of R, L, and E gives

$$i(t) = 0.01\left(1 - e^{-100{,}000t}\right)$$

Initial current value ($t = 0$) can be obtained by substituting $t = 0$ in this equation.

$$i(0) = \frac{E}{R}\left(1 - e^{-0}\right)$$

$$i(0) = \frac{E}{R}(1 - 1) = 0$$

Similarly, the final (steady-state) current value ($t \to \infty$) is obtained by taking the limit of the equation

$$i(\infty) = \frac{E}{R}\left(1 - e^{\infty}\right)$$

$$i(\infty) = \frac{E}{R}(1 - 0) = \frac{E}{R} = \frac{10}{1000} = 10 \text{ mA}$$

EXAMPLE 3.32

The following differential equation describes the response of a car's suspension system when subjected to a unit impulse force. Find the equation describing the movement $x(t)$.

$$\frac{d^2x}{dt^2} + 2\frac{dx}{dt} + x = \delta(t)$$

Solution

The value $x(t)$ can be obtained by first taking the Laplace transform, isolating $X(s)$, and then performing an inverse transformation. Taking the Laplace transformation of both sides of the equation gives

$$\mathbf{L}\left[\frac{d^2x}{dt^2} + 2\frac{dx}{dt} + x\right] = \mathbf{L}[\delta(t)]$$

$$\mathbf{L}\left[\frac{d^2x}{dt^2}\right] + \mathbf{L}\left[2\frac{dx}{dt}\right] + \mathbf{L}[x] = \mathbf{L}[\delta(t)]$$

$$\mathbf{L}\left[\frac{d^2x}{dt^2}\right] + 2\mathbf{L}\left[\frac{dx}{dt}\right] + \mathbf{L}[x] = \mathbf{L}[\delta(t)]$$

$$s^2X(s) + 2sX(s) + X(s) = 1$$

$$X(s)(s^2 + 2s + 1) = 1$$

$$X(s) = \frac{1}{(s^2 + 2s + 1)}$$

$$X(s) = \frac{1}{(s + 1)^2}$$

Taking the inverse transformation (entry 10) gives

$$x(t) = \mathbf{L}^{-1}[X(s)] = \mathbf{L}^{-1}\left[\frac{1}{(s + 1)^2}\right]$$

$$x(t) = te^{-t}$$

EXAMPLE 3.33

A resistor, capacitor, and inductor are connected in series. A dc voltage of 10 V is applied through a battery with low internal impedance. The current i in the circuit is given by the following differential equation:

$$v = L\frac{di}{dt} + Ri + \frac{1}{c}\int_0^t i\,dt$$

All initial conditions are zero; that is, at time $t = 0$, there is no magnetic field across the inductor, no charge on the capacitor, and no current through the resistor. Determine a general expression for current, which is valid for all values of time greater than or equal to 0 s. See Figure 3.1.

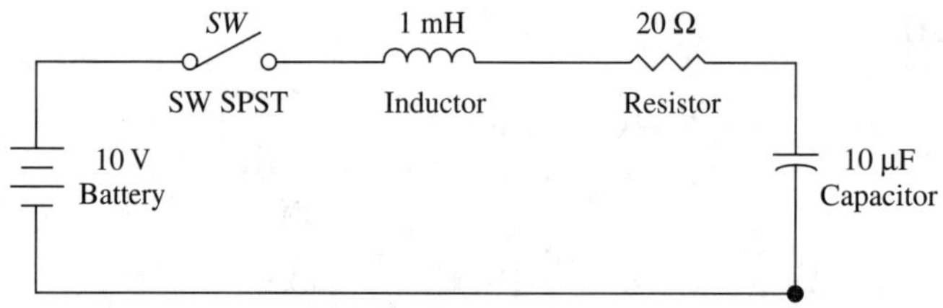

Figure 3.1 Schematic diagram for Example 3.33.

Solution

Because current in the circuit is expressed as a differential equation, it needs to be isolated through Laplace transformation. First, substitute the component values and battery voltage in the differential equation:

$$10 = 0.001\frac{di}{dt} + 20i + \frac{1}{0.00001}\int_0^t i\,dt$$

Taking the Laplace transform of both sides of the equation gives

$$\mathbf{L}[10] = L\left[0.001\frac{di}{dt} + 20i + \frac{1}{0.00001}\int_0^t i\,dt\right]$$

$$\frac{10}{s} = 0.001sI(s) + 20I(s) + \frac{10^5 I(s)}{s}$$

$$\frac{10}{s} = \left[\frac{10^{-3}s^2 + 20s + 10^5}{s}\right] I(s)$$

Isolating the current term $I(s)$, we have

$$I(s) = \left[\frac{s}{10^{-3}s^2 + 20s + 10^5}\right]\frac{10}{s}$$

$$I(s) = \frac{10^4}{s^2 + 2\times 10^4 s + 10^8}$$

$$I(s) = \frac{10^4}{(s + 10^4)}$$

The expression for the current in real time can now be determined by performing the inverse Laplace transformation of the equation:

$$L^{-1}[I(s)] = L^{-1}\left[\frac{10^4}{(s+10^4)^{\text{[illegible]}}}\right]$$

$$i(t) = 416.67t^4e^{-10000t}$$

The plot in Figure 3.2 shows the current flow in the circuit immediately after the switch is closed. It shows that current initially rises rapidly, because the capacitor has zero initial charge. Current flow reduces with the charging of the capacitor, eventually reaching zero.

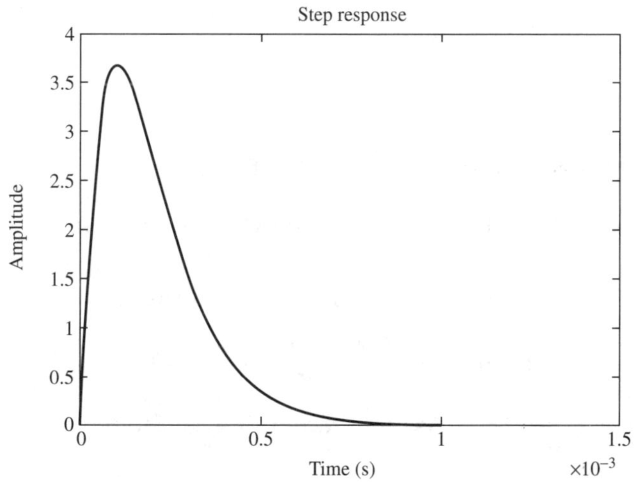

Figure 3.2 Step response for Example 3.33.

PROBLEMS

1. Identify the real-time term (time-domain variable) in each of the following.
   a. 10.256
   b. $\frac{1}{5t}$
   c. sin 10
   d. $x(t)$
   e. e
   f. E

2. Identify the s-domain term (Laplace-transformed quantity) in each of the following.

a. v

b. $10s$

c. $\dfrac{1}{s+1}$

d. $\dfrac{1}{s^2+10s+1}$

e. I

f. $i(t)$

3. Represent each of the following in s-domain (take the Laplace transform).

a. 0.12

b. t

c. $e^{2.5}$

d. $2.5+E$

e. $10+v$

f. $i(t)$

g. $5\delta(t)$

4. Take Laplace transformation of each of the following.

a. $\dfrac{dx(t)}{dt}+4x(t)+1$ (Given $\dfrac{dx(t)}{dt}=0$ and $x(t)=0$ at $t=0$.)

b. $3\dfrac{d^2x(t)}{dt^2}+x(t)=10$ (Assume zero initial conditions.)

c. $2\dfrac{dv(t)}{dt}+v(t)=100$ (Given $\dfrac{dv(t)}{dt}=0$ and $v(t)=1$ at $t=0$.)

d. $x+2.5\int x dt+\dfrac{dx}{dt}=y$ (Assume zero initial conditions.)

e. $10\sin 5t$

f. $\sin t+\cos 10t$

g. $2(\sin 100t+\cos 10t)$

h. $\dfrac{10}{t}$

i. e^{-5t}

j. $2te^{-t}$

5. Take the inverse Laplace transform of each of the following.

a. 1.25

b. $\dfrac{1}{10s}$

c. $\dfrac{1}{s+3}$

d. $\dfrac{4}{(s+2)^2}$

e. $\dfrac{12.5}{2s+1}$

f. $\dfrac{1}{s^2+100}$

g. $\dfrac{10}{s(5s+1)}$

h. $\dfrac{0.5s}{s^2+5}$

i. $\dfrac{1000}{s^2+0.15s+16}$

6. Using partial fractions, determine the inverse Laplace transform of the following.

a. $\dfrac{120}{s(s+1)(10s+2)}$

b. $\dfrac{10}{s(s+2)^2}$

7. Find an expression for voltage $v(t)$ in a circuit, expressed by the following differential equation. Assume zero initial conditions.

$$0.1\frac{d^2v}{dt^2} + 8\frac{dv}{dt} + 160 = 10$$

8. Find an expression for displacement $x(t)$, given

$$\frac{dx}{dt} + 8x = 2 \qquad \text{(All initial conditions are zero.)}$$

9. Output $x(t)$ in a control system is given by the following s-domain expression:

$$X(s) = \frac{2}{s+0.4}$$

Determine the output at time $t = 1$ s.

10. Displacement $x(t)$ in a spring, mass, and damper system is given by the following differential equation. Assuming zero initial conditions, determine an expression for displacement $x(t)$.

$$1.25\frac{d^2x}{dt^2} + 5\frac{dx}{dt} + 10x = 3$$

11. Current $i(t)$ in an RC circuit is related to the applied voltage $e(t)$ expressed by the following differential equation:

$$Ri(t) + \frac{1}{C}\int i(t)\,dt = e(t)$$

Using the given circuit parameters, determine an expression for the current $i(t)$. Assume zero initial conditions.

$$\begin{aligned} \text{resistor } R &= 10\ \text{k}\Omega \\ \text{capacitor } C &= 0.22\ \mu\text{F} \\ \text{applied voltage } e(t) &= 10\ \text{V dc} \end{aligned}$$

12. Linear displacement m in a position-control system is given by the following s-domain expression:

$$1.2 = \frac{d^2m}{dt^2} + 0.8\frac{dm}{dt} + m$$

Determine the steady-state output (output m as $t \to \infty$).

4

MATHEMATICAL MODELING

INTRODUCTION

Architects and structural engineers carry out extensive stress analysis on proposed designs and create architectural models. Automobile body designers work with clay models. Hydraulic engineers and shipbuilders carry out extensive modeling of a proposed design, followed by physical model testing.

A control system is conceived to ensure that some dynamic variable maintains a desired state with respect to time. Before this control system is constructed or assembled, it has to be designed and analyzed for operation in the field. After a control system is in operation, it continues to require tuning and operational analysis from time to time.

It is thus necessary to develop a technique for the purpose of analysis, design, and tuning of control systems. Transfer functions are used for analyzing the operation and performance of closed-loop control systems. The transfer function of a control system depends on the characteristics of its components (or subsystems) and also on the way these components are connected together.

Behavior of some of these components, such as electrical resistors and mechanical springs, can be described by linear algebraic expressions, whereas the majority of the components, such as inductors and capacitors, require integral or derivative terms to properly model their behavior. A differential equation is required to model a simple series electrical circuit containing a resistor, a capacitor, and an inductor. From this differential equation, a transfer function can be developed, which completely describes its dynamic behavior.

MATHEMATICAL MODEL

Mathematical Model

The **mathematical model** for an element or a system is an equation or set of equations that define the relationship between the input and output (variables).

Operational behavior of an element depends on its characteristics. For a system, its behavior is dependent not only on its components, but also on how these components are linked together.

EXAMPLE 4.1

Ohm's law describes the relationship between current and voltage for a resistor. In other words, it can be stated that Ohm's law is a mathematical model for a resistor.

$$\text{voltage} = \text{resistance} \times \text{current}$$
$$V = RI$$

Nonlinear Behavior

Even though a number of physical components exhibit linear relationships between input and output, when viewed over a wide operating range these components exhibit nonlinear relationships.

Consider the case of a mechanical spring. The relationship between applied spring force and resultant deflection (extension or compression) is expressed as a linear equation.

$$\text{force} = \text{spring stiffness} \times \text{deflection}$$
$$F = K \times \Delta L$$

But when the spring is subjected to a large amount of compression force, the spring seems to reach the limit of compression, with coils coming close to each other. As this happens, the linear relationship between the applied force and resultant deflection no longer exists. If the force is further increased, the coils are squeezed together with no further increase in deflection. It can be said that the spring has reached the saturation limit. Graphically, this can be shown as in Figure 4.1.

In our study of (classical) control systems, we assume that all elements, subsystems, and even complete systems can be described by linear algebraic and/or differential equations. This will be true only if every element is operated over a relatively narrow range of its entire span. Thus, the preceding spring, even though it exhibits nonlinear

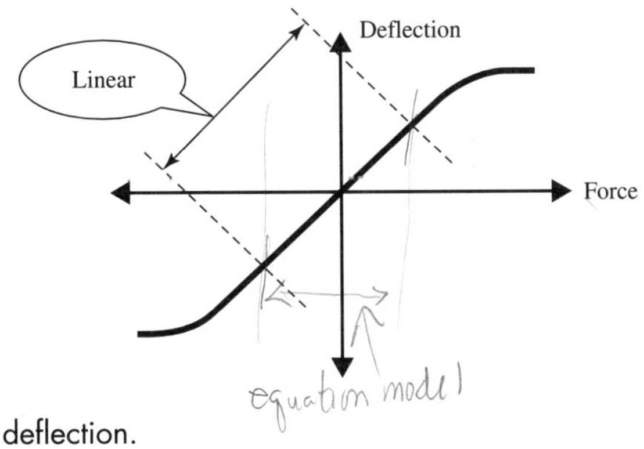

Figure 4.1 Spring deflection.

behavior at the extreme deflection, can be considered to be linear over the middle (smaller) range.

Operational amplifiers also exhibit a similar behavior. Output voltage of an amplifier (say a noninverting amplifier) is directly proportional to the applied input voltage. This is true for all small-signal applications. On the other hand, if the input-signal amplitude is sufficiently large or if the amplifier gain is set too high, the output voltage can reach the saturation limit (dictated by power supply), and a linear relationship no longer exists.

It is assumed throughout the rest of the text that every element can be described by a linear model. This may be an inherent characteristic of the element, or it may be substantiated by an assumption that operation is over a small and linear range of the element.

TRANSFER FUNCTIONS

Transfer Function

The **transfer function** of an element or a system is an s-domain expression describing the relationship that exists between input and output variables, assuming that all initial conditions are zero.

A mathematical model of a resistor element describing the relationship between current flow through it and the voltage applied across it is fairly simple and is a numerical constant (R). Now consider a series RC circuit, where it is desired to express the relation between capacitor voltage v_C (output) and voltage applied to circuit e (input). This relation is expressed through a differential equation,

$$RC\frac{dv_c}{dt} + v_c = e$$

where both v_C and e are functions of time.

It is not possible to isolate the capacitor voltage v_c from the differential equation, and thus it is not feasible to come up with a mathematical model directly linking the output and input variables using time-variable terms $v_c(t)$ and $e(t)$. But this relationship can be easily expressed using the s-domain terms $V_c(s)$ and $E(s)$.

$$V_c(s) \frac{1}{RCs + 1} \times E(s)$$

In other words, output (capacitor) voltage in a series RC circuit is related to input voltage through the model $1/(RCs + 1)$; also known as the transfer function of the circuit (Figure 4.2).

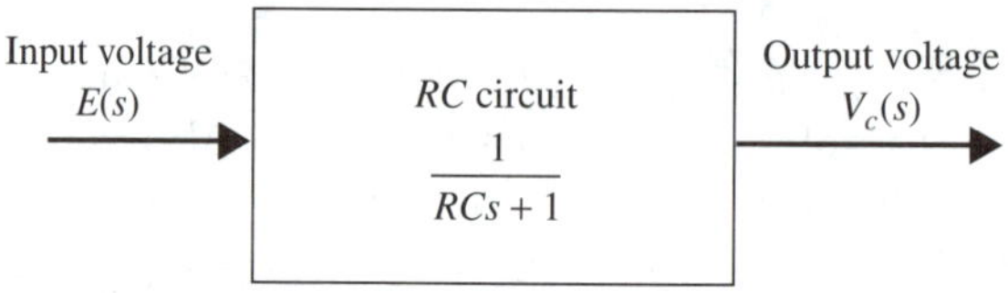

Figure 4.2 Transfer function of an RC circuit.

In a circuit (or a system), there are several variables that may be of interest. Before proceeding further, *one and only one* output variable has to be decided upon. A mathematical expression (usually a differential equation) can then be set up. Laplace transforms are applied to this differential equation, and a transfer function linking the desired output variable and the input variable can then be obtained. The derivation of a transfer function for a system depends on the desired output variable. It is possible to obtain multiple transfer functions for a system relating different input and output variables. Fortunately, in many control systems, only one output variable is of concern; hence a single transfer function is needed.

ELECTRICAL NETWORKS

A passive electrical system can contain a number of resistors, capacitors, and inductors. If possible, all similar components can be lumped together as a single equivalent component for the purpose of analysis. It is important that circuit operation remain unaffected by the lumped-parameter approach; otherwise the simplified model is incorrect and the components cannot be lumped together.

Resistor

Current i (amperes) and voltage v (volts) through a resistor R (ohms) are governed by Ohm's law.

$$v = Ri$$
$$i = \frac{v}{R}$$

The corresponding transfer functions are as shown in Figure 4.3.

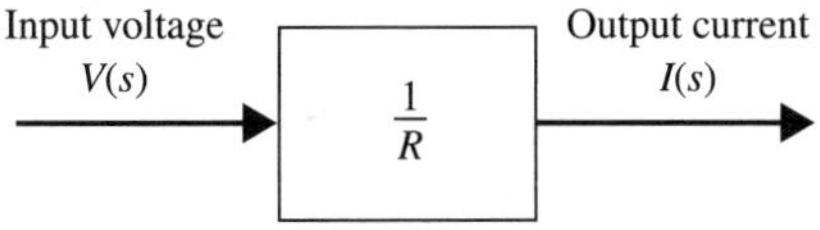

(a) Resistor TF current versus voltage.

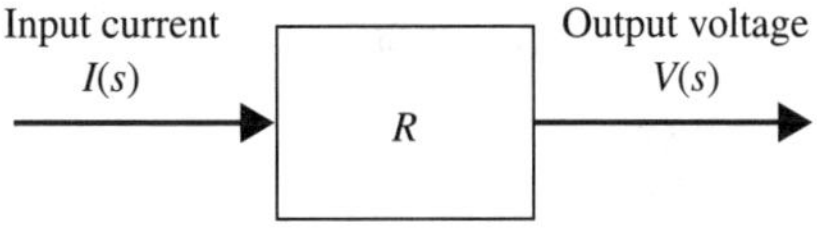

(b) Resistor TF voltage versus current.

Figure 4.3 Resistor transfer functions.

Capacitor

Current i (amperes) and voltage v (volts) through a capacitor C (farads) are governed by the following:

$$v = \frac{1}{C}\int i\,dt$$
$$i = C\frac{dv}{dt}$$

Because integral and derivative terms are involved, Laplace transformation is applied.

$$V(s) = \frac{i}{Cs} \times I(s), \quad \text{or} \quad I(s) = Cs \times V(s)$$

Capacitor transfer functions are shown in Figure 4.4.

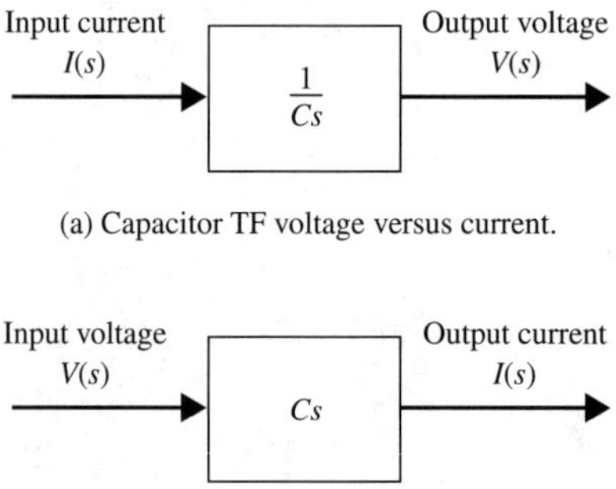

(a) Capacitor TF voltage versus current.

(b) Capacitor TF current versus voltage.

Figure 4.4 Capacitor transfer functions.

Inductor

Current i (amperes) and voltage v (volts) through an inductor L (henrys) are governed by the following:

$$v = L\frac{di}{dt}$$

$$i = \frac{1}{L}\int v\,dt$$

Applying Laplace transformation to eliminate integral and derivative terms gives

$$V(s) = L\,s \times I(s), \quad \text{or} \quad I(s) = \frac{1}{L\,s} \times V(s)$$

Inductor transfer functions are shown in Figure 4.5.

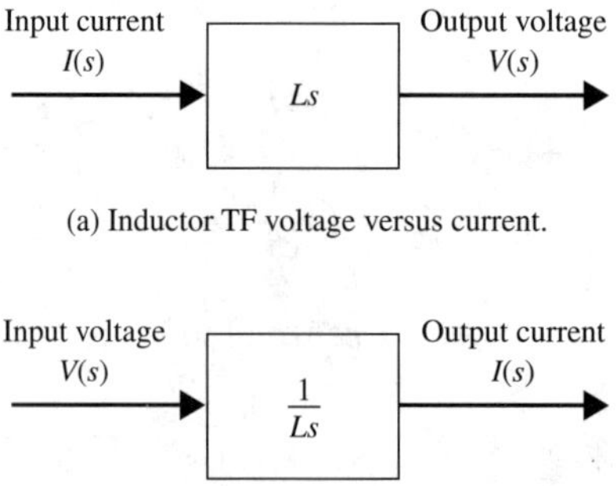

(a) Inductor TF voltage versus current.

(b) Inductor TF current versus voltage.

Figure 4.5 Inductor transfer functions.

Series *RC* Circuit

Figure 4.6 shows a series RC circuit. Initially, there is no current flow in the circuit and the switch is in the open position, as shown. At time $t = 0$, the switch is placed in the closed position and voltage e is applied. Current starts to build up in the circuit and starts charging the capacitor. Eventually, the capacitor is fully charged and full voltage appears across it. An equation can be developed that fully describes the capacitor voltage (or current in the circuit) from start to final steady-state value.

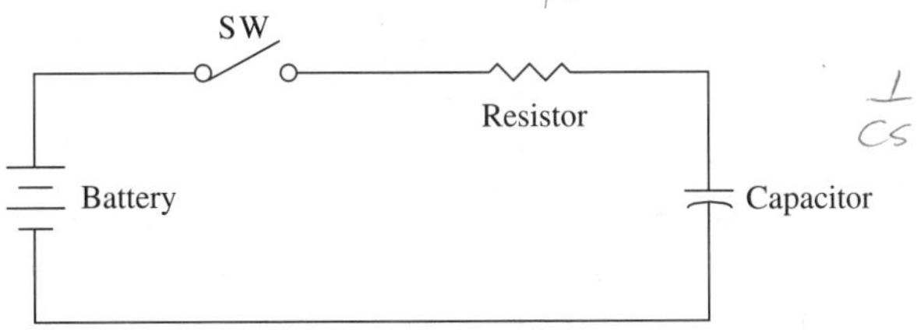

Figure 4.6 Series RC circuit.

The voltage in the circuit at time t is given by

$$\text{capacitor voltage} + \text{resistor voltage} = \text{applied voltage}$$

$$v_C + v_R = e$$

The transfer function depends upon the desired output variable. Two separate transfer functions will be developed: one for capacitor voltage and the other for resistor current.

Capacitor Voltage Because voltage across resistor is not a desired quantity, it needs to be replaced with a term containing a reference to capacitor voltage. Expanding the resistor voltage in terms of current gives

$$v_C + Ri = e$$

Capacitor current i is same as the current through the resistor. Capacitor current is related to capacitor voltage through the relation $i = C dv_C/dt$. Replacing the current variable i,

$$v_C + R\left(C\frac{dv_C}{dt}\right) = e$$

$$RC\frac{dv_C}{dt} + v_C = e$$

This differential equation fully describes the behavior of capacitor voltage upon application of an external voltage e.

Transfer Function: By taking the Laplace transform of both sides of the equation and simplifying, a transfer function of the RC circuit can be obtained.

$$\mathbf{L}\left[RC\frac{dv_C}{dt} + v_C\right] = \mathbf{L}[e]$$

$$RCsV_C(s) + V_C(s) = E(s)$$

$$V_C(s)(RCs + 1) = E(s)$$

$$V_C(s) = \frac{1}{(RCs + 1)}E(s)$$

$$\frac{\text{output}}{\text{input}} = \frac{V_C(s)}{E(s)} = \frac{1}{(RCs + 1)} = \text{TF}$$

Using the transfer function, a black box approach can be developed, where output (capacitor) voltage can be determined for any applied input voltage (Figure 4.7).

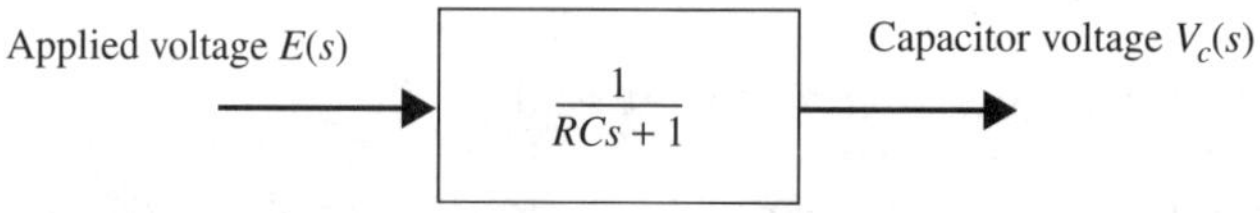

Figure 4.7 TF: capacitor voltage versus applied voltage.

Resistor Current The transfer function related to current in the circuit can be developed similarly. Again starting with an equation for voltage across the circuit,

$$v_C + v_R = e$$

Here the variable of interest is the current through the resistor. Neither the resistor voltage across v_R nor the capacitor voltage, v_C, is the desired variable; both need to be replaced with terms containing circuit current. Capacitor voltage is related to capacitor current through

$$v_C = \frac{1}{C}\int i\,dt$$

Substituting it in the equation,

$$\frac{1}{C}\int i\,dt + Ri = e$$

This differential equation fully describes the current flow in the circuit from time $t = 0$ to steady-state condition upon application of an external voltage e.

Transfer Function: Taking the Laplace transform of both sides of the equation and simplifying to develop the transfer function gives

$$\mathbf{L}\left[\frac{1}{C}\int i\,dt + Ri\right] = \mathbf{L}[e]$$

$$\frac{I(s)}{Cs} + RI(s) = E(s)$$

$$I(s)\frac{(1 + RCs)}{Cs} = E(s)$$

$$I(s) = \frac{Cs}{(RCs + 1)}E(s)$$

$$\frac{\text{output}}{\text{input}} = \frac{I(s)}{E(s)} = \frac{Cs}{(RCs + 1)} = \text{TF}$$

This transfer function allows the output current to be determined at any instant as a result of an applied (input) voltage (Figure 4.8).

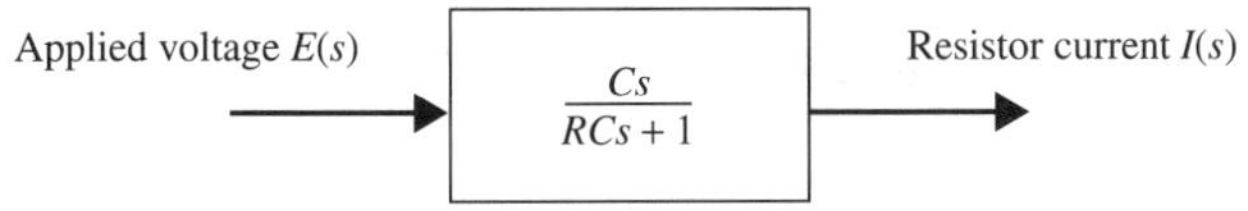

Figure 4.8 Series *RC* TF resistor current versus applied voltage.

Series *RL* Circuit

Figure 4.9 shows a series *RL* circuit. Initially, there is no current flow in the circuit and the switch is in open position as shown. At time $t = 0$, the switch is placed in the closed position and voltage *e* is applied. Current starts to increase from an initial condition of 0 A. It eventually reaches a maximum value and remains in this steady-state condition. An equation relating inductor voltage (or current in the circuit), from start to final steady-state value, can be developed.

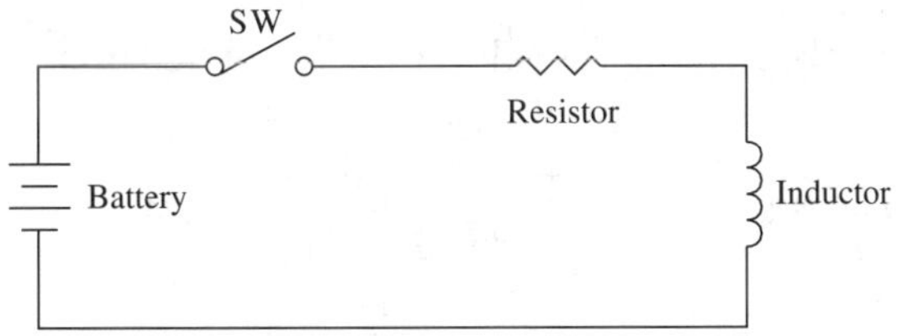

Figure 4.9 Series *RL* circuit.

Equating the voltage drops in the circuit gives

$$\text{inductor voltage} + \text{resistor voltage} = \text{applied voltage}$$
$$v_L + v_R = e$$

Inductor Voltage The term v_L will be retained, whereas the v_R term will be changed to relate to inductor voltage. Expanding the v_R term,

$$v_L + Ri = e$$

Since the resistor current is same as the inductor current and inductor current is related to inductor voltage through $i = \frac{1}{L}\int v_L\,dt$, substituting the value of current i gives

$$v_L + R\left(\frac{1}{L}\int v_L\,dt\right) = e$$
$$v_L + \frac{R}{L}\int v_L\,dt = e$$

This integral equation describes the behavior of inductor voltage upon application of an external voltage e.

Transfer Function: By taking the Laplace transform of both sides of the equation and simplifying,

$$\mathbf{L}\left[v_L + \frac{R}{L}\int v_L\,dt\right] = \mathbf{L}[e]$$
$$V_L(s) + \frac{R}{L}\frac{V_L(s)}{s} = E(s)$$
$$V_L(s)\left(1 + \frac{R}{Ls}\right) = E(s)$$
$$V_L(s) = \left(\frac{Ls + R}{Ls}\right)E(s)$$
$$\frac{\text{output}}{\text{input}} = \frac{V_L(s)}{E(s)} = \left(\frac{Ls + R}{Ls}\right) = \text{TF}$$

This transfer function relates inductor voltage to the applied input voltage (Figure 4.10).

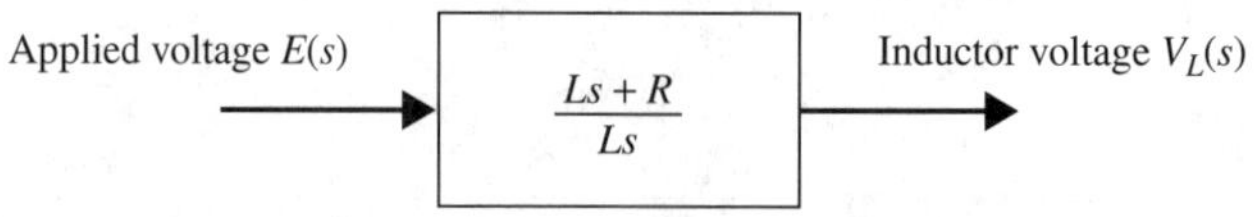

Figure 4.10 Series RL TF inductor voltage versus applied voltage.

Resistor Current A similar approach can be taken if a relationship between resistor current and applied voltage is desired. Starting with the basic voltage equation,

$$v_L + v_R = e$$

Inductor voltage is related to inductor current through $v_L = L\frac{di}{dt}$, so

$$L\frac{di}{dt} + Ri = e$$

This is the differential equation for current flow in the circuit.

Transfer Function Taking the Laplace transform of both sides of the differential equation and simplifying,

$$L\,sI(s) + RI(s) = E(s)$$

$$I(s)(L\,s + R) = E(s)$$

$$I(s) = \frac{1}{(L\,s + R)}E(s)$$

$$\frac{\text{output}}{\text{input}} = \frac{I(s)}{E(s)} = \frac{1}{(L\,s + R)} = \text{TF}$$

See Figure 4.11.

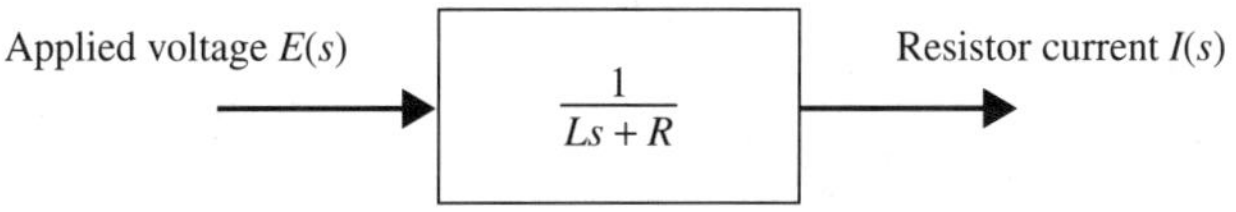

Figure 4.11 Series RL TF current versus applied voltage.

Series *RLC* Circuit

Figure 4.12 shows a series *RLC* circuit. A mathematical model relating the circuit current to the applied voltage will be developed. Equating the voltage drops in the circuit,

$$v_L + v_C + v_R = e$$

Replacing voltages across *R*, *L*, and *C* with common circuit current,

$$L\frac{di}{dt} + \frac{1}{C}\int i\,dt + Ri = e$$

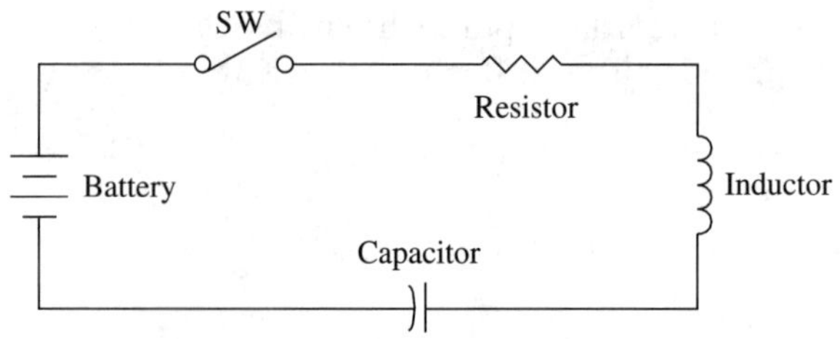

Figure 4.12 Series RLC circuit.

Taking the Laplace transform of both sides gives

$$L\,s\,I(s) + \frac{1}{Cs} I(s) + R\,I(s) = E(s)$$

$$I(s)\left(\frac{LCs^2 + RCs + 1}{Cs}\right) = E(s)$$

$$I(s) = \left(\frac{Cs}{LCs^2 + RCs + 1}\right) E(s)$$

$$\frac{\text{output } I(s)}{\text{input } E(s)} = \frac{Cs}{LCs^2 + RCs + 1} = \text{TF}$$

Circuit current at any instant can be determined through this transfer function (Figure 4.13).

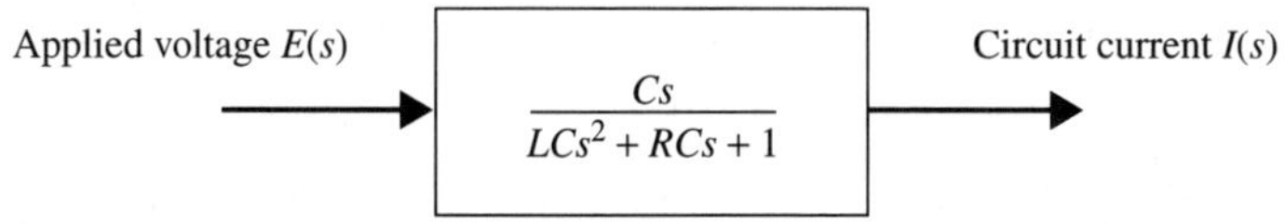

Figure 4.13 Series RLC circuit TF current versus applied voltage.

EXAMPLE 4.2

Determine the transfer function of a series *RLC* circuit relating circuit current to applied voltage, when subjected to an input voltage of 10 V dc. Given that $R = 2.2\,\text{k}\Omega$, $C = 4.7\,\mu\text{F}$, and $L = 1$ mH.

Solution

The transfer function ($V_{in} \rightarrow$ current) of a series *RLC* circuit is given by the following:

$$\frac{I(s)}{E(s)} = \frac{Cs}{LCs^2 + RCs + 1}$$

Substituting the values of *R*, *L*, and *C* in this equation gives

$$\text{TF} = \left(\frac{4.7 \times 10^{-6}s}{4.7 \times 10^{-9}s^2 + 1.034 \times 10^{-2}s + 1}\right)\text{A/V}$$

EXAMPLE 4.3

Determine the new transfer function if the circuit is subjected to an input voltage of 12 V instead of 10.

Solution

Transfer function of an *RLC* circuit depends only on the value of its components, *R*, *L*, and *C*, and not on applied input voltage. Thus, changing the applied voltage to 12 V instead of 10 V will not change the transfer function.

EXAMPLE 4.4

Determine the transfer function of the circuit in Figure 4.14 for output versus input voltage. Output voltage is measured across the *RC* filter R_2C.

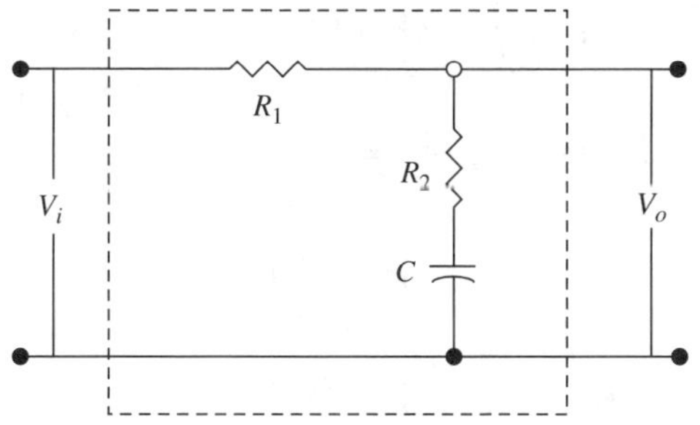

Figure 4.14 Circuit for Example 4.4.

Solution

Let the current through the circuit at any instant be given by i. Writing the equation for voltage drop in the circuit,

$$v_{R1} + v_{R2} + v_c = v_i$$

$$R_1 i + R_2 i + \frac{1}{C}\int i\,dt = v_i \qquad \text{(a)}$$

Similarly, the equation for output voltage is

$$v_{R2} + v_c = v_o$$

$$R_2 i + \frac{1}{C}\int i\,dt = v_o \qquad \text{(b)}$$

Taking the Laplace transform of Equation (a),

$$L\left[R_1 i + R_2 i + \frac{1}{C}\int i\,dt\right] = L\,[v_i]$$

$$R_1 I(s) + R_2 I(s) + \frac{1}{C}\frac{I(s)}{s} = V_i(s)$$

$$I(s)\left[R_1 + R_2 + \frac{1}{Cs}\right] = V_i(s)$$

$$I(s)\left[\frac{(R_1 + R_2)Cs + 1}{Cs}\right] = V_i(s)$$

$$I(s) = \frac{Cs}{(R_1 + R_2)Cs + 1} V_i(s) \qquad \text{(c)}$$

Similarly, taking the Laplace transform of Equation (b),

$$L\left[R_2 i + \frac{1}{C}\int i\,dt\right] = L\,[v_o]$$

$$R_2 I(s) + \frac{1}{C}\frac{I(s)}{s} = V_o(s)$$

$$I(s)\left[R_2 + \frac{1}{Cs}\right] = V_o(s)$$

$$I(s)\left[\frac{R_2 Cs + 1}{Cs}\right] = V_o(s)$$

$$I(s) = \frac{Cs}{R_2 Cs + 1} V_o(s) \qquad \text{(d)}$$

Using Equations (c) and (d), variable $I(s)$ can be eliminated.

$$\frac{Cs}{R_2Cs+1}V_o(s) = \frac{Cs}{(R_1+R_2)Cs+1}V_i(s)$$

Simplifying,

$$V_o(s) = \frac{R_2Cs+1}{(R_1+R_2)Cs+1}V_i(s)$$

Thus the transfer function relating output voltage to input voltage is

$$\text{TF} = \frac{V_o(s)}{V_i(s)} = \frac{R_2Cs+1}{(R_1+R_2)Cs+1}$$

EXAMPLE 4.5

a. Determine the transfer function of the preceding circuit with the following component values: $R_1 = 1\ \text{k}\Omega$, $R_2 = 680\ \Omega$, $C = 0.22\ \mu\text{F}$.
b. Determine an expression for output voltage for a step input of 10 V dc.

Solution

a. The circuit transfer function as obtained above,

$$\text{TF} = \frac{V_o(s)}{V_i(s)} = \frac{R_2Cs+1}{(R_1+R_2)Cs+1}$$

Substituting the component values,

$$\text{TF} = \frac{V_o(s)}{V_i(s)} = \frac{680 \times 0.22 \times 10^{-6}s+1}{(1000+680) \times 0.22 \times 10^{-6}s+1}$$

$$\text{TF} = \frac{V_o(s)}{V_i(s)} = \frac{1.5 \times 10^{-4}s+1}{3.7 \times 10^{-4}s+1}$$

b. Output voltage in the s-domain is

$$\text{output} = \text{TF} \times \text{input}$$

$$V_o(s) = \text{TF} \times V_i(s)$$

$$V_o(s) = \frac{1.5 \times 10^{-4}s+1}{3.7 \times 10^{-4}s+1} \times V_i(s)$$

This expression is valid for any type of input voltage. Since an input of 10 V dc is applied,

$$v_I = 10\text{ V}$$
$$V_i(s) = L[v_i]$$
$$V_i(s) = L[10]$$
$$V_i(s) = \frac{10}{s}$$

Substituting the value of $V_i(s)$ in the expression for output voltage,

$$V_o(s) = \frac{1.5 \times 10^{-4}s + 1}{3.7 \times 10^{-4}s + 1} \times \frac{10}{s}$$

The expression in real time (time domain) is obtained by taking the inverse transform:

$$v_o = L^{-1}[V_o(s)]$$
$$v_o = L^{-1}\left[\frac{1.5 \times 10^{-4}s + 1}{3.7 \times 10^{-4}s + 1} \times \frac{10}{s}\right]$$
$$v_i = 10 - 5.95e^{-2706t}$$

MECHANICAL SYSTEMS

A linear-translation, passive mechanical system has three main components: spring, mass, and damper. A car's suspension system consists of a wheel (mass), spring, and a shock absorber (damper). For the purpose of analysis, the components are assumed to contain only pure elements, that is, a spring is considered to be a pure spring with zero mass and is assumed to dissipate no energy during compression or extension.

Mass

According to Newton's second law of motion, force applied to a mass results in acceleration of mass given by

$$\text{applied force} = \text{mass} \times \text{resultant acceleration}$$
$$f = Ma$$

This results in a transfer function (Figure 4.15):

$$\text{TF} = \frac{\text{output acceleration } A(s)}{\text{applied force } F(s)} = \frac{1}{M}$$

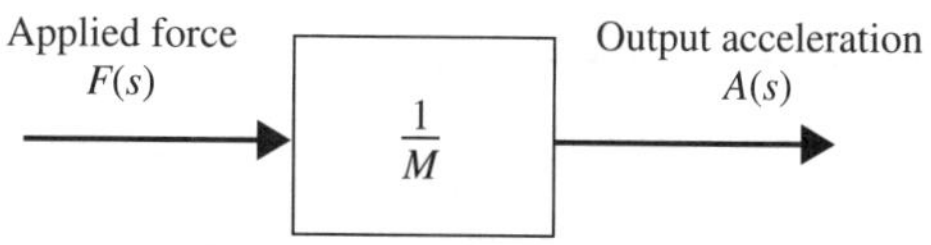

Figure 4.15 TF of mass acceleration versus force.

Mass can also be related to resultant change in velocity by

$$\text{applied force} = \text{mass} \times \text{rate of change of velocity}$$

$$f = M\frac{dv}{dt}$$

Taking the Laplace transform of both sides,

$$F(s) = Ms \cdot V(s)$$

This leads to a transfer function of the form

$$\text{TF} = \frac{\text{output velocity } V(s)}{\text{applied force } F(s)} = \frac{1}{Ms}$$

Similarly, from the mass and displacement relation,

$$f = M\frac{d^2x}{dt^2}$$

$$\text{TF} = \frac{\text{output displacement } X(s)}{\text{applied force } F(s)} = \frac{1}{Ms^2}$$

Spring

When a spring is stretched or compressed, it develops a force related to its spring stiffness (K) and total displacement from the equilibrium position:

$$f = Kx$$

The transfer function for a spring is thus

$$\text{TF} = \frac{\text{output displacement } X(s)}{\text{applied force } F(s)} = \frac{1}{K}$$

Damper

A damper is a mechanical component that provides damping in the motion control systems. Some form of damping is usually desirable in most motion control systems, because damping reduces vibrations and the tendency to hunt (back-and-forth movement about the target position). This type of damper (viscous) provides a resistance only to movement. Force (resistance) developed by a viscous damper is directly proportional to velocity.

$$f = Bv$$

$$f = B\frac{dx}{dt}$$

where B = damping coefficient.

Taking the Laplace transform of both sides,

$$F(s) = Bs \cdot X(s)$$

Thus the damper transfer function is

$$\mathrm{TF} = \frac{\text{output displacement } X(s)}{\text{applied force } F(s)} = \frac{1}{Bs}$$

Spring Mass Damper System

A spring mass damper system is shown in Figure 4.16. The system is fixed at ends between two supports. Mass is supported by the spring and is free to oscillate about the position of rest. The frequency at which these oscillations take place is a function of the spring stiffness and the inertial mass and is known as the natural frequency (ω_n) of the system. An equation of motion relating vertical motion of mass to applied force will be developed.

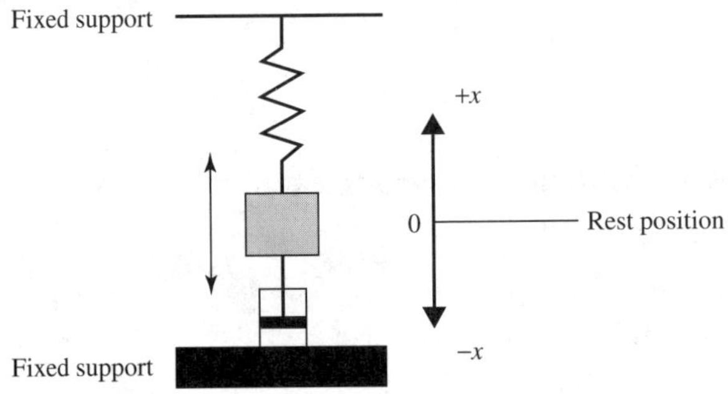

Figure 4.16 Spring mass damper system.

When not in motion, mass rests at the equilibrium position (zero position). The weight of the mass is balanced by an extension of the spring. Let a downward force F be applied to the mass. It will start to move from rest position (accelerate), and in doing so it will resist the movement. The spring and the damper will also resist the motion.

Four different forces are at work here:

- Spring force (force = stiffness × displacement)
- Mass (inertia) force (resistance = mass × acceleration)
- Damping force (resistance = damping coefficient × velocity)
- External force F

Equating all the forces,

$$\text{mass force} + \text{damping force} + \text{spring force} = F$$

$$M\frac{d^2x}{dt^2} + B\frac{dx}{dt} + Kx = F$$

This differential equation fully describes the behavior of this spring, mass, and damper system.

Transfer Function By taking the Laplace transformation of both sides and simplifying, a transfer function can be developed that relates the movement of mass to an external force:

$$\mathbf{L}\left[M\frac{d^2x}{dt^2} + B\frac{dx}{dt} + Kx\right] = \mathbf{L}[F]$$

$$Ms^2X(s) + BsX(s) + KX(s) = F(s)$$

Because the exact nature of the external force is not known, $F(s)$ is used to indicate the transform of the forcing function (Figure 4.17).

$$X(s)(Ms^2 + Bs + K) = F(s)$$

$$X(s) = \frac{1}{(Ms^2 + Bs + K)}F(s)$$

$$X(s) = TFxF(s)$$

$$\text{TF} = \frac{1}{(Ms^2 + Bs + K)}$$

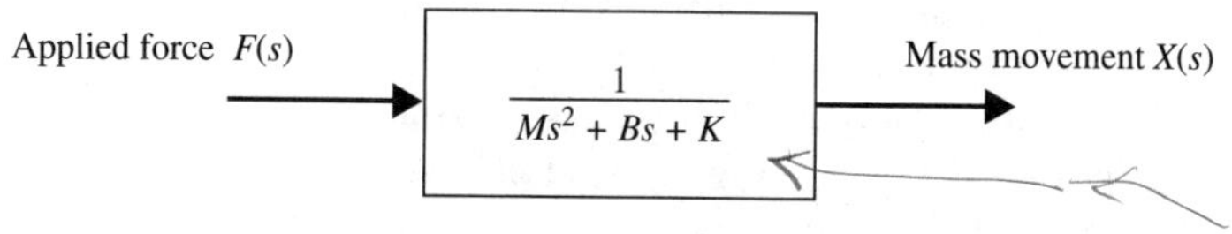

Figure 4.17 TF of spring mass damper system.

ANALOGY BETWEEN PHYSICAL SYSTEMS

Components from different physical systems exhibit similar behavior (Table 4.1) and are modeled by almost identical differential equations. A resistor from the electrical system, a damper from the mechanical system, and a thermal conductor (insulator) all show similar behavior.

- A resistor dissipates electrical energy; a damper dissipates mechanical energy; a thermal conductor dissipates thermal energy.
- The voltage drop across a resistor is given by a linear product of current and its resistance value ($V = RI$). Force across a damper is given by linear product of velocity and damping coefficient ($F = BV$); the temperature differential across a thermal conductor is given by the linear product of thermal resistance and heat-flow rate.

Table 4.1 Analogy Between Physical Systems

Element Type	***Electrical***	***Mechanical***	***Thermal***
Energy dissipater	Resistor $i = \frac{1}{R}v$	Damper $f = Bv$	Thermal conductor $q = \frac{1}{R}\Delta t$
Energy storage	Capacitor $i = C\frac{dv}{dt}$	Mass $f = M\frac{dv}{dt}$	Heat-storage capacity $q = C\frac{d(\Delta t)}{dt}$
Energy storage	Inductor $i = \frac{1}{L}\int v\,dt$	Spring $f = K\int u\,dt$	(No equivalent element)
	i = Current v = Voltage R = Resistance C = Capacitance L = Inductance	f = Force v = Velocity u = Displacement B = Damping coefficient M = Mass K = Spring stiffness	q = Heat quantity t = Temperature R = Thermal resistance C = Heat capacity

Once a system has been replaced by its transfer function, the apparent attributes related to its physical nature are lost. It is not possible to determine the physical nature of a system from the knowledge of its transfer function. But the transfer function retains all the characteristics (pertinent to control system operation) of its components and fully describes the operating behavior, which is of concern in analysis. Once the overall transfer function of a control system is obtained, it can be drawn as a black box, with just one input and one output. This is true even for a control system, which may incorporate components and subsystems from different disciplines.

FLUID SYSTEMS

Fluid systems contain tanks and reservoirs connected through pipes, fittings, valves, and flow-measuring elements. All these elements can be broadly classified as falling in one of the two categories.

Flow Element: Fluid Resistance Element Fluid flow can be either *laminar* or *turbulent*. A dimensional number known as the *Reynolds number*, R_e, indicates the type of flow that may exist. The Reynolds number takes into consideration the operating parameters of pipe diameter, fluid viscosity, and fluid velocity. Flow is considered to be laminar when the value of R_e is below 2000 and turbulent for numbers greater than 5000. For a number between 2000 and 5000, the flow type lies in between the two states, or is what is known as *transient flow*.

For laminar flow conditions, the flow rate is directly proportional to the *pressure head* (pressure).

$$\text{flow rate } q = \text{pressure head } h\,/\,\text{flow resistance } R$$

For turbulent flow conditions, the flow rate is a nonlinear function of the pressure head. For simplified analysis of fluid systems, it can be assumed that flow conditions are laminar, even though it is not entirely true. The unit of fluid resistance is N·s /m^3.

Tank: Fluid Capacitance Element A hydraulic tank behaves very much like an electrical capacitor. An electrical capacitor stores electrical charge (integral of current flow over time), but a fluid capacitor stores fluid volume (integral of flow rate over time). Electrical capacitance increases with an increase in plate size, and so does fluid capacitance with an increase in tank dimensions. The unit of fluid capacitance is m^3/N.

Fluid capacitance C relates to pressure head h developed at the base and net tank inflow rate q through the equation

$$h = \frac{1}{C}\int q\,dt$$

or

$$C\frac{dh}{dt} = q$$

Single-Tank Fluid-Level System

Tank fluid-level control is very common in the process industry. A model will be developed for pressure at the base of the tank related to flow rate into the tank (Figure 4.18). Fluid flows into the tank at an inflow rate of q_i. The fluid flow rate out of the tank is q_o.

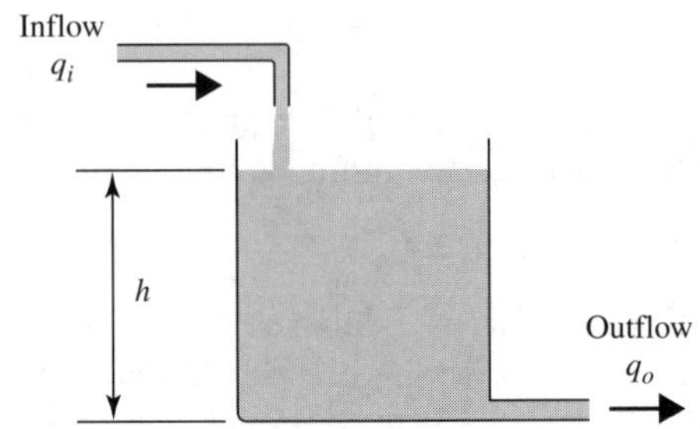

Figure 4.18 Tank-level control.

Thus, the net inflow rate into the tank is given as

$$\text{net in flow rate} = q_i - q_o$$

From the expression for tank capacitance, it is related to the rate of change of the pressure head (pressure at the base of the tank):

$$C\frac{dh}{dt} = q_i - q_o$$

Rearranging the terms,

$$C\frac{dh}{dt} + q_o = q_i$$

Outflow, q_o, is dependent upon the tank pressure head, h, and outflow resistance, R. Thus, replacing the outflow term, q_o, gives

$$C\frac{dh}{dt} + \frac{h}{R} = q_i$$

This differential equation describes the mathematical relationship between the applied inflow rate and the resultant tank pressure head. Taking the Laplace transform of the equation,

$$CsH(s) + \frac{H(s)}{R} = Q_i(s)$$

$$H(s)\left[Cs + \frac{1}{R}\right] = Q_i(s)$$

$$\text{TF}: \quad \frac{H(s)}{Q_i(s)} = \frac{R}{RCs + 1}$$

This is the transfer function of a tank-level system relating the applied inflow rate and the resultant tank pressure head.

Two-Tank Fluid-Level System

In a process industry, several process vessels are interconnected, and output from one system feeds the input to the next. It is very common to find the output of a tank-level system feeding to a downstream tank. In such a case, the level in the downstream tank is going to be affected by the outflow from upstream tank. But outflow from upstream tank is a function of its tank level. Thus level of upstream tank affects the level in the downstream tank. Consequently, two tank-level systems can be classified into two categories.

Noninteracting Tanks If two tanks are set up such that fluid level in the downstream tank does not affect the fluid-level dynamics of the upstream tank, the system is known as a noninteracting system. In this case, the fluid level of the upstream tank can be determined without any consideration of the downstream tank (Figure 4.19).

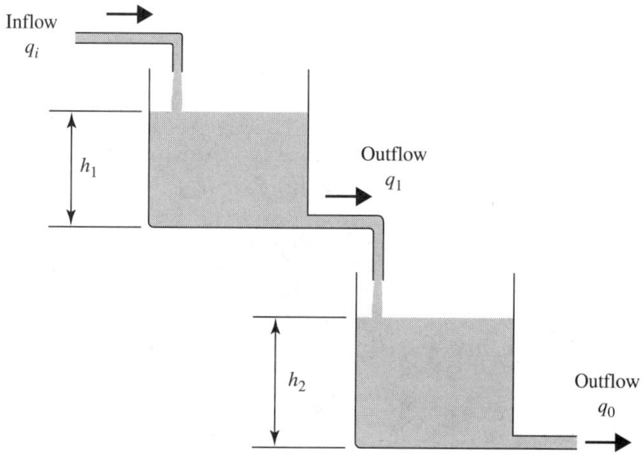

Figure 4.19 Noninteracting tanks.

Following are the variables used in the analysis:

- h_1 Fluid level in the upstream tank (1)
- h_2 Fluid level in the downstream tank (2)
- q_i Fluid flow rate into tank 1
- q_1 Fluid flow rate out of tank 1 and into tank 2
- q_2 Fluid flow rate out of tank 2
- R_1 Resistance of the tank 1 outlet
- R_2 Resistance of the tank 2 outlet

Tank 1:

$$\text{net inflow into the tank} = q_i - q_1$$

But this is related to the rate of change of the fluid level:

$$C_1 \frac{dh_1}{dt} = q_i - q_1$$

$$C_1 \frac{dh_1}{dt} + q_1 = q_i$$

q_1 is related to the pressure head (fluid level) and outlet resistance:

$$q_1 = \frac{h_1}{R_1}$$

Substituting the value into the flow equation,

$$C_1 \frac{dh_1}{dt} + \frac{h_1}{R_1} = q_i$$

Tank 2: The flow equation for tank 2 can be similarly obtained, keeping in mind that flow rate into tank 2 is same as the outflow from tank 1 and is specified as q_1.

$$C_2 \frac{dh_2}{dt} + \frac{h_2}{R_2} = q_1$$

These two equations specify the relationship between the flow rate and level for the two tanks.

Overall Transfer Function: It should be pointed out that outflow from tank 1 (q_1) and fluid level in tank 1 (h_1) are both intermediate variables. It is desired to find a relationship between inflow q_i (into tank 1) and the level of tank 2, and the two intermediate variables are of no concern.

Variable q_1 can be eliminated first. Replacing it in terms of tank 1 pressure head and outlet resistance, we have

$$C_2 \frac{dh_2}{dt} + \frac{h_2}{R_2} = \frac{h_1}{R_1}$$

Taking the Laplace transform of the first equation gives

$$C_1 s H_1(s) + \frac{H_1(s)}{R_1} = Q_i(s)$$

Simplifying,

$$H_1(s)\frac{(R_1C_1s+1)}{R_1} = Q_i(s)$$

$$H_1(s) = \frac{R_1}{(R_1C_1s+1)} \times Q_i(s)$$

Similarly, taking the Laplace transform of the equation for tank 2 and simplifying,

$$H_2(s) = \frac{R_2}{(R_2C_2s+1)} \times \frac{H_1(s)}{R_1}$$

Substituting the expression for $H_1(s)$ into the preceding equation yields

$$H_2(s) = \frac{R_2}{(R_2C_2s+1)} \times \frac{1}{R_1} \times \frac{R_1}{(R_1C_1s+1)} \times Q_1(s)$$

$$H_2(s) = \frac{R_2}{(R_2C_2s+1)(R_1C_1s+1)} \times Q_1(s)$$

This expression relates the final level in tank 2 to inflow q_i into the system. The transfer function can now be developed:

$$\frac{H_2(s)}{Q_1(s)} = \frac{R_2}{(R_2C_2s+1)(R_1C_1s+1)}$$

This is the overall transfer function of two noninteracting tanks. The expression contains all the characteristics (C_1, C_2, R_1, and R_2) of both tanks but does not retain any intermediate variables (q_1 and h_1).

Interacting Tanks If fluid level in tank 2 affects the fluid level in tank 1, the system is known as an interacting system. It is not possible to determine fluid level in one tank without consideration of the flow dynamics of the second tank (Figure 4.20).

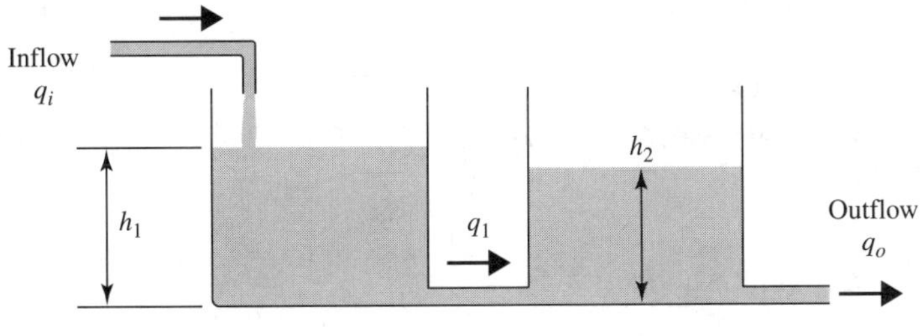

Figure 4.20 Interacting tanks.

The overall transfer function between level h_2 in tank 2 and the inflow q_i can be evaluated; it is

$$\frac{H_2(s)}{Q_1(s)} = \frac{1}{(R_2 R_1 C_2 C_1 s^2 + R_2 C_2 s + R_1 C_1 s + R_2 C_1 s + 1)}$$

THERMAL SYSTEMS

Thermal systems are concerned with transfer of heat from one body to another. These systems are governed by laws of thermodynamics and heat transfer. Heat flows from a body at a higher temperature to a body at a lower temperature. As in fluid systems, there are two dynamic components: thermal capacitance and thermal resistance.

Thermal Capacitance

Thermal capacitance is the capacity of an object to store heat.

$$C = M \times c$$

where

$$C = \text{thermal capacitance}$$
$$M = \text{mass of object}$$
$$c = \text{specific heat of object.}$$

Thermal capacitance is also related to the rise in temperature with heat transfer

$$C = \frac{dT}{dt} = q$$

where

$$C = \text{thermal capacitance}$$
$$\frac{dT}{dt} = \text{rate of change in temperature}$$
$$q = \text{heat transfer rate.}$$

Thermal capacitance is analogous to fluid capacitance as well as to electrical capacitance. Its units are BTU/°F and Kcal/°C.

Thermal Resistance

Thermal resistance is the change in temperature for a unit change in heat-flow rate.

$$\text{thermal resistance} = \frac{\text{change in temperature}}{\text{heat-flow rate}}$$

$$R = \frac{\Delta T}{q}$$

Thermal resistance is analogous to electrical resistance and fluid resistance. Its units are °F/BTU/s and °C/Kcal/s (Figure 4.21).

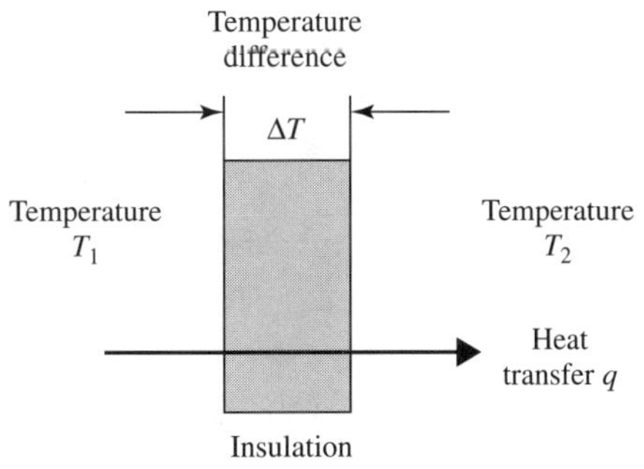

Figure 4.21 Thermal resistance.

Heating Systems

Consider the case of an object being heated by an external energy source. For the sake of simplicity, it can be assumed that the object is a fluid mass completely filling the internal cavity. The heater unit can be electric, steam, or hot water (Figure 4.22). The object is covered on all sides by a layer of insulating material. Because the object temperature is higher than that of outside, there is going to be a loss of heat. This loss will have a tendency to lower the temperature of the object.

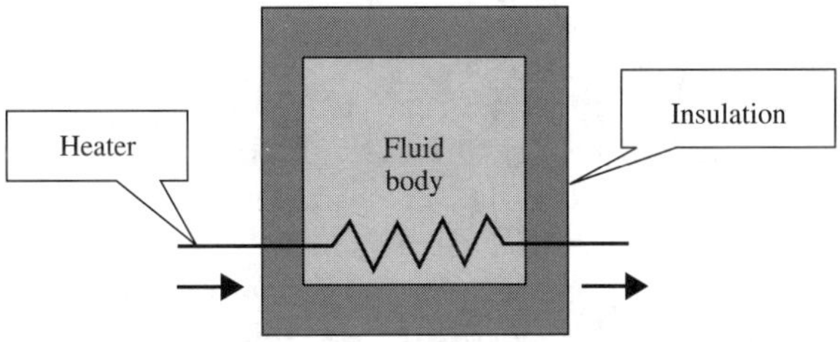

Figure 4.22 Heating system.

We assume the following variables:

- q_i Rate at which heat is added to the object by the heater
- q_o Rate at which heat is lost to ambient air through insulation
- T_1 Temperature of object
- T_2 Ambient temperature
- R Thermal resistance of insulation

- C Thermal capacity of object
- M Mass of object
- c Specific heat of object (fluid)

The thermal capacity of the object is

$$C = M \cdot c$$

Heat loss through insulation is

$$q_o = \frac{T_1 - T_2}{R}$$

The net rate at which heat is gained by the object is

$$\text{rate of heat addition} - \text{rate of heat loss} = q_i - q_o$$

As a result of heat gained by the object, its temperature is going to rise; this change is related to its heat capacity and given by the equation

$$C\frac{dT_1}{dt} = q_i - q_o$$

Substituting the value of the heat-loss rate, q_o,

$$C\frac{dT_1}{dt} = q_i - \frac{T_1 - T_2}{R}$$

$$RC\frac{dT_1}{dt} + T_1 = R \cdot q_i + T_2$$

This differential equation describes the heat transfer for a simple system. It should be noted that the equation contains two input variables, q_i, the heat added to system, and T_2, the temperature of the external environment. It is not possible to write a single transfer function with two inputs. Classical control system theory permits only one input and one output. *Failure to handle more than one input and output is one of the weaknesses of classical control theory, and this is where modern control system theory (state variable analysis) and even fuzzy logic control stand out.* Fortunately, the principle of superposition can be applied and two separate transfer functions can be developed, one with q_i as input and the other with T_2 as the input variable.

PROBLEMS

1. The input/output relationship of a control system component is described by a nonlinear equation. Explain if Laplace transforms can be used to develop a transfer function for it.

2. A series *RLC* circuit is subjected to an undefined input voltage. Write down the differential equation for capacitor voltage.
3. Express the equation in Problem 2 as a transfer function.
4. Modify the differential equation in Problem 2 for an input voltage of 10 sin 100t.
5. Develop a transfer function for capacitor voltage V_c and applied input V_1 for the following circuit.

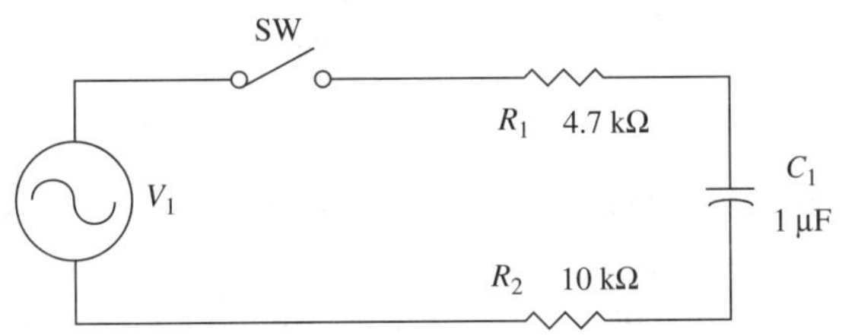

6. Develop a transfer function for the following noninteracting circuit.

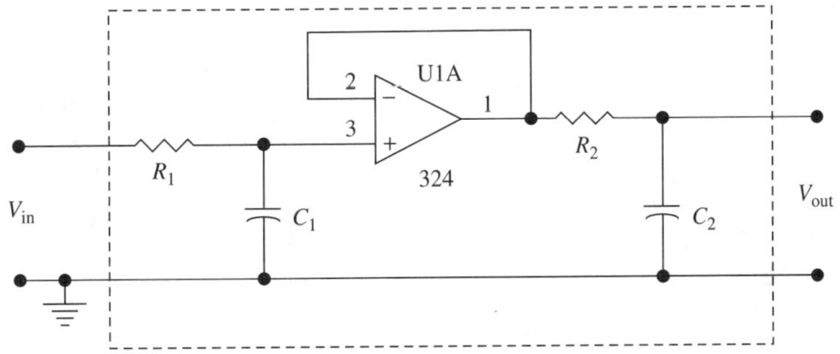

7. Identify the electrical equivalent of each of the following.
   a. Fluid capacitance
   b. Fluid resistance
   c. Fluid pressure
   d. Mechanical spring
   e. Mechanical mass
   f. Mechanical damper
8. Develop a transfer function for the vertical displacement of the following spring (K) and mass (M) system. Assume that there is no aerodynamic drag and the

amount of damping is zero. What is the behavior of the mass when a unit impulse input is applied to the mass in the downward direction?

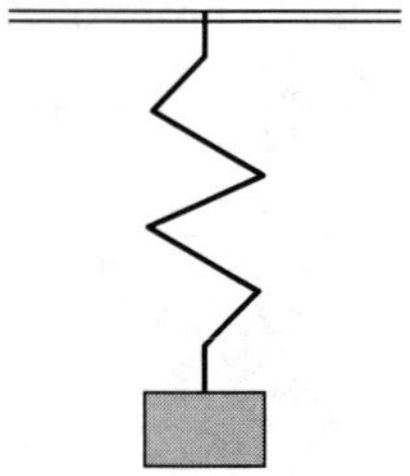

9. Explain if it is possible to develop a transfer function for a system that is being subjected to two inputs.

10. What is the significance of the analogy between electrical systems and other systems, such as mechanical, thermal, and fluid?

5

TRANSIENT RESPONSE

INTRODUCTION

Once a mathematical model has been derived for a control system, a number of techniques can be applied to analyze the system performance. Transient response (time response) analysis and frequency response analysis are two of the most popular methods of analysis. This section introduces the basics of transient response; the frequency response analysis is covered in Chapter 6. Characteristic equations, s-planes, and pole-zero mappings are also introduced. The time response of a system when subjected to an input is dependent upon its characteristics, which are, in turn, completely described by its transfer function. (Real) time response provides the ultimate test of system performance, because the (real-world) events take place as functions of time. Through transient response analysis, the real-time behavior can be determined without ever subjecting the actual system to a real-time input. Stability (Chapter 8) of a system can also be related to its transient response, the characteristic equation and pole-zero map.

TIME RESPONSE

Time Response

Time response is the output response (in time) of a system when subjected to an input.

A control system is always subjected to a dynamic input (set point) and load (disturbance) conditions. In some control systems the input is changing continuously, e.g., a servomotor controlling the movement of a pen in a plotter or a sun-tracking system in a modern solar-power plant. In some other systems, changes in input may be less frequent, but the load on the system may be changing all the time, as in a home heating system.

The time response (Figure 5.1) of a system can be broken down as follows.

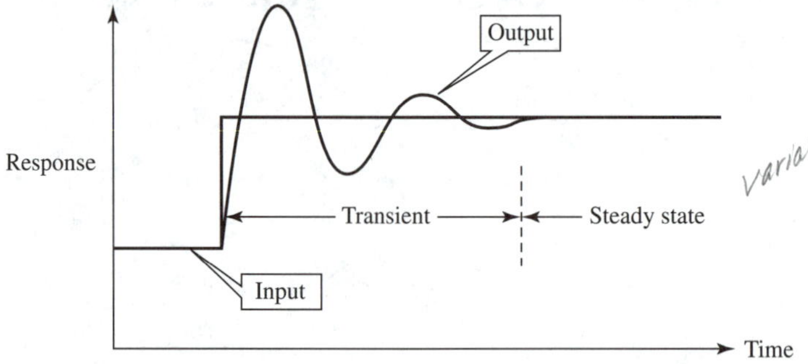

Figure 5.1 Control system: time response.

Total Response

Total response is the response of a system from the time of application of an input to the point when time approaches infinity.

Total response consists of two parts: the transient response and steady-state response.

Transient Response

Transient response is the response of a system from the time of application of input to the time when output has reached steady state.

The transient response of a system is a function of its transfer function.

Steady-State Response

The **steady-state response** is the system response after the transients have died and output has settled (time response after transient response).

Steady-state response is also known as *forced response*. Forced response is a function of applied input.

Sometimes control system output never settles to a steady-state value. The output may either oscillate continuously or, worse still, may continue to increase without any bounds. Such uncontrolled behavior is not at all desirable. A nuclear reactor's temperature control system is normally a stable control system, but under certain conditions, when commanded for increased output, temperature (reactivity) may increase without bound, resulting in a meltdown (Chernobyl).

In the study of control systems, transient response is more significant than steady-state response. It dictates the manner in which the steady-state condition is reached. At the same time, steady-state response indicates the residual error (offset) in the controlled variable. *In this text, the term transient response is used synonymously with time response, or total response.*

INPUT FUNCTIONS

The input signal (or, rather, the range of signals) to which a control system is going to be subjected during its operation in the field is not known in advance. For the purpose of analysis and comparison of various systems, a set of standard signals has been chosen. These input signals are completely described as mathematical functions. These functions are the

- Step function
- Pulse function →more often a disturbance
- Impulse function
- Ramp function
- Sinusoidal function

These signals (functions) can be easily generated in the field as well as in laboratory practice and are used for testing, analysis, and design of control systems.

Step Function

> ***Step Function***
> The **step function** is a function of time, which has a zero value before $t = 0$ and has a constant value for all time $t \geq 0$.

Mathematically speaking,

$$f(t) = 0 \quad \text{for} \quad t < 0$$
$$f(t) = K \quad \text{for} \quad t \geq 0$$

K = 1

See Figure 5.2.

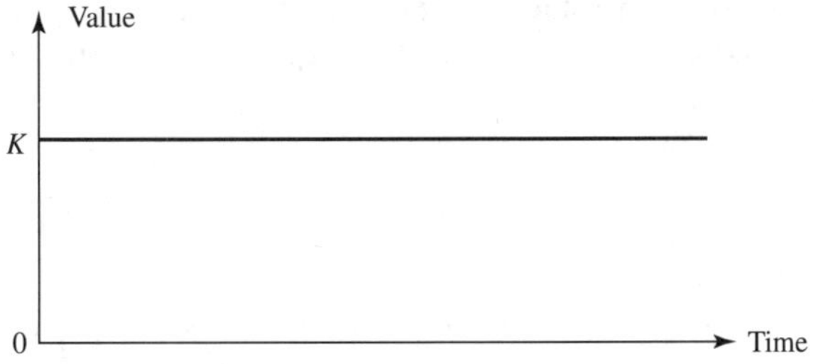

Figure 5.2 Step function.

The step function commonly occurs in control systems. Flicking of a light switch, a quick change of a thermostat setting, and turning on an electric motor are a few examples of step functions.

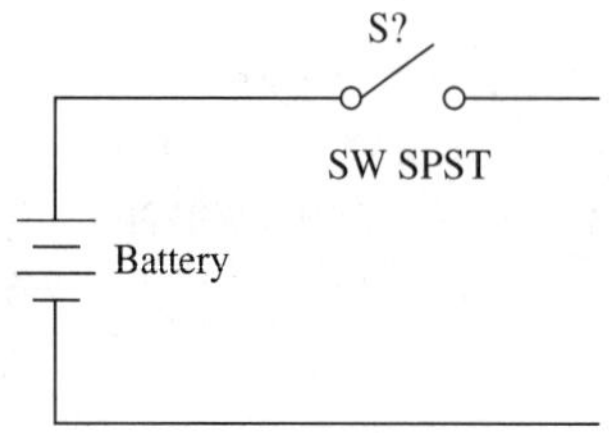

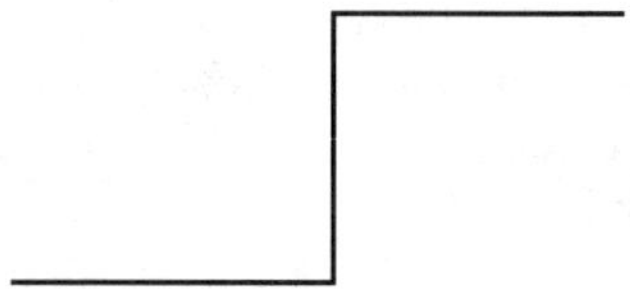

Unit Step Function

A **unit step function** is a function of time that has zero magnitude before time $t = 0$, and unit magnitude after that.

It is commonly represented by the symbol $u(t)$ (Figure 5.3).

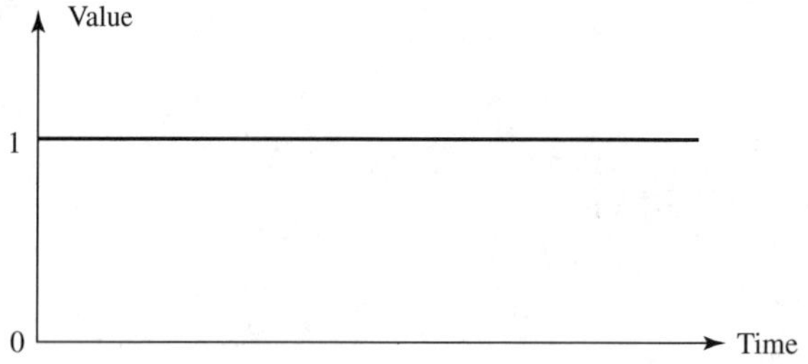

$$f(t) = 0 \quad \text{for} \quad t < 0$$
$$f(t) = 1 \quad \text{for} \quad t \geq 0$$

Figure 5.3 Unit step function.

Delayed Step Function

A **delayed step function,** $F(t - a)$, is a function of time that has a zero magnitude before $t = 0$ and a constant amplitude after that.

See Figure 5.4.

EXAMPLE 5.1

A step function has zero amplitude before $t = 0$ and 1.25 after that. Express this function mathematically.

Solution

$$f(t) = 0 \quad \text{for} \quad t < 0$$
$$f(t) - 1.25 \quad \text{for} \quad t \geq 0$$

$$f(t-a) = 0 \quad \text{for} \quad t < a$$
$$f(t-a) = K \quad \text{for} \quad t \geq a$$

Figure 5.4 Delayed step function.

EXERCISES

5.1 Draw the unit step function that has a zero value until $t = 2$ s.

5.2 A step function has an amplitude of zero until $t = 1$ s and an amplitude of 2.5 after that. Express this function mathematically.

Pulse Function

A pulse function is derived from a step function.

> ***Pulse Function***
> A **pulse function** is the difference between a step function and a delayed step function.

See Figure 5.5.

EXERCISE

5.3 A pulse function has an amplitude of 1.5 between $t = 0$ s and $t = 5$ s. Express this function mathematically, and draw a graph representing this function.

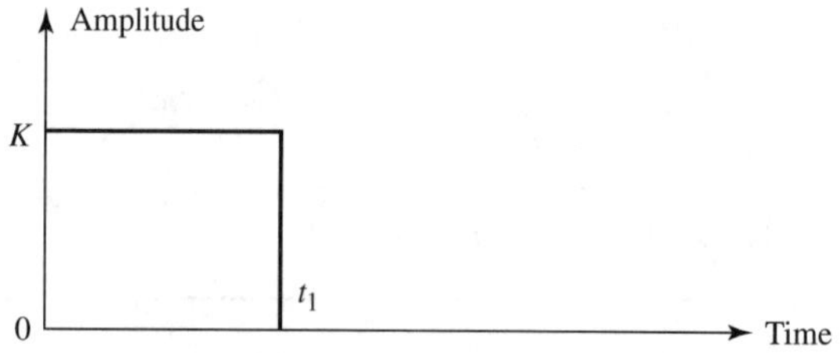

$f(t) = 0 \quad t < 0$
$f(t) = K \quad 0 \le t < t_1$
$f(t) = 0 \quad t \ge t_1$

Figure 5.5 Pulse function.

Impulse Function

Impulse Function $\delta(t)$
A function which has zero amplitude everywhere, except at the origin, where it has infinite amplitude [or where the amplitude is infinity].

Delayed Impulse Function $\delta(t-\tau)$
A function which has zero amplitude everywhere except at time $t = \tau$, where the amplitude is infinity.

An impulse function, $\delta(t)$, is also known as a Dirac delta function. It is zero everywhere except at time $t = 0$ (Figure 5.6).

$$\delta(t) = 0 \qquad t \neq 0$$
$$\delta(t) = \infty \qquad t = 0$$
$$\int_{-\infty}^{\infty} \delta(t)\, dt = 1$$

It is essentially a pulse function with unity area (Figure 5.7) but with a zero width (time). It does not exist in nature in pure form, but lightning strikes and sudden impacts can be approximated to fall in this category.

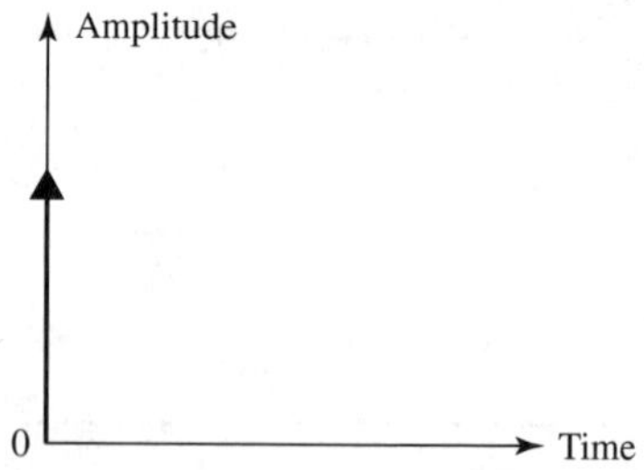

Figure 5.6 Impulse function.

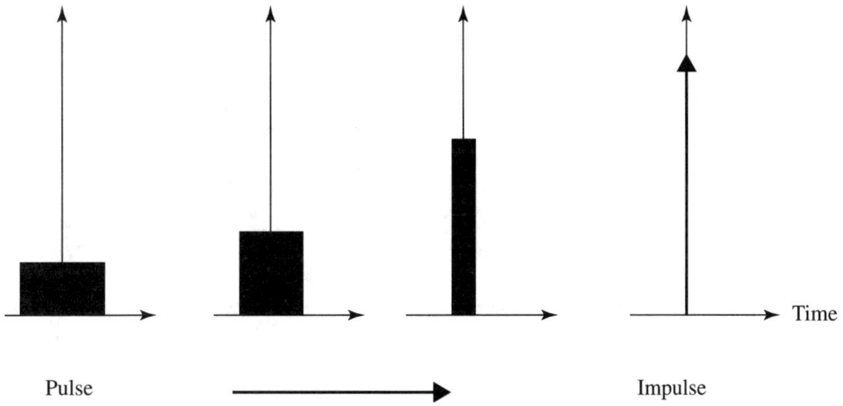

Figure 5.7 Pulse to impulse function.

Delayed Impulse Function

A delayed impulse function, $\delta(t - a)$, occurs only at the instant when time $t = a$ and is zero everywhere else (Figure 5.8).

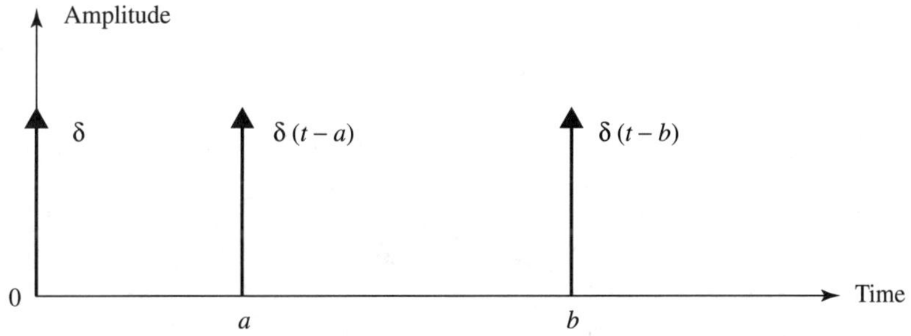

Figure 5.8 Delayed impulse function.

EXAMPLE 5.2

Identify the times at which the following functions exist.

a. $\delta(t)$
b. $\delta(t - 10)$
c. $\delta(t + 2)$

Solution

a. It exists at $t = 0$.
b. It exists at $t = 10$.
c. It exists at $t = -2$.

EXERCISE

5.4 Draw a graph for each of the following impulse functions.

a. $\delta(t - 1.50)$
b. $\delta(t - 0.10)$

Ramp Function

Ramp Function
A **ramp function** is a function whose amplitude increases linearly with time.

In metal processing, the ingot soaking-pit temperature is increased in a ramplike fashion (Figure 5.9).

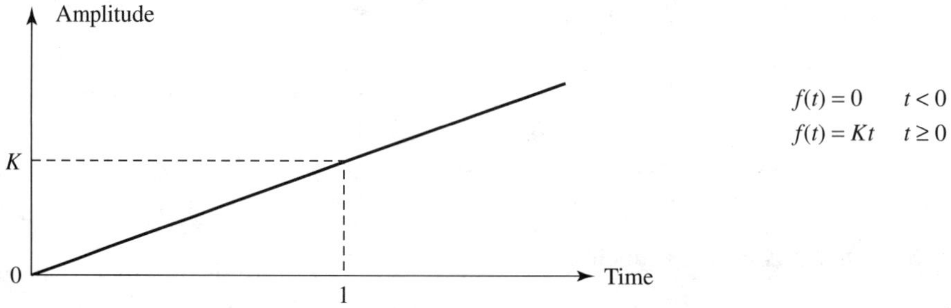

Figure 5.9 Ramp function.

Sinusoidal Function

Sinusoidal functions are commonly used for testing the response of electrical networks and audio, video, and industrial control systems. They are also the input signals used for frequency response analysis (Figure 5.10).

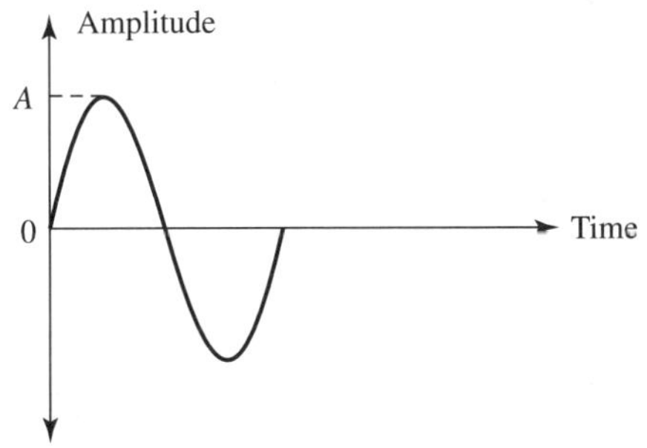

Figure 5.10 Sinusoidal function.

Sinusoidal Function

A **sinusoidal function** is a function of time, which is periodically changing and is described by

$$f(t) = 0 \qquad t < 0$$
$$f(t) = A \sin \omega t \qquad t \geq 0$$

where

$A =$ peak amplitude
$\omega = 2\pi f$
$f =$ frequency

TYPES OF TRANSIENT RESPONSES

In spite of the enormous variation in transfer functions and physical natures of control systems, the transient response falls in one of six categories. The output responses from six different systems are shown here. Because overall response is also dependent upon the *forcing* (input) function, it is assumed that every system is subjected to a unit step input.

Overdamped Response

In an overdamped response, as the system is subjected to a step input, the output rises slowly and reaches the final (steady-state) value after a long period of time (Figure 5.11). There are no oscillations, and the output never overshoots the final value. In a milling machine, even a small amount of overshoot will result in tool chattering or gouging of material, ruining the surface finish. This type of behavior is typical of control systems used in machine tools, printers, and plotters (*stable response*).

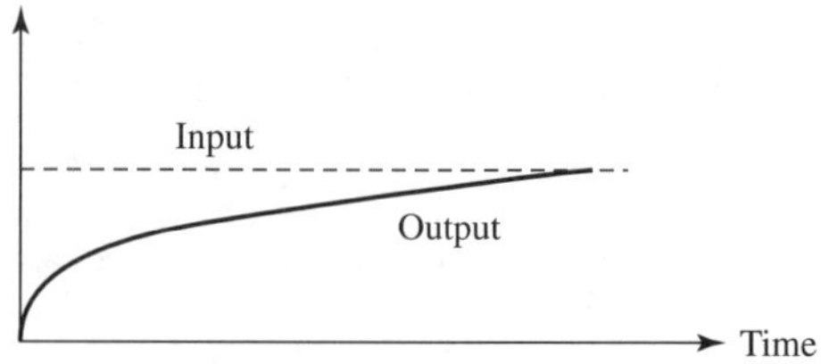

Figure 5.11 Overdamped response.

Critically Damped Response

Critically damped response is almost identical to overdamped response, but with one subtle difference. The time needed to reach the final value is shorter than that for an overdamped system. Steady-state output is reached in the shortest possible time without any oscillation or overshoot (Figure 5.12). Critically damped response is highly desirable in a number of applications but it is difficult to achieve in practice. Theoretically speaking, a system can be designed and tuned for critically damped behavior, but there is a possibility that with passage of time, system characteristics may change. As a result, the system, which initially operated as a critically damped unit, may later exhibit underdamped behavior (with some overshoot). Thus, to ensure oscillation-free operation, a slightly overdamped response is preferred (*stable response*).

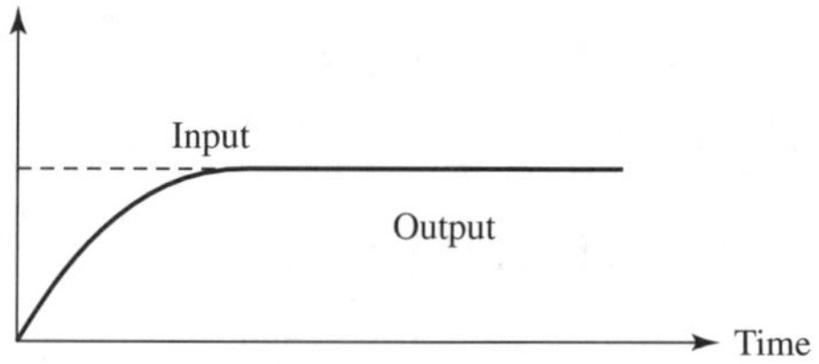

Figure 5.12 Critically damped response.

Underdamped Response

An underdamped type of response is characterized by fast rise, overshoot, undershoot, and continued oscillations before the output settles down to the final value (Figure 5.13). This type of response results in the shortest time to *first* reach the target value, but the amount of time taken to settle down may be much longer than that for an underdamped or even an overdamped system. This is perhaps the most popular type of response, especially in the process industry. The amount of maximum overshoot that occurs at the first peak of oscillation and the settling time are some of the criteria for system design and controller tuning (*stable response*).

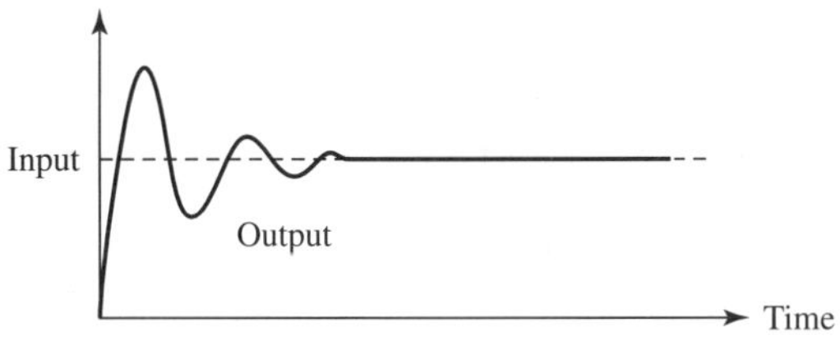

Figure 5.13 Underdamped response.

Constant Oscillations

Sometimes when a system is subjected to a step input, the output will rise quickly, overshoot the final (target) value, and continue to oscillate back and forth about this final value without ever settling down (Figure 5.14). This is more or less a hypothetical behavior, because most of the physical systems contain some amount of damping to cause the oscillations to eventually die out.

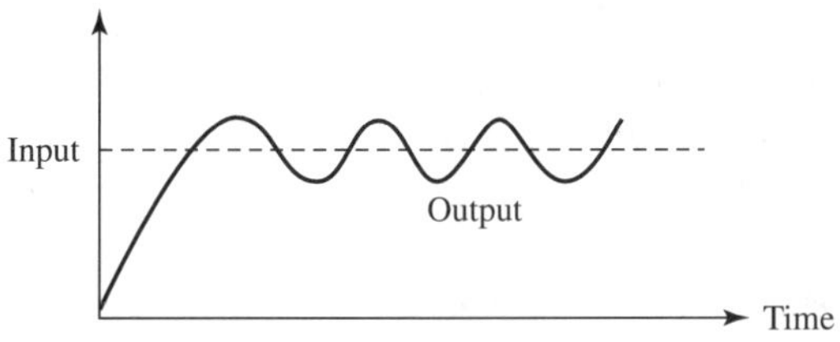

Figure 5.14 Constant oscillations.

Theoretically, a combination of a pure spring and a pure mass will exhibit this behavior. Once set in motion, the mass will continue to oscillate forever in the absence of any resistance. In practice, as the spring stretches and compresses, energy is dissipated as heat. The aerodynamic drag on the moving mass and spring coils further robs the spring mass system of energy, and the overall system response tends to be underdamped rather than with constant oscillations. A similar type of response is exhibited by an electrical system containing a pure inductor and a pure capacitor. This type of response is obviously not desirable, except in vibrators and oscillators, where lost energy is replenished through a continuous intake from an external energy source (*quasi-stable response*).

Increasing Oscillations

In the case of increasing oscillations, the system output rises quickly, overshoots the final value, and continues to oscillate back and forth (Figure 5.15). The output

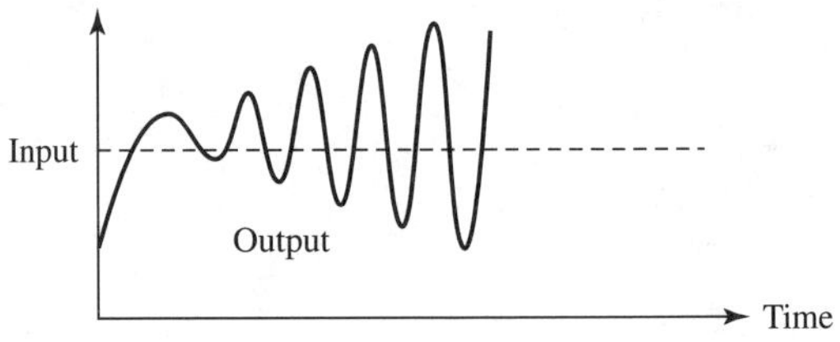

Figure 5.15 Increasing oscillations.

oscillations continue to increase in amplitude with each successive cycle. Eventually, the system is brought to a halt either through a failure of the physical system or through some catastrophic event. This type of behavior is highly undesirable (*unstable response*).

Exponential Rise

The exponential rise is the most severe type of system response (Figure 5.16). The output continues to rise in an exponential manner without ever reaching a final value. For a physical system, the system is brought to a halt either through a failure of the physical system or through some catastrophic event. Some chemical (explosions) and thermonuclear (nuclear reactor meltdown, nuclear bomb) operations fall in this category. This type of response is totally undesirable in control systems (*unstable response*).

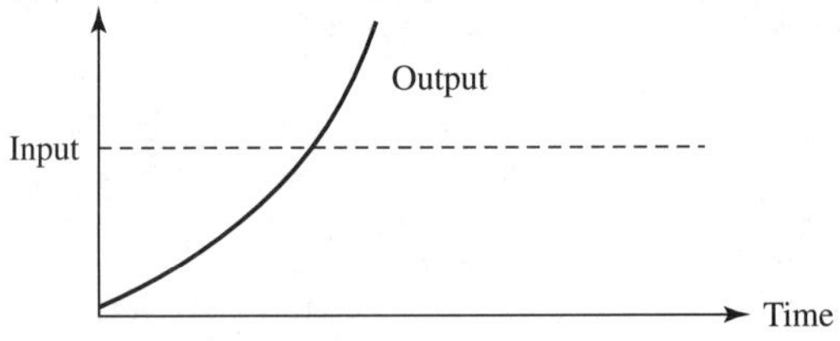

Figure 5.16 Exponential rise of response.

TRANSIENT RESPONSE THROUGH LAPLACE TRANSFORMATION

Laplace transformation can be applied to determine the transient response of a linear system. The process is fairly simple if the system transfer function is known. Otherwise, transformation can be applied to the system linear differential equation along with the forcing function, and the transient response can be extracted from that. Both these possibilities are considered here.

Transfer Function not Known

- Obtain (or derive) the differential equation describing the system behavior.
- Include the forcing (test/input) function in the equation.
- Take the Laplace transform of the equation.
- Isolate the desired variable.
- Take the inverse transform, which gives the time response.

EXAMPLE 5.3

Determine the output of the given integrator when subjected to a dc input voltage of 1.5 V ($R = 100\,\text{k}\Omega$ and $C = 4.7\,\mu\text{F}$). See Figure 5.17.

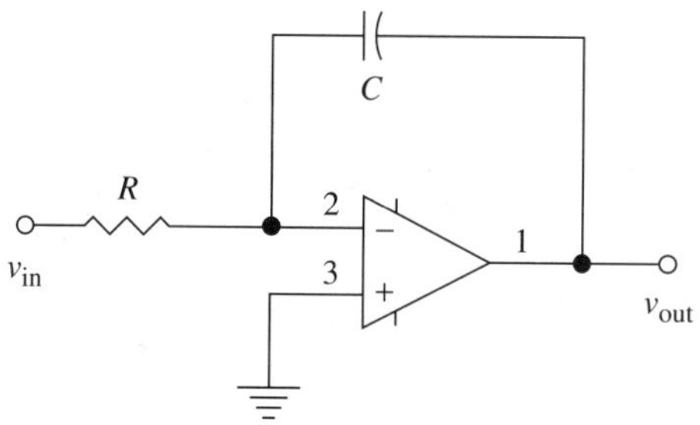

Figure 5.17 Schematic diagram for Example 5.3.

Solution

Integrator output is given by

$$v_{out} = \frac{1}{RC}\int v_{in}\,dt$$

Substituting the values,

$$v_{out} = \frac{1}{100 \times 10^3 \times 4.7 \times 10^{-6}}\int 1.5\,dt$$

$$v_{out} = 2.128\int 1.5\,dt$$

Taking the Laplace transform of the equation,

$$\mathbf{L}[v_{out}] = \mathbf{L}\left[2.128 \int 1.5\, dt\right]$$

$$\mathbf{L}[v_{out}] = 3.192\mathbf{L}\left[\int 1\, dt\right]$$

$$V_{out}(s) = 3.192\frac{1}{s^2}$$

$$V_{out}(s) = \frac{3.192}{s^2}$$

Taking the inverse transform,

$$v_{out} = \mathbf{L}^{-1}\left[\frac{3.192}{s^2}\right]$$

$$v_{out} = 3.192\,t$$

A constant (step) input voltage to the integrator results in a continuously rising (ramp) output (Figure 5.18). This is as expected from an integrator.

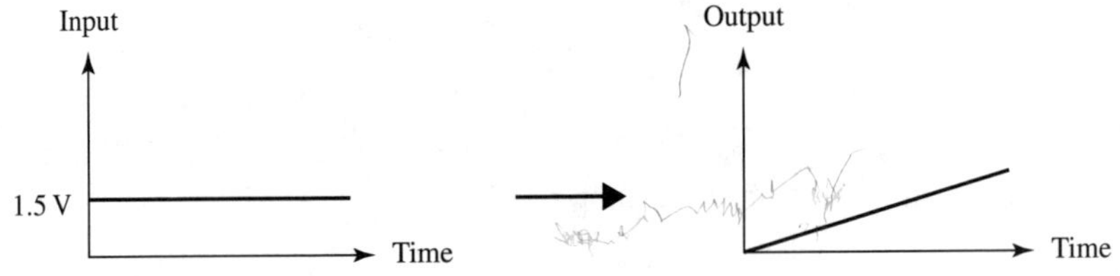

Figure 5.18 Integrator response to a step input.

Transfer Function Known

- Take the Laplace transform $R(s)$ of the forcing (test/input) function $r(t)$:

$$R(s) = \mathbf{L}[r(t)]$$

- Multiply the system transfer function, TF, by the transform of the forcing function, $R(s)$, to get the (controlled) output response, $C(s)$, in the s-domain.

$$C(s) = \text{TF} \times R(s)$$

- Take the inverse transform of the product. It provides the desired time response, $c(t)$:

$$c(t) = \mathbf{L}^{-1}[\text{TF} \times R(s)]$$

EXAMPLE 5.4

Current (i) in a series RL circuit, when subjected to an input voltage (e), is given by the following transfer function:

$$\frac{I(s)}{E(s)} = \frac{1}{L(s + R/L)}$$

Determine an equation for the current i when a dc voltage of 10 V is applied to the circuit ($R = 100\,\text{k}\Omega$ and $L = 1\,\text{mH}$).

Solution

Taking the transform of the input voltage, (1)

$$E(s) = \mathbf{L}[e(t)]$$

But $e(t) = 10\,\text{V}$; hence,

$$E(s) = \mathbf{L}[10]$$

$$E(s) = \frac{10}{s}$$

output voltage (in s-domain) = transfer function × input function (in s-domain)

$$I(s) = \text{TF} \times \text{input function}$$

$$I(s) = \frac{1}{L(s + R/L)}\,\frac{10}{s}$$

$$I(s) = \frac{10}{L\,s(s + R/L)}$$

Substituting the values of L and R, (2)

$$I(s) = \frac{10}{0.1\,s(s + 10{,}000)}$$

Taking the inverse transform gives the expression for current i in real time.

$$i(t) = \mathbf{L}^{-1}\left[\frac{100}{s(s+10{,}000)}\right]$$

$$i(t) = 100\mathbf{L}^{-1}\left[\frac{1}{s(s+10{,}000)}\right]$$

$$i(t) = \frac{100}{10{,}000}(1 - e^{-10000t})$$

$$i(t) = 0.01(1 - e^{-10000t})$$

EXAMPLE 5.5

Rotational velocity n of a dc servomotor, when subjected to a voltage e, is given by the following transfer function:

$$\frac{N(s)}{E(s)} = \frac{100}{s+20}$$

a. Determine the time equation for velocity when the motor is subjected to an input voltage of 90 V dc.
b. Determine the final (steady-state) velocity.

Solution

a. First, determine the s-domain equation; then take its inverse transform to determine the time equation for velocity.

$$N(s) = \text{TF} \times \text{input function}$$

$$N(s) = \frac{100}{s+20} \times E(s)$$

But $E(s) = \mathbf{L}[90]$, so

$$E(s) = \frac{90}{s}$$

Substituting the value of $E(s)$,

$$N(s) = \frac{100}{s+20} \times \frac{90}{s}$$

$$N(s) = \frac{9000}{s(s+20)}$$

The real-time expression for $n(t)$ is derived by taking the inverse transform.

$$n(t) = \mathbf{L}^{-1}\left[\frac{9000}{s(s+20)}\right]$$

$$n(t) = 450(1 - e^{-20t})$$

b. The steady-state velocity can be found in two possible ways.

- *Using the real-time equation*. Because the time equation is available, it can be solved for the case of t approaching infinity.

$$n(t \to \infty) = 450(1 - e^{-20 \times \infty})$$

$$n(\infty) = 450$$

- *Using the final value theorem*. This theorem allows the final value to be determined without taking the inverse transform.

$$n(t \to \infty) = \lim_{s \to 0}[sN(s)]$$

$$n(\infty) = \lim_{s \to 0}\left[s\frac{9000}{s(s+20)}\right]$$

$$n(\infty) = \lim_{s \to 0}\left[\frac{9000}{(s+20)}\right]$$

$$n(\infty) = \frac{9000}{(0+20)}$$

$$n(\infty) = 450$$

which is an identical result.

IMPULSE FUNCTION RESPONSE

The transient response $C(s)$ of a linear system, when subjected to an input function $R(s)$, is given as

$$C(s) = \text{system TF} \times R(s)$$

If the input function is an impulse function, $\delta(t)$, then

$$R(s) = \mathbf{L}[\delta(t)]$$
$$R(s) = 1$$

Substituting the value of $R(s)$ to get the final response gives

$$C(s) = \text{TF} \times 1$$
$$C(s) = \text{TF}$$

Thus, *the transfer function of a system is the same as its response to a unit impulse function in the s-domain.*

EXAMPLE 5.6

Determine the time response of the system in Figure 5.19 when it is subjected to a unit impulse.

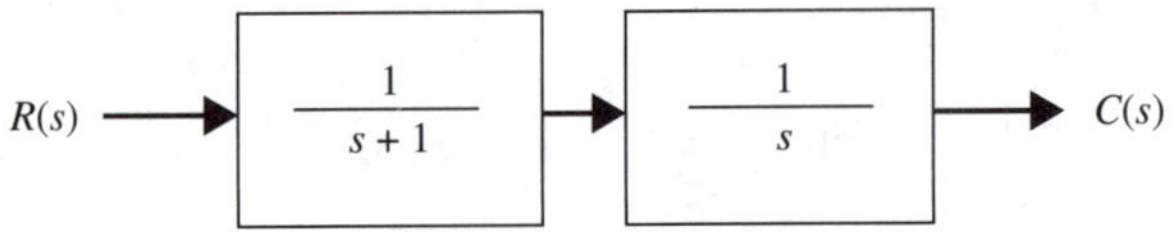

Figure 5.19 Diagram for Example 5.6.

Solution

The system time response can be determined from its response in the s-domain.

$$s\text{-domain response } C(s) = \text{TF}(s) \times R(s)$$

$$\text{TF}(s) = \frac{1}{(s+1)} \times \frac{1}{s}$$

$$\text{TF}(s) = \frac{1}{s(s+1)}$$

The input function is $R(s) = \mathbf{L}[\delta(t)] = 1$. Substituting the values,

$$C(s) = \frac{1}{s(s+1)} \times 1$$

$$C(s) = \frac{1}{s(s+1)} \qquad \text{(same as TF(s))}$$

Taking the inverse Laplace transform to obtain the real-time response gives

$$c(t) = \mathbf{L}^{-1}\left[\frac{1}{s(s+1)}\right]$$

$$c(t) = 1 - e^{-t}$$

See Figure 5.20.

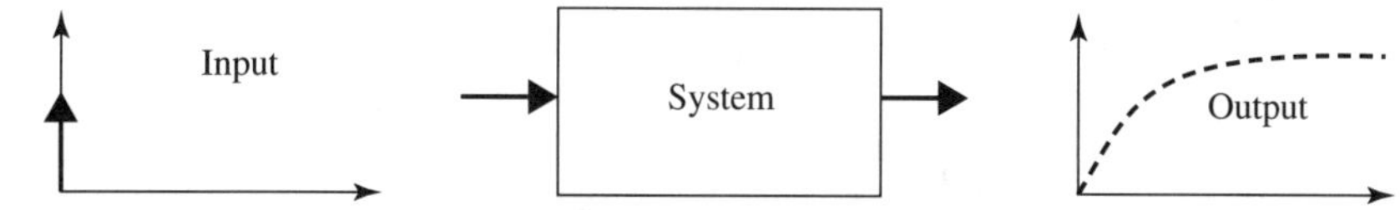

Figure 5.20 Time response, Example 5.6.

EXAMPLE 5.7

Determine the response of the closed-loop control system in Figure 5.21 to a unit step input.

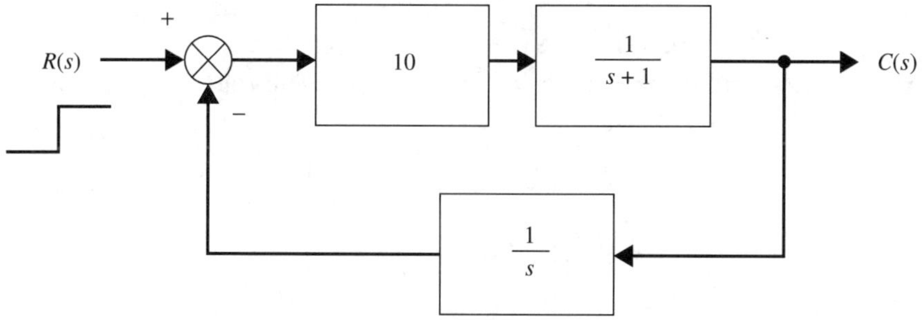

Figure 5.21 Block diagram for Example 5.7.

Solution

$$\text{system response } C(s) = \text{TF}(s) \times R(s)$$

$$\text{input function } R(s) = \mathbf{L}[r(t)]$$

$$= \frac{1}{s} \qquad (r(t) \text{ is a unit step input})$$

The control system is a negative-feedback closed-loop system, and its TF is given as

$$\text{TF}(s) = \frac{\text{forward gain}}{1 + \text{feedback gain} \times \text{forward gain}}$$

$$\text{TF}(s) = \frac{10\dfrac{1}{(s+1)}}{1 + 10\dfrac{1}{(s+1)}\dfrac{1}{s}}$$

$$\text{TF}(s) = \frac{10s}{s(s+1) + 10}$$

$$\text{TF}(s) = \frac{10s}{s^2 + s + 10}$$

Substituting the values,

$$C(s) = \text{TF}(s)R(s)$$

$$C(s) = \frac{10s}{(s^2 + s + 10)}\frac{1}{s}$$

$$C(s) = \frac{10}{(s^2 + s + 10)}$$

Taking the inverse Laplace transform of $C(s)$ to obtain the real-time function $c(t)$,

$$c(t) = \mathbf{L}^{-1}\left[\frac{10}{(s^2 + s + 10)}\right]$$

$$c(t) = 3.2e^{-0.5t} \sin 3.12\,t$$

The unit step response of the control system is shown in Figure 5.22. It is an underdamped response with a large overshoot and oscillations.

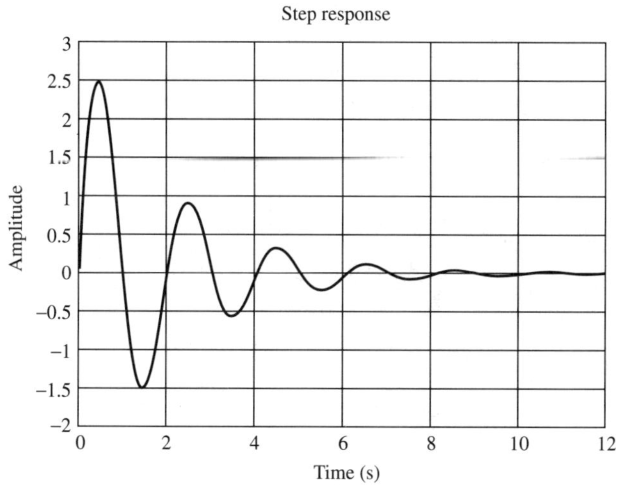

Figure 5.22 Unit step response, Example 5.7.

EXAMPLE 5.8

A dc motor–speed control system regulates the motor velocity based upon the control voltage applied to it. It is applied with a control voltage of 10 V dc. Determine the motor velocity at the following intervals:

a. time $t = 1\,\text{ms}$
b. time $t = 10\,\text{ms}$
c. time $t = 100\,\text{ms}$
d. time $t = 1\,\text{s}$
e. time t approaching infinity (steady-state velocity)

The motor transfer function (velocity/control voltage) is

$$\text{TF}(s) = \frac{200}{s+5}$$

Solution

Motor velocity at different time intervals can be determined from the real-time equation of motor velocity. The real-time expression for time response, in turn, can be obtained from its s-domain response.

$$\text{velocity}(s) = \text{TF}(s) \times \text{control voltage}(s)$$

$$\text{control voltage}(s) = \mathbf{L}[10]$$

$$\text{control voltage}(s) = \frac{10}{s}$$

Therefore,

$$\text{velocity}(s) = \frac{200}{s+5}\frac{10}{s}$$

$$\text{velocity}(s) = \frac{2000}{s(s+5)}$$

The real-time response is provided by the inverse transformation:

$$\text{velocity}(t) = \mathbf{L}^{-1}[\text{velocity}(s)]$$

$$\text{velocity}(t) = \mathbf{L}^{-1}\left[\frac{2000}{s(s+5)}\right]$$

$$\text{velocity}(t) = \frac{2000}{5}(1 - e^{-5t})$$

$$\text{velocity}(t) = 400(1 - e^{-5t})$$

a. The velocity at 1 ms after the application of control voltage can be obtained by substituting the value of t as 0.001 in this expression:

$$\text{velocity}(t = 0.001\ \text{s}) = 1.99\ \text{rev/min}$$

b. velocity ($t = 0.010$ s) = 19.51 rev/min
c. velocity ($t = 0.1$ s) = 157.39 rev/min
d. velocity ($t = 1$ s) = 397.30 rev/min
e. steady-state velocity (t approaches infinity) = 400 rev/min

The last result can also be derived from the final value theorem. Figure 5.23 shows the complete step response.

CHARACTERISTIC EQUATION

The transfer function of a system can be expressed as a ratio of two polynomials of s.

$$\text{transfer function} = \frac{N(s)}{D(s)}$$

Characteristic Equation

The **characteristic equation** is obtained by equating the denominator of the transfer function to zero.

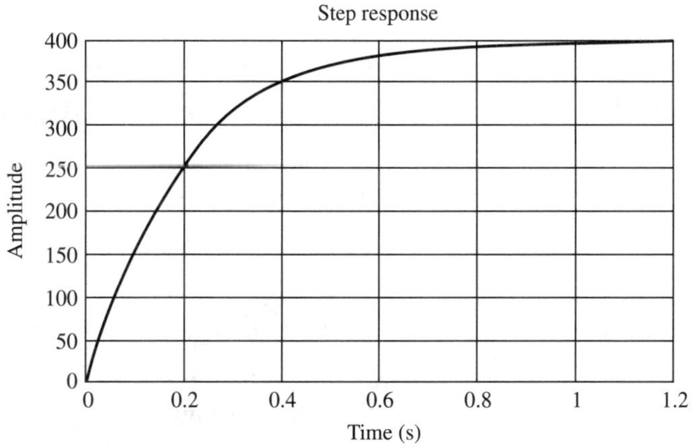

Figure 5.23 Step response for Example 5.8.

For the system with the preceding transfer function, the characteristic equation is

$$D(s) = 0$$

For a negative-feedback closed-loop system with forward gain of $G(s)$ and feedback gain of $H(s)$, the transfer function is

$$\text{CLTF} = \frac{G(s)}{1 + GH(s)}$$

The characteristic equation of the system is, therefore,

$$1 + GH(s) = 0$$

EXAMPLE 5.9

The transfer function of a low-pass electrical filter is given here. Determine the characteristic equation.

$$\text{TF} - \frac{12}{5s + 1}$$

Solution

The characteristic equation is obtained by equating the denominator of the transfer function to zero. Thus, the characteristic equation for the low-pass filter is

$$5s + 1 = 0$$

EXAMPLE 5.10

Determine the characteristic equation of a closed-loop system, given the following:

$$G(s) = \frac{s}{s+10} \quad \text{and} \quad H(s) = \frac{1}{0.5s+1}$$

Solution

Two different approaches can be taken to determine the characteristic equation.

Approach I: The closed-loop transfer function is first determined, and the characteristic equation is then determined by equating the denominator to zero.

$$\mathrm{TF} = \frac{G(s)}{1+GH(s)}$$

$$\mathrm{TF} = \frac{\dfrac{s}{s+10}}{1+\dfrac{s}{s+10}\dfrac{1}{0.5s+1}}$$

$$\mathrm{TF} = \frac{0.5s^2+s}{0.5s^2+11.5s+10}$$

Equating the denominator to zero, the characteristic equation is obtained as

$$0.5s^2+11.5s+10=0$$

Approach II: Without determining the closed-loop transfer function, the characteristic equation can be determined from the following:

$$1+GH(s)=0$$

Substituting the values of $G(s)$ and $H(s)$,

$$1+\frac{s}{s+10}\frac{1}{0.5s+1}=0$$

$$\frac{(s+10)(0.5s+1)+s}{(s+10)(0.5s+1)}=0$$

$$(s+10)(0.5s+1)+s=0$$

$$0.5s^2+11.5s+10=0$$

which is the same expression, obtained before.

POLES AND ZEROS OF THE TRANSFER FUNCTION

The terms *poles* and *zeros* have been adopted from algebraic equations and are used extensively in control system analysis and design. These represent the values of complex variable s that make the value of a transfer function infinite or zero, respectively.

Pole of a Transfer Function

- The **pole** is the root (solution) of the (characteristic) equation, obtained by setting the denominator polynomial of the transfer function equal to 0.
- The value of s that makes (the value of) the transfer function approach infinity (hence the term *pole* (rising to infinity)).
- A complex pole (pole with imaginary component) always appears as a complex conjugate pair.

Zero of a Transfer Function

- The **zero** is the root (solution) of the equation, obtained by setting the numerator polynomial of the transfer function equal to 0.
- The value of s that makes (the value of) the transfer function equal to zero (hence the term *zero*).

EXAMPLE 5.11

Determine the poles and zeros of the given transfer function:

$$\frac{s+1}{s(s+5)}$$

Solution

Poles: With $s = 0$ and $s = -5$, the transfer function approaches infinity, so there are two poles of transfer function ($s = 0, s = -5$).

These are also the solutions to the characteristic equation, which is

$$s(s+5) = 0$$

Clearly, there are two roots of this equation ($s = 0$ and $s = -5$)

Zeros: With $s = -1$, the numerator of the transfer function becomes zero, setting the value of the transfer function to zero as well. Thus, there is one (*finite*) zero at $s = -1$.

Note: From a theoretician's point of view, the number of zeros of a function should match the number of poles. Because in this example there is only one finite zero but two poles, it can be said that the remaining zero exists at infinity. This concept may seem abstract at this point, but it is an important one, and it is helpful in visualizing the path traversed by the poles of the system (root locus), because the location of these plays an important role in system performance.

EXAMPLE 5.12

The characteristic equation of a system is given here. Identify the poles of the system.

$$10s^2 + 4s + 15 = 0$$

Solution

The characteristic equation is of second degree, and there will be two roots (poles). Applying the quadratic formula, the roots are

$$s = \frac{-4 \pm \sqrt{4^2 - 4 \times 10 \times 15}}{2 \times 10}$$

$$s = 2 + j3.317 \quad \text{and} \quad s = 2 - j3.317$$

The poles are complex and appear as a complex conjugate pair.

EXAMPLE 5.13

Determine the poles and zeros of the closed-loop system in Figure 5.24, given

$$G_1(s) = \frac{0.1s + 1}{s} \quad \text{and} \quad G_2(s) = \frac{s + 1}{s^2 + 2s + 4}$$

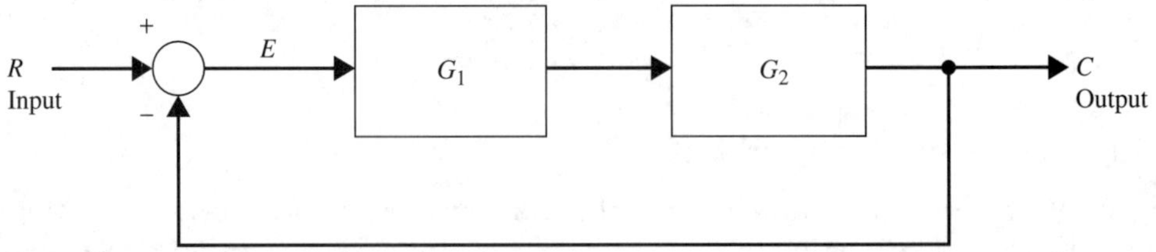

Figure 5.24 Block diagram for Example 5.13.

Solution

The system transfer function is

$$\mathrm{TF} = \frac{\dfrac{0.1s+1}{s}\,\dfrac{s+1}{s^2+2s+4}}{1+\dfrac{0.1s+1}{s}\,\dfrac{s+1}{s^2+2s+4}}$$

$$\mathrm{TF} = \frac{(0.1s+1)(s+1)}{s(s^2+2s+4)+1}$$

Zeros: The numerator of the transfer function is already in factored form, and the zeros can be directly identified. The two zeros are given by $0.1s+1=0$ and $s+1=0$. Thus, the zeros are $s=-10$ and $s=-1$.

Poles: The poles can be determined from the roots of the characteristic equation, which is

$$s(s^2+2s+4)+1=0$$
$$s^3+2s^2+4s+1=0$$

The characteristic equation is of third degree and thus there will be three roots (poles). Computer programs such as MATLAB and MathCad, programmable calculators, or numerical approximation techniques can be used.

The roots of the characteristic equation, and thus the poles of this system, are

$$s=-0.8576+j1.6661$$
$$s=-0.8576-j1.6661$$
$$s=-0.2848$$

---

s-PLANE AND POLE-ZERO MAP

s-Plane

An **s-plane** is a special complex plane used to plot the poles and zeros of a transfer function. The poles' locations on the *s*-plane indicate the transient performance of the system.

The imaginary axis component relates to the natural frequency of oscillation, and the real-axis component relates to the exponential term (Figure 5.25).

Pole-Zero Map

A **pole-zero map** is an *s*-plane diagram showing the location of poles and zeros of a system transfer function.

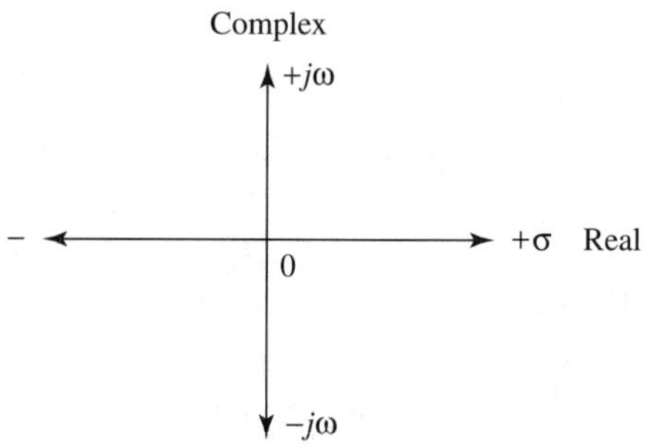

Figure 5.25 s-Plane.

EXAMPLE 5.14

Draw the pole-zero map of a unity gain closed-loop system, given

$$G(s) = \frac{s+1}{s^2+2s+10}$$

Solution

The system transfer function is

$$\mathrm{TF} = \frac{\dfrac{s+1}{s^2+2s+10}}{1+\dfrac{s+1}{s^2+2s+10}\times 1}$$

$$\mathrm{TF} = \frac{s+1}{s^2+3s+11}$$

Zeros: There is only one zero, at $s = -1$.

Poles: The poles are determined from the roots of the characteristic equation, which is

$$s^2+3s+11=0$$

The roots are $s = -1.5 + j2.96$ and $s = -1.5 - j2.96$.

The pole-zero map is shown in Figure 5.26. Zero is identified with a circle and poles are indicated with x's.

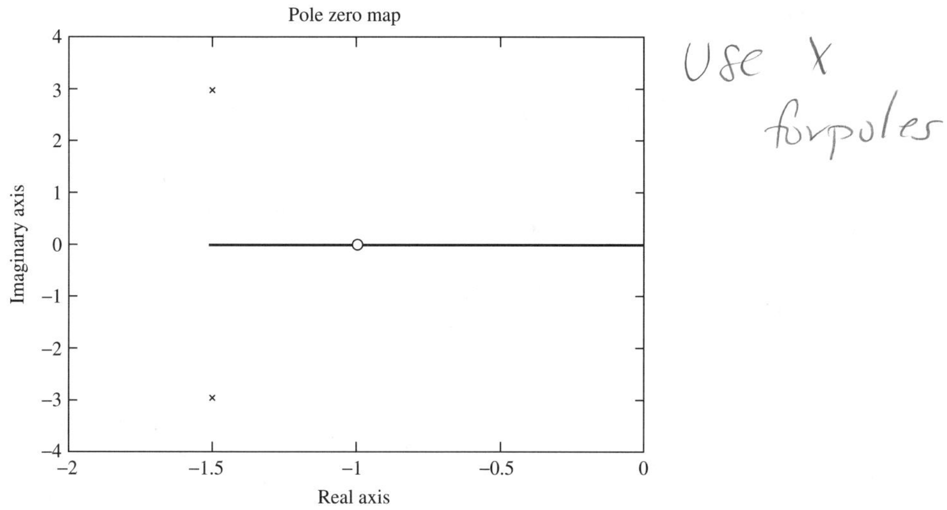

Figure 5.26 Pole-zero map for Example 5.14.

Pole Location and Transient Response

The location of a pole on an s-plane specifies the type of response contributed by that pole. The number of poles present is directly related to the order (order of the highest-power s term) of the characteristic equation. In the simplest case (first-order), there will be only one pole. A second-order system has two poles, both of which may be real or a complex conjugate pair. Poles can be of two types.

Distinct (single or not associated with other) Pole A distinct pole has no imaginary component and thus always lies on the real axis.

Pole on LHS of s-Plane

- A unit impulse input results in an output that exponentially decays to zero (*stable* operation).
- A step input results in an output that exponentially settles to a finite value (*stable* operation).
- The distance from the origin is related to the inverse of the time constant.
- A pole far away from the origin has a small time constant.
- A pole closer to the origin has a large time constant and is known as a *dominant pole*.

Pole on RHS of s-Plane

- A step input results in an output with a continuous exponential rise (*unstable* operation).

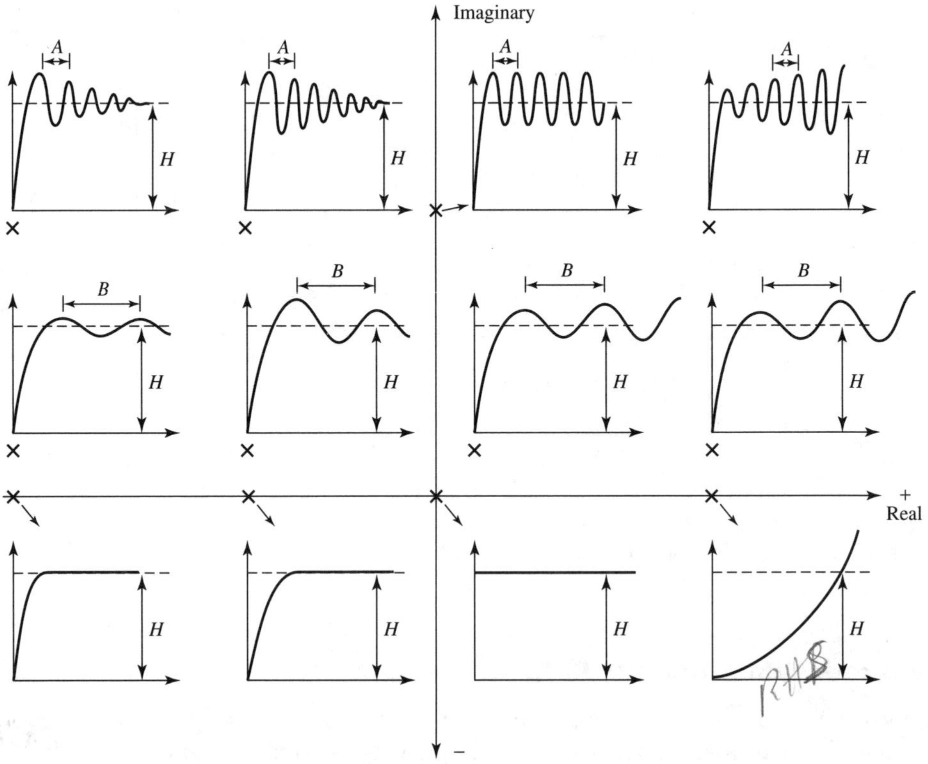

Figure 5.27 Pole location and transient response.

Complex Conjugate Poles

- Complex poles appear as complex conjugate pairs and thus do not lie on the real axis.
- It is possible for a complex pair to have a real component of zero, and in that case both poles lie on the imaginary axis.

Poles on the LHS of the s-Plane

- A step input results in an oscillatory output that decays exponentially and settles down to a finite value (*stable* operation).
- The distance from the origin is directly related to the natural frequency of the oscillations.
- The ratio of the real component to the imaginary relates to the amount of damping.
- Poles closer to the imaginary axis are known as *dominant poles*.
- With complex poles on the imaginary axis (zero real component), a step input results in an oscillatory output that neither decays nor rises (*quasi-stable* operation).

Poles on the RHS of the s-Plane

- A step input results in an oscillatory output that does not decay but continues to rise (*unstable* operation).

Figure 5.27 shows the significance of pole location on output response. The diagram shows the output response for a step input in each case.

EXAMPLE 5.15

The transfer function of a system is

$$\text{TF} = \frac{20}{(s+1)(s+10)(s+12)}$$

Determine the following.

a. Characteristic equation
b. Zeros and poles of the system
c. The type of output response when subjected to a unit step input

Solution

a. The characteristic equation is

$$(s+1)(s+10)(s+12) = 0$$
$$s^3 + 23s^2 + 142s + 120 = 0$$

The system transfer function is already in factored form, so poles and zeros can be directly determined.

b. There are no zeros in the system transfer function. There are three poles, located at $s = -1, s = -10$, and $s = -12$.

c. The output in the s-domain is

$$\text{output }(s) = \text{TF} \times \text{input }(s)$$
$$\text{output }(s) = \frac{20}{(s+1)(s+10)(s+12)} \times \mathbf{L}[u(t)]$$
$$\text{output }(s) = \frac{20}{(s+1)(s+10)(s+12)} \times \frac{1}{s}$$
$$\text{output }(s) = \frac{20}{s(s+1)(s+10)(s+12)}$$

The output in real time is obtained by inverse transformation:

$$\text{output}(t) = \mathbf{L}^{-1}[\text{output}(s)]$$

$$\text{output}(t) = \mathbf{L}^{-1}\left[\frac{20}{s(s+1)(s+10)(s+12)}\right]$$

The s-domain expression can be simplified through partial fractions. Because the denominator of the output is already in factored form, the factored form of the output is

$$\text{output}(s) = \frac{A}{s} + \frac{B}{s+1} + \frac{C}{s+10} + \frac{D}{s+12}$$

$$A = \frac{20}{s(s+1)(s+10)(s+12)} \times s\bigg|_{s=0}$$

$$A = \frac{20}{(s+1)(s+10)(s+12)}\bigg|_{s=0}$$

$$A = \frac{20}{(0+1)(0+10)(0+12)}$$

$$A = 0.167$$

$$B = \frac{20}{s(s+1)(s+10)(s+12)} \times (s+1)\bigg|_{s=-1}$$

$$B = -0.202$$

$$C = \frac{20}{s(s+1)(s+10)(s+12)} \times (s+10)\bigg|_{s=-10}$$

$$C = 0.111$$

$$D = \frac{20}{s(s+1)(s+10)(s+12)} \times (s+12)\bigg|_{s=-12}$$

$$D = -0.076$$

Substituting the coefficient values gives

$$\text{output}(s) = \frac{0.167}{s} - \frac{0.202}{s+1} + \frac{0.111}{s+10} - \frac{0.076}{s+12}$$

The corresponding system response in the time domain is

$$\text{output}(t) = 0.167 - 0.202e^{-t} + 0.111e^{-10t} - 0.076e^{-12t}$$

EXAMPLE 5.16

A first-order system with transfer function TF = $10/(s+2)$ is subjected to the following inputs. Determine the output response in each case.

a. Unit step $u(t)$

b. Unit ramp t

Solution

The output V_o in the s-domain is

$$V_o(s) = \text{TF} \times \text{input }(s)$$

$$V_o(s) = \frac{10}{s+2} \times \text{input }(s)$$

a. The unit step input is

$$V_o(s) = \frac{10}{s+2} \times \mathbf{L}[u(t)]$$

$$V_o(s) = \frac{10}{s+2} \times \frac{1}{s}$$

$$V_o(s) = \frac{10}{s(s+2)}$$

Taking the inverse transform to get the time response yields

$$v_o(t) = \mathbf{L}^{-1}\left[\frac{10}{s(s+2)}\right]$$

$$v_o(t) = 5(1 - e^{-2t})$$

b. The unit ramp input is

$$V_o(s) = \frac{10}{s+2} \times L\,[t]$$

$$V_o(s) = \frac{10}{s^2(s+2)}$$

$$v_o(t) = \mathbf{L}^{-1}\left[\frac{10}{s^2(s+2)}\right]$$

$$v_o(t) = 5t - 2.5 + 2.5e^{-2t}$$

PROBLEMS

1. Output voltage $v(t)$ in a circuit, when subjected to an input $e(t)$, is given by the differential equation

$$10.5\frac{d^2v}{dt^2} + 4\frac{dv}{dt} + v = e$$

Determine the transfer function $V(s)/E(s)$.

2. Displacement x in a mass and damper assembly, when subjected to a force $f(t)$, is given by a first-order differential equation

$$\frac{dx}{dt} + 10x = f(t)$$

Determine its transfer function, $X(s)/F(s)$.

3. Determine an expression (in time) for the velocity of mass in Problem 2 if the mass is subjected to a constant unit force. Assume that initially mass is at rest.

4. The displacement $x(t)$ of mass in a suspension system, when subjected to a force $f(t)$, is given by the transfer function

$$\frac{X(s)}{F(s)} = \frac{10}{s^2 + 5s + 10}$$

Determine the time response if the system is subjected to a cyclic force of $5 \sin 2t$.

5. A position control system for a robotic arm is designed to accept a voltage (volts dc) signal from an external source and move the arm correspondingly. Displacement is measured in meters. The transfer function for output displacement $x(t)$ and input voltage $e(t)$ is given as

$$\frac{X(s)}{E(s)} = \frac{0.1}{0.1s^2 + 0.25s + 1}$$

Determine the output displacement (m) if a signal of 10 V dc is applied to it.

6. What is the (final) steady-state displacement of the system in Example 5? Verify the answer by applying the final value theorem.

7. Determine the steady-state value of the following systems when subjected to a unit impulse

   a. $\dfrac{1}{(s+10)}$

b. $\dfrac{(s+1)}{(s^2+3s+5)}$

c. $\dfrac{5s+1}{s(s+2)(s^2+0.5s+1)}$

8. The current $i(t)$ in an *RLC* circuit, when subjected to an input voltage $e(t)$, is given by the *s*-domain expression

$$\frac{Cs}{LCs^2+RCs+1}\times E(s)$$

Determine the current at time $t = 0.01$ ms when a voltage of 2 V dc is applied. The component values are $R = 50\ \Omega$, $C = 4.7\ \mu$F, and L = 1 mH.

9. Determine the steady-state error in the following unity feedback closed-loop systems when they are subjected to a unity step input.

a. $\dfrac{10}{s+11}$

b. $\dfrac{10}{s^2+2s+10}$

c. $\dfrac{10}{s(s^2+2s+10)}$

10. A reference voltage of 5 V dc is applied to a servomotor position control system. Develop an expression for motor position if the transfer function for position $Z(s)$ and input voltage $E(s)$ is given as

$$\text{TF:}\quad \frac{Z(s)}{E(s)} = \frac{25}{(s+3)(s+6)}$$

6

FREQUENCY RESPONSE

INTRODUCTION

Frequency response techniques are used extensively to analyze the characteristics and performance of control systems. They are also used in specifications of numerous systems and components such as audio and video amplifiers, speakers, sound cards, and servomotors. A significant portion of the earlier work in control system theory was based on frequency response techniques. These techniques also lend themselves to simple, yet fruitful, procedure for experimental plant testing and model verification. This chapter discusses the principles of frequency response testing and analysis. The numerical method for obtaining frequency response of a system from its transfer function is explained. The use of a MATLAB program for determination of frequency response is illustrated through examples.

Frequency Response

Frequency response is the response of a system, when it is subjected to sinusoidal signal, over a range of frequencies.

Theoretically speaking, frequency response covers the entire frequency spectrum extending from near zero to near infinity. In practice, only a limited frequency range significant to the system under consideration is used. Consider a system (continually) subjected to a sinusoidal input waveform (Figure 6.1).

$$r(t) = A \sin \omega t$$

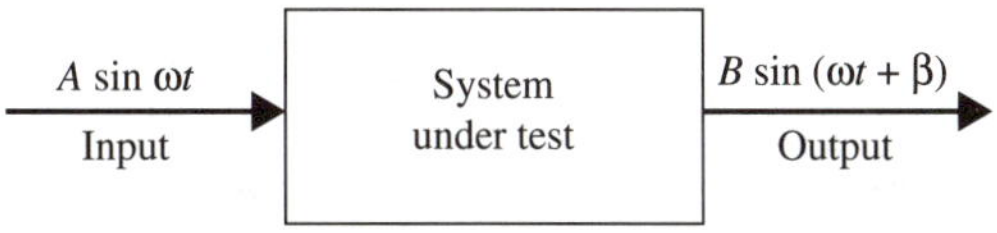

Figure 6.1 Frequency response.

where $A =$ input peak amplitude
$\omega =$ input signal frequency

When all the initial transients have died out, a stable sinusoidal output response will be obtained, which can be expressed as:

$$c(t) = B \sin(\omega t + \beta)$$

where $B =$ output peak amplitude
$\omega =$ output signal frequency
$\beta =$ phase difference between the output and input signals

The frequency of the output waveform will be identical to the input-signal frequency, but the two signals may not be in phase. This phase difference, β, may be positive (output leading input) or negative (output lagging input). At the same time there may be a difference in the amplitude of the input and output waveforms, resulting in signal gain.

$$\text{gain} = \frac{\text{peak output signal amplitude } B}{\text{peak input signal amplitude } A}$$

Again, the gain may be less than 1 (attenuation), equal to 1 (no amplification), or greater than 1 (amplification). Both the phase difference and the signal gain are characteristics of the system under consideration. Thus it is possible to set up system classifications based upon the frequency response characteristics (Chapter 7). System characteristics, performance, and stability analysis (Chapter 8) dictate the range of frequencies to be used. Frequency response forms the basis of a number of criteria used for specifying the performance of control systems.

When this information about the signal gain, and phase difference over frequency range is expressed in a proper form, it is referred to as the *frequency response* of the given system. Gain is commonly expressed as *magnitude*.[1]

$$\text{magnitude } M = 20 \log \text{ (gain)}$$

[1] The term *gain* is used to represent the ratio of output and input amplitudes and is not used to imply that amplification is taking place.

where

$$\text{gain} = \frac{\text{output signal amplitude}}{\text{input signal amplitude}}$$

Methods Used for Frequency Response Representation

Frequency response data about signal gain and phase difference can be presented either in a tabular form or graphical form.

Tabular Form Tabular form is useful when the data about phase difference and gain need to be declared to a fair degree of precision or the data set is relatively small, as in the case of experimental measurement (Table 6.1).

Graphical Form Graphical form provides a convenient way to view the frequency response data. It is more commonly used than the tabular form of presentation. Again, there are several forms used for graphical presentation. The two most common, which are discussed here, are

- Bode plot (diagram), or logarithmic plot
- Nyquist plot, or polar plot

Bode Plot: Logarithmic plots are more common of the two. This plot is commonly referred to as a *Bode plot* in honor of H. W. Bode, who first demonstrated it in the 1930s. A Bode plot uses semilog paper and actually consists of two plots, the magnitude plot and the phase-angle plot.

Bode Plot

A Bode plot is the plot of magnitude and phase angle versus frequency on a semilog paper.

Magnitude versus frequency: Linear scale is used for plotting the magnitude, which is a logarithmic quantity (dB). A linear scale is used for the y-axis. Frequency is displayed in rad/s and is plotted on the x-axis. A logarithmic scale is used for the x-axis because a large frequency range can be covered.

Phase angle versus frequency: Phase angle is calculated in degrees and is plotted on the y-axis. A linear scale is used for the y-axis. Similar to magnitude plot, frequency is in rad/s and is plotted on a logarithmic x-axis.

Because frequency (x-axis) is common to both the plots and a linear scale is used for both the magnitude and phase angle, it is more convenient to use the same semilog paper for both the plots. Conventionally, the top section of paper is used for the magnitude plot and the bottom section is used for the phase-angle plot. Figure 6.2 shows the Bode plot of the same system shown in Table 6.1.

Table 6.1 Frequency Response Data TF = 1/(s + 1)

Gain	***Magnitude (dB)***	***Phase Angle (degrees)***
0.9950	0.0432	5.7106
0.9932	0.0593	6.6851
0.9907	−0.0812	−7.8225
0.9873	−0.1112	−9.1476
0.9827	−0.1520	−10.6882
0.9764	−0.2075	−12.4742
0.9680	−0.2826	−14.5363
0.9568	−0.3837	−16.9051
0.9420	−0.5188	−19.6072
0.9228	−0.6979	−22.6621
0.8982	−0.9325	−26.0767
0.8674	−1.2354	−29.8392
0.8299	−1.6198	−33.9144
0.7854	−2.0979	−38.2397
0.7346	−2.6791	−42.7278
0.6785	−3.3688	−47.2722
0.6190	−4.1668	−51.7603
0.5580	−5.0680	−56.0856
0.4976	−6.0629	−60.1608
0.4396	−7.1394	−63.9233
0.3853	−8.2841	−67.3379
0.3356	−9.4844	−70.3928
0.2908	−10.7285	−73.0949
0.2510	−12.0067	−75.4637
0.2160	−13.3110	−77.5258
0.1855	−14.6348	−79.3118
0.1590	−15.9732	−80.8524
0.1361	−17.3226	−82.1775
0.1164	−18.6799	−83.3149
0.0995	−20.0432	−84.2894

Polar Plot: Although not as common as a Bode plot, a polar plot provides a concise form of presenting a complete frequency response (from zero frequency to infinity) of a system. It is also known as a *Nyquist plot* in honor of H. Nyquist, who first presented this idea back in the early 1930s.

A polar plot is a single plot between the system gain and the phase angle. Complex, yet linear, scales are used for the polar plot. Gain and phase angle are calculated for a wide frequency range (ideally from 0 rad/s to infinity). A vector is used to represent gain and phase-angle data. The vector is not drawn in entirety, but only as a point on the graph representing the tip of the vector, with the origin serving as the base

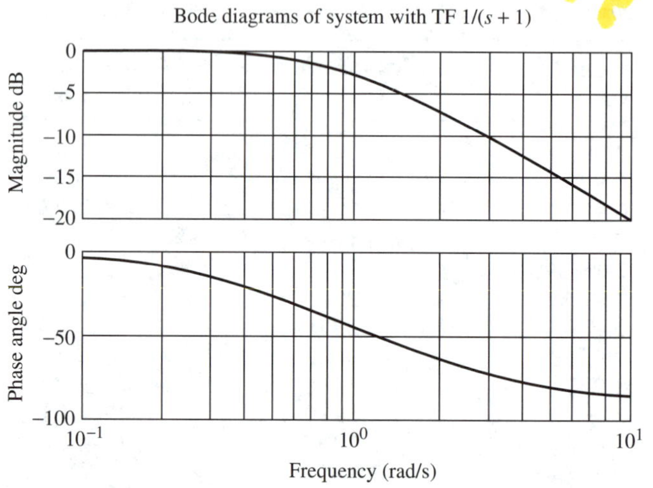

Figure 6.2 Bode plot of TF: $1/(s+1)$.

of the vector. Phase angle is considered positive when measured in a counterclockwise direction. When repeated for various frequencies, all these points generate a continuous curve, known as a polar plot.

Polar Plot

A ***polar plot*** is the loci of vectors drawn for a frequency range extending from zero to infinity, with each vector representing system gain and the phase angle at a certain frequency.

Figure 6.3 shows the polar plot of the same system ($1/(s+1)$) used earlier. Compare it with tabular form (Table 6.1) and Bode diagrams.

The graphical form provides a lot more information about the trend of the frequency response. Numerical values can be obtained for any frequency within the plot and can also be extrapolated for frequency points outside the plot, although it is difficult to accurately read the numerical values from a graph.

Frequency Response Determination

The frequency response of a system can be determined by one of the three methods: experimentally, mathematically, or through computer software such as MATLAB.

Experimental Determination of the Frequency Response If system transfer function is not known, experimental determination is the only recourse available. The experimental method is used for plant testing and verification of the plant model. A mathematical model can be extracted from experimental data. The transfer function

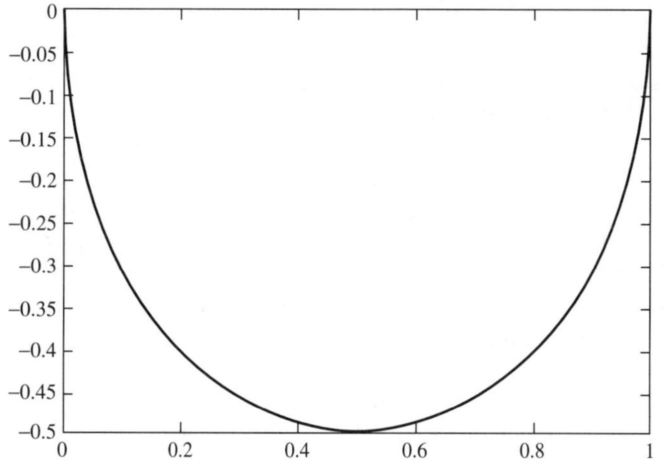

Figure 6.3 Polar plot of TF: $1/(s + 1)$.

obtained from experimental data is somewhat approximate, because it is difficult to collect experimental data in a field with precision and over an extended period of time. At the same time, it is not very difficult to experimentally determine the frequency response of a small system in a laboratory environment.

A typical setup for the experimental determination of frequency response is shown in Figure 6.4. A test of electrical systems requires a power supply, a function generator, and a dual-trace oscilloscope or a chart recorder. For a test of pneumatic systems, pneumatic sine-wave generators are available. Mechanical cycle generators are used for testing of mechanical systems. In all experimental testing, appropriate transducers

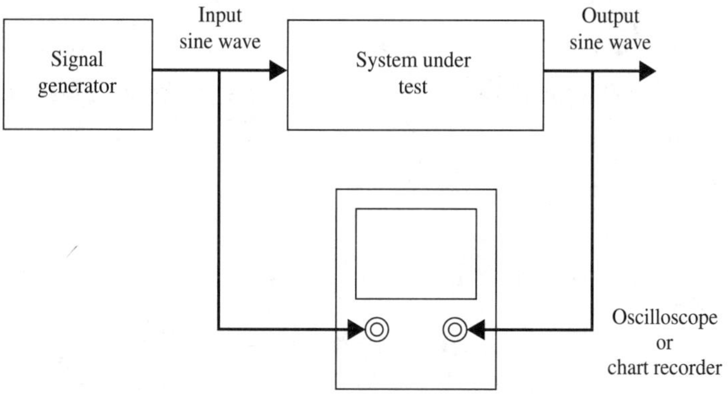

Figure 6.4 Experimental determination of frequency response.

are needed to convert pressure and/or displacement information into electrical signals so that these can be recorded.

Before a test can be conducted, the test frequency range and the input signal amplitude are established. The sinusoidal signal with the lowest frequency is first applied, and sufficient time is allowed for all the transients to settle down. Once a stable sinewave output is reached, the output amplitude and the phase difference between input and output waveform are recorded. The ratio of output to input amplitude provides the system gain. The input signal frequency is increased and output amplitude and phase shift are recorded after the output has settled down. This process is repeated until the entire frequency range (of interest) is covered. Experimental data can now be tabulated and displayed in graphical form.

Mathematical Evaluation of the Frequency Response System gain and phase angle can be easily calculated with the help of complex numbers. The transfer function can be considered to be a function of frequency, which is treated as a complex variable. The magnitude and phase angle of this complex function provide the system gain and phase difference at that frequency.

Simple First-Order Transfer Function: The following procedure can be used:

- Establish the frequency range of interest.
- Replace the s term with $j\omega$.
- Represent the resulting complex number in polar form.
- Substitute the first frequency (rad/s) in place of ω.
- Calculate the complex magnitude. This yields the system gain.
- Calculate the complex angle. This yields phase difference.
- Repeat for the entire frequency range.

EXAMPLE 6.1

Determine the gain and phase angle of the following system at a frequency of 1 rad/s.

$$G(s) = (s + 1)$$

Solution

Replacing the s term with $j\omega$ in the transfer function,

$$G(j\omega) = (j\omega + 1)$$

Converting to polar form gives

$$G(j\omega) = M\angle\beta$$

where

$$M = \text{system gain at frequency } \omega = \sqrt{(1+\omega^2)}$$

$$\beta = \text{phase difference at frequency } \omega = \tan^{-1}(\omega)$$

Substituting the value of frequency ($\omega = 1$ rad/s),

$$M = \sqrt{(1+1^2)} = 1.414$$

$$\beta = \tan^{-1}(1) = 45^\circ$$

Substituting these values, G (1 rad/s) = 1.414 $\angle 45$. Therefore,

$$\text{system gain} = 1.41 = 20 \log(1.414) = 3.01 \text{ dB}$$

$$\text{phase angle} = 45^\circ$$

EXAMPLE 6.2

Determine the frequency response of the following transfer function over a frequency range of 0.1 to 10 rad/s.

$$TF = \frac{1}{2s+1}$$

Solution

For the given system, the gain and phase angle data will be calculated at the following frequencies.

a. 0.1 rad/s
b. 0.5 rad/s
c. 1.0 rad/s
d. 2 rad/s
e. 10 rad/s

Replacing the s term with $j\omega$ in the transfer function gives

$$G(j\omega) = \frac{1}{2j\omega + 1}$$

Converting to polar form,

$$\begin{aligned} G(j\omega) &= \frac{1}{1+2j\omega}\frac{1-2j\omega}{1-2j\omega} \\ &= \frac{1-2j\omega}{1+4\omega^2} \\ &= \frac{1}{1+4\omega^2} - \frac{2j\omega}{1+4\omega^2} \\ &= M\angle\beta \end{aligned}$$

where

$$\begin{aligned} M &= \text{system gain at frequency } \omega \\ &= \sqrt{\left(\frac{1}{1+4\omega^2}\right)^2 + \left(\frac{2\omega}{1+4\omega^2}\right)^2} \\ \beta &= \text{phase difference at frequency } \omega \\ &= \tan^{-1}(-2\omega) \end{aligned}$$

a. Frequency $\omega = 0.1$ rad/s:

$$\begin{aligned} \text{system gain } M &= \sqrt{\left(\frac{1}{1+4(0.1)^2}\right)^2 + \left(\frac{2\times 0.1}{1+4(0.1)^2}\right)^2} \\ &= 0.9806 \\ &= 20\log(0.9806) \\ &= -0.122\text{ dB} \\ \text{phase difference} &= \tan^{-1}(-2\times 0.1) \\ &= -11.3^\circ \end{aligned}$$

b. Frequency $\omega = 0.5$ rad/s:

$$\begin{aligned} \text{system gain } M &= 0.7071 \\ &= -3.01\text{ dB} \\ \text{phase difference} &= -45^\circ \end{aligned}$$

c. Frequency $\omega = 1.0\,\text{rad/s}$:

$$\begin{aligned}\text{system gain } M &= 0.4472\\ &= -6.99\,\text{dB}\\ \text{phase difference} &= -63.43^\circ\end{aligned}$$

d. Frequency $\omega = 2.0\,\text{rad/s}$:

$$\begin{aligned}\text{system gain } M &= 0.2425\\ &= -12.31\,\text{dB}\\ \text{phase difference} &= -75.96^\circ\end{aligned}$$

e. Frequency $\omega = 10.0\,\text{rad/s}$:

$$\begin{aligned}\text{system gain } M &= 0.0499\\ &= -26.04\,\text{dB}\\ \text{phase difference} &= -87.14^\circ\end{aligned}$$

Gain and phase angle can thus be determined for any other value of the frequency.

Higher-Order Transfer Function (Multiple Factors) When multiple factors are involved, each factor can be represented in polar form. The complex numbers in polar form allow easier multiplication and division. Thus, an overall complex result can be easily obtained. The following procedure can be applied:

- Separate the complex function into a product of lower-order terms.
- Establish the frequency range of interest.
- Replace every s term with $j\omega$.
- Represent every function in polar form.

- For every function:
  - Substitute the first frequency (rad/s) in place of ω.
  - Calculate the complex magnitude. This yields the system gain.
  - Calculate the complex angle. This yields phase difference.

- Obtain overall magnitude by multiplying all magnitude terms.
- Obtain the overall phase angle by adding the phase-angle terms.
- Repeat for the entire frequency range.

EXAMPLE 6.3

Determine the gain and phase angle of the given system at a frequency of 0.1 Hz.

$$G(s) = (10s + 1)(2s + 5)$$

Solution

The transfer function contains two first-order terms in cascade.

$$G(s) = G_1(s)G_2(s)$$

where

$$G_1(s) = (10s + 1)$$
$$G_2(s) = (2s + 5)$$

Frequency is expressed in Hz and needs to be converted into rad/s.

$$\text{frequency } \omega = 0.1 \times 2 \times \pi \text{ rad/s}$$
$$= 0.2\pi \text{ rad/s}$$

$G_1(s) = (10s + 1)$: Replacing the s term with $j\omega$,

$$G_1(j\omega) = (10j\omega + 1)$$

Converting to polar form gives

$$G_1(j\omega) = M_1 \angle \beta_1$$

where

$$M_1 = \sqrt{(100\omega^2 + 1)}$$
$$\beta_1 = \tan^{-1}(10\omega)$$

$G_2(s) = (2s + 5)$: Similarly,

$$G_2(j\omega) = M_2\angle\beta_2$$

where

$$M_2 = \sqrt{(4\omega^2 + 25)}$$
$$\beta_2 = \tan^{-1}\left(\frac{2\omega}{5}\right)$$

Therefore, the overall complex function is

$$\begin{aligned} G(j\omega) &= M_1\angle\beta_1 \times M_2\angle\beta_2 \\ &= M_1 M_2\angle(\beta_1 + \beta_2) \end{aligned}$$
$$G(j\omega) = \sqrt{(100\omega^2 + 1)} \times \sqrt{(4\omega^2 + 25)}\ \angle\left(\tan^{-1}(10\omega) + \tan^{-1}\left(\frac{2\omega}{5}\right)\right)$$

Substituting the value of the frequency ($\omega = 200\ \pi$ rad/s),

$$G(j\omega) = 32.8\angle 95.06$$
$$\text{system gain} = 32.8(30.318\ \text{dB})$$
$$\text{phase angle} = 95.06^\circ$$

EXAMPLE 6.4

Calculate the frequency response of the following system over a frequency range of 0.01 to 10 rad/s:

$$G(s) = \frac{2.5(s+1)}{(s+2)(s+3)}$$

Solution

Separating the function into simpler terms gives

$$G(s) = 2.5 \times (s+1) \times \frac{1}{(s+2)} \times \frac{1}{(s+3)}$$

Replacing every s term with $j\omega$,

$$G(j\omega) = 2.5 \times (j\omega + 1) \times \frac{1}{(j\omega + 2)} \times \frac{1}{(j\omega + 3)}$$

Converting to polar form gives

$$\begin{aligned} G(j\omega) &= M_1 \angle\beta_1 M_2 \angle\beta_2 M_3 \angle\beta_3 M_4 \angle\beta_4 \\ M_1 &= 2.5 \\ \beta_1 &= 0^\circ \end{aligned}$$

The constant term 2.5 can also be treated either as a scalar (2.5) or as a complex quantity with a 0° phase angle.

$$\begin{aligned} M_2 &= \sqrt{\omega^2 + 1} \\ \beta_2 &= \arctan\,(\omega) \\ M_3 &= \sqrt{\omega^2 + 4} \\ \beta_3 &= -\arctan\,(\omega/2) \\ M_4 &= \sqrt{\omega^2 + 9} \\ \beta_4 &= -\arctan\,(\omega/3) \end{aligned}$$

Each of these terms can be evaluated by substituting the frequency (ω) data. A few of the selected values are shown in Table 6.2.

Table 6.2 Frequency Response Data, Example 6.4

Frequency (rad/s)	Gain	Phase Angle (degrees)
0.1000	0.4180	0.9390
0.1668	0.4203	1.5200
0.2783	0.4265	2.3298
0.4642	0.4422	3.0379
0.7743	0.4758	2.1148
1.2915	0.5251	−3.8954
2.1544	0.5469	−17.7115
3.5938	0.4844	−36.5990
5.9948	0.3587	−54.4358
10.0000	0.2360	−67.7014

Frequency Response Through MATLAB MATLAB contains several powerful commands for frequency response determination. It further allows the frequency response to be displayed either in a table or in graphical form. The following section describes the procedure required by MATLAB for representation of the system transfer function, frequency range, and the type of output (tabular or graphical) needed.

Polynomial Representation: The system transfer function $G(s)$ is expressed as a ratio of two polynomials of s: Num(s) and Den(s).

$$G(s) = \frac{\text{Num}(s)}{\text{Den}(s)}$$

EXAMPLE 6.5

Express the following transfer function in a proper form suitable for MATLAB programming.

$$G(s) = \frac{5(s+1)}{(2s+1)(s+3)}$$

Solution

Each of the numerator and denominator need to be expressed as polynomials of s. Multiplying out the terms,

$$G(s) = \frac{5s+5}{2s^2+7s+3}$$

The transfer function is now in suitable form for MATLAB programming.

Polynomial Entry: Both the numerator and denominator polynomials are entered with s terms arranged in descending order. Only the coefficients of the s terms are entered (s, s^2, s^3, etc., are not entered). The first term (leftmost) is the highest-power term of s and the last term (rightmost) is the s term with power zero (constant term). The total number of terms is always one more than the order of the polynomial.

EXAMPLE 6.6

Enter the following transfer function:

$$G(s) = \frac{s+10}{3s^2+5s+2}$$

Solution

Let the transfer function be represented as

$$G(s) = \frac{\text{Num}(s)}{\text{Den}(s)}$$

where

$$\text{Num}(s) = s + 10$$
$$\text{Den}(s) = 3s^2 + 5s + 2$$

In MATLAB, these are entered as

```
Num = [1 10]    or    Num = [1,10]
Den = [3 5 2]         Den = [3,5,2]
```

The coefficients can be separated from each other either by blank spaces or by commas.

EXAMPLE 6.7

Enter the following transfer function:

$$G(s) = \frac{s}{5 + 3s^3}$$

Solution

Let the transfer function be represented as

$$G(s) = \frac{\text{Num}(s)}{\text{Den}(s)}$$

It should be remembered that MATLAB is case-sensitive.

Both numerator and denominator are arranged in descending order:

$$\text{Num}(s) = s$$
$$\text{Den}(s) = 3s^3 + 5$$

The highest power of a numerator is 1; thus, there should be two terms, with the last term being the s^0 term. Thus, rewriting gives

$$\text{Num}(s) = s + 0$$

Similarly, the denominator is a polynomial of third order, so rewriting yields

$$\text{Den}(s) = 3s^3 + 0s^2 + 0s + 5$$

In MATLAB, these are entered as

```
Num = [1 0]         or    Num = [1, 0]
Den = [3 0 0 5]     or    Den = [3, 0, 0, 5]
```

Graphical Output (Bode Plot): A bode plot is obtained with the following command:

```
bode(num,den)
```

where

$$\text{num} = \text{numerator polynomial}$$
$$\text{den} = \text{denominator polynomial}$$

The frequency range is automatically selected by the MATLAB program. Frequency response is displayed as a semilogarithmic plot. Scaling for magnitude, phase angle, and frequency is again automatically set.

EXAMPLE 6.8

Draw the Bode plot of the following system:

$$G(s) = \frac{2.5s + 10}{3s^2 + 8s}$$

Solution

Rewriting the transfer function,

$$G(s) = \frac{n}{m}$$

where

$$n = 2.5\ s + 10$$
$$m = 3s^2 + 8s + 0$$

Entering the data in MATLAB,

```
n = [2.5  10]
m = [3 8 0]
bode(n,m)
```

The last command results in the Bode plot of the system being drawn (Figure 6.5).

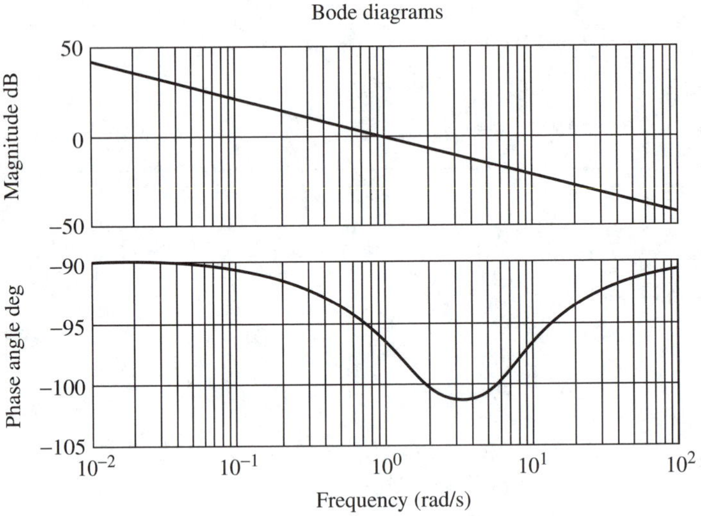

Figure 6.5 Bode plot for Example 6.8.

Tabular Output: Sometimes it is desirable to obtain the numerical values of magnitude and phase angle and display the data in a tabular form. MATLAB allows such information to be obtained through the following command:

```
[g,p] = bode(num,den)
```

where g = vector that contains gain data
p = vector that contains phase-angle data

Numerical values are displayed in tabular form by the following command:

```
[g,p]
```

It should be pointed out that in graphical output, magnitude (gain in dB) is plotted, whereas in numerical output, the gain (amplification factor) information is provided. However, it is extremely simple to determine magnitude data.

$$m = 20 \cdot \log_{10} g$$

EXAMPLE 6.9

Obtain the magnitude and gain value for the preceding transfer function.

Solution

Here the numerator is $n = [2.5\ 10]$ and the denominator is $m = [3\ 8\ 0]$.
Gain and phase angle are obtained by invoking the *bode* function with the statement

```
[g,p] = bode(n,m)
```

Gain and phase angle data can be checked by entering

```
[g,p]
```

This command displays the following values (gain and phase angle).

```
124.9995   -90.0716
 90.9863   -90.0984
 66.2280   -90.1352
 48.2065   -90.1857
 35.0885   -90.2551
 25.5396   -90.3504
 18.5886   -90.4813
 13.5285   -90.6610
  9.8444   -90.9074
  7.1617   -91.2448
  5.2075   -91.7056
  3.7831   -92.3316
  2.7437   -93.1734
  1.9839   -94.2845
  1.4268   -95.7011
  1.0174   -97.3954
  0.7163   -99.1976
  0.4970 -100.7335
  0.3407 -101.5079
  0.3125 -101.5370
  0.2329 -101.2062
  0.1606   -99.9561
  0.1123   -98.2164
  0.0797   -96.4405
  0.0571   -94.8881
  0.0412   -93.6406
  0.0299   -92.6831
  0.0217   -91.9660
  0.0158   -91.4361
  0.0115   -91.0473
  0.0083   -90.7631
```

Magnitude information can be added to the preceding with

```
m = 20 * log10(g)
[m,g,p]
```

The magnitude information is now added to the table (magnitude, gain, phase angle).

41.9382	124.9995	-90.0716
39.1795	90.9863	-90.0984
36.4208	66.2280	-90.1352
33.6621	48.2065	-90.1857
30.9033	35.0885	-90.2551
28.1443	25.5396	-90.3504
25.3849	18.5886	-90.4813
22.6250	13.5285	-90.6610
19.8638	9.8444	-90.9074
17.1003	7.1617	-91.2448
14.3326	5.2075	-91.7056
11.5570	3.7831	-92.3316
8.7668	2.7437	-93.1734
5.9502	1.9839	-94.2845
3.0875	1.4268	-95.7011
0.1495	1.0174	-97.3954
-2.8976	0.7163	-99.1976
-6.0736	0.4970	-100.7335
-9.3531	0.3407	-101.5079
-10.1030	0.3125	-101.5370
-12.6552	0.2329	-101.2062
-15.8848	0.1606	-99.9561
-18.9899	0.1123	-98.2164
-21.9734	0.0797	-96.4405
-24.8660	0.0571	-94.8881
-27.7006	0.0412	-93.6406
-30.5009	0.0299	-92.6831
-33.2820	0.0217	-91.9660
-36.0527	0.0158	-91.4361
-38.8177	0.0115	-91.0473
-41.5798	0.0083	-90.7631

Determination of Frequency Points: Information can be obtained about the frequency points used in calculation of gain and phase angle. It can be achieved through the following command,

```
[g,p,w] = bode(num.den)
```

The vector **w** will contain the values of frequencies at which magnitude and phase angle are calculated.

EXAMPLE 6.10

Obtain the frequency response of the following system in a tabular form. The result should include magnitude, gain, phase angle, and frequency.

$$G(s) = \frac{s(s+10)}{0.2s^2 + 3s + 1.25}$$

Solution

Rewriting the numerator in polynomial form gives

$$G(s) = \frac{s^2 + 10s}{0.2s^2 + 3s + 1.25}$$

Enter the transfer function data into MATLAB:

```
num = [1 10 0]
den = [0.2 3 1.25]
```

Gain, phase angle, and frequency information can now be determined,

```
[g,p,ω] = bode(num,den)
```

Magnitude data are calculated from

$$m = 20 \cdot log_{10} g$$

Final value can now be displayed in tabular form,

```
[m,g,p,ω]
```

The preceding command results in following data to be displayed.

```
-21.9406    0.0800   88.6824    0.0100
-19.1840    0.1098   88.1902    0.0137
-16.4294    0.1508   87.5144    0.0189
-13.6782    0.2071   86.5872    0.0259
-10.9335    0.2840   85.3166    0.0356
 -8.2012    0.3890   83.5789    0.0489
 -5.4918    0.5314   81.2124    0.0672
 -2.8245    0.7224   78.0134    0.0924
 -0.2332    0.9735   73.7476    0.1269
  2.2261    1.2921   68.1948    0.1743
  4.4707    1.6732   61.2526    0.2395
  6.3994    2.0891   53.0991    0.3290
  7.9258    2.4905   44.3091    0.4520
  9.0244    2.8263   35.7457    0.6210
  9.7475    3.0717   28.2166    0.8532
```

```
10.1950     3.2341    22.1863     1.1721
10.4703     3.3382    17.7571     1.6103
10.6578     3.4111    14.8144     2.2122
10.8227     3.4764    13.1567     3.0392
11.0192     3.5560    12.5380     4.1753
11.0919     3.5859    12.5090     4.5938
11.2950     3.6707    12.6278     5.7362
11.6803     3.8372    12.9494     7.8805
11.9109     3.9404    13.0156     9.2275
12.1625     4.0562    12.9286    10.8264
12.6736     4.3021    12.1490    14.8735
13.1245     4.5313    10.6183    20.4336
13.4612     4.7104     8.7001    28.0722
13.6820     4.8317     6.7984    38.5662
13.8145     4.9060     5.1526    52.9832
13.8898     4.9487     3.8351    72.7895
13.9313     4.9724     2.8254   100.0000
```

Specifying a Single Frequency: Sometimes it is desired to calculate the gain and phase angle at a specific value of a frequency. The following command calculates the gain and phase angle at the specified value of frequency.

```
[g,p] = bode(num,den,ω)
```

where $\omega =$ single frequency of interest in rad/s

EXAMPLE 6.11

Calculate the magnitude, gain, and phase angle for the given system at a frequency of 10 Hz:

$$G(s) = \frac{10}{0.002s^2 + 0.05s + 1}$$

Solution

The frequency value is specified in Hz, but MATLAB requires the value to be in rad/s. So converting,

$$\begin{aligned} \text{frequency } \omega &= 2\pi f \\ &= 2 \cdot \pi \cdot 10 \\ &= 20 \text{ pi rad/s} \end{aligned}$$

Note: pi is a defined constant in MATLAB. It allows a more precise value of pi to be used in calculations.

Specify the numerator:

```
num = 10
```

Specify the denominator:

```
den = [0.002 0.05 1]
```

Gain and phase-angle values can now be calculated at the desired frequency with the following command.

```
[g,p] = bode(num,den,ω)
```

Magnitude can also be calculated and displayed along with gain and phase angle.

```
[20 * log10(g), g, p]
```

It yields the following result:

```
2.4094  1.3197  -155.5065
```

Specifying the Frequency Range: When the *bode* command is given without specifying any frequency value, MATLAB determines a suitable range for the frequency. But it is possible to specify any desired frequency range for a graphical plot or tabular data. This can be achieved in one of two ways.

- *Specifying individual frequency points:* Individual frequency points can be specified, such as

```
ω = [0.1 2 8 100 500].
```

- *Specifying frequency range:* This requires specifying the terminal (minimum and maximum) values. A terminal value can be specified only as an integer power of 10; e.g., 2 implies a frequency of 10^2, and -3 implies a frequency of 10^{-3}.

Because a logarithmic frequency scale is used for Bode diagrams, it is more convenient to specify the frequency range through minimum and maximum frequency values. The program can establish the frequency values, which are spaced apart logarithmically. The command to specify the frequency range is as follows:

```
ω = logspace(n1,n2)
```

where n_1 = minimum frequency expressed as an exponent of 10
n_2 = maximum frequency expressed as an exponent of 10

Fifty frequency values logarithmically spaced between the maximum and minimum frequency are calculated. These values are logarithmically spaced for ease in plotting on a logarithmic scale. Once the frequency range is established, frequency response can be obtained through one of the following commands:

Data table: [g, p] = bode (num, den, ω)
Bode diagram: bode (num, den, ω)

EXAMPLE 6.12

Set up a frequency range of 0.01 rad/s to 100 rad/s.

Solution

The minimum frequency is 0.01 rad/s. Expressing it as a power of 10, minimum frequency = 10^{-2} rad/s. Similarly, the maximum frequency of 100 rad/s is expressed as 10^2 rad/s.

The frequency range command can now be issued:

```
ω = logspace(-2, 2)
```

Fifty frequency values logarithmically spaced between 0.01 rad/s and 100 rad/s are calculated as shown in Table 6.3.

Table 6.3 Frequency Response Data for Example 6.12

0.0100	0.0655	1.3257	2.8118	18.4207
0.0121	0.0791	1.5999	3.3932	22.2300
0.0146	0.0954	1.9307	4.0949	26.8270
0.0176	0.1151	2.3300	4.9417	32.3746
0.0212	0.1389	0.4292	5.9636	39.0694
0.0256	0.1677	0.5179	7.1969	47.1487
0.0309	0.2024	0.6251	8.6851	56.8987
0.0373	0.2442	0.7543	10.4811	68.6649
0.0450	0.2947	0.9103	12.6486	82.8643
0.0543	0.3556	1.0985	15.2642	100.0000

EXAMPLE 6.13

Investigate the frequency response of the following system. Obtain precise numerical data on gain and frequency in the frequency region where the phase angle drops below −180°.

$$G(s) = \frac{s+1}{s(5s^2 + s + 100)}$$

Solution

A Bode diagram of the given system can be drawn first. From the diagram, the pertinent frequency range over which the signal phase angle crosses the -180° value can be approximately determined. A new plot can then be obtained, spanning the new and reduced frequency range, or tabular data can be obtained for this new range of frequencies.

The Bode diagram is obtained through the following commands. Specify the numerator of the system transfer function:

```
num = [10 1]
```

Specify the denominator:

```
den = [5 1 100 0]
```

Draw the Bode plot (Figure 6.6):

```
bode(num,den)
```

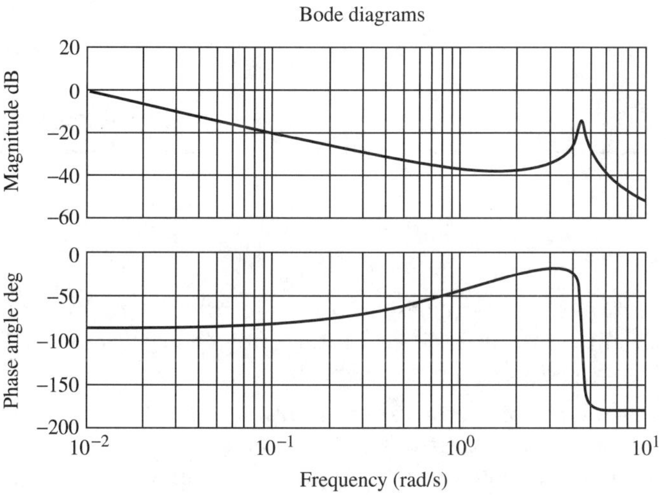

Figure 6.6 Bode plot for Example 6.13.

Table 6.4 Results from Example 6.13

Gain	*Phase Angle (degree)*	*Frequency (rad/s)*
0.0149	−45.6031	1.0000
0.0139	−39.8110	1.2328
0.0135	−34.3267	1.5199
0.0137	−29.3895	1.8738
0.0149	−25.2114	2.3101
0.0178	−22.0902	2.8480
0.0270	−21.1273	3.5112
0.1341	−47.4623	4.3288
0.0238	−183.4396	5.3367
0.0087	−185.4082	6.5793
0.0044	−184.9993	8.1113
0.0025	−184.2785	10.0000

From the plot, it is observed that phase angle crosses the -180° line somewhere between a frequency range of 4 and 5 rad/s. Thus, a suitable range will be 1 to 10 rad/s. The frequency response data over a narrow frequency range can now be obtained. Specify the frequency range:

```
ω = logspace(0, 1)
```

Obtain the tabular data:

```
[g,p,ω] = bode(num,den,ω)
```

Selected values are shown in Table 6.4.

PROBLEMS

1. Explain the procedure for experimentally determining the frequency response of an electric circuit containing several interconnected components. Draw a complete schematic diagram of the experimental setup.
2. A voltage to pressure converter takes an input of 0–10 V and generates a corresponding pressure output in the range of 3–15 psig.
   a. Devise a scheme to measure the frequency response.
   b. Draw a block diagram of the experimental setup.
   c. If the converter response is to be tested only for a constant input voltage, what is a good voltage setting to use for the test? Explain your reasoning.

3. Express the following frequency ranges in decades.
a. 0.1 Hz to 100 Hz
b. 0.1 rad/s to 100 rad/s
c. 10 Hz to 100 KHz
d. 2.5 Hz to 15 KHz

4. What are the lowest and highest frequencies that can be plotted on the following frequency response plots?
a. Bode plot
b. Polar plot

5. Express the following quantities in decibels.
a. 100
b. 120
c. 0.25
d. 0.0001
e. 1000000

6. Express the following decibel values as standard numbers.
a. 6 dB
b. −6 dB
c. 20 dB
d. −40 dB
e. 120 dB

7. Determine the gain, magnitude, and phase angle of each of the following transfer functions at the specified frequencies.
a. $\frac{10}{s+1}$ at $\omega = 1$ rad/s and 100 rad/s
b. $0.02s$ at $\omega = 10$ rad/s and 125 rad/s
c. $\frac{s}{(s^2+2s+10)}$ at $\omega = 1$ Hz and 100 Hz
d. $\frac{(4s+1)}{s(s+2)(s+3)}$ at $\omega = 0.1$ rad/s and 400 rad/s

8. Calculate and plot the frequency response of the system with given transfer function over a frequency range of 0.1 to 100 rad/s:

10^{-1} to 10^{2}

$$\frac{s+3}{s(s+10)}$$

ω = logspace(-1, 2)

9. The transfer function of an electric circuit is given here. Determine
a. DC gain and phase angle
b. Gain and phase angle at a frequency of 1 KHz

$$\frac{20}{10s^2+4s+1}$$

10. An amplifier is subjected to a sinusoidal voltage given as $v = 1.25 \sin 100t$. Determine the amplitude of output signal and the phase difference between output and input. The amplifier transfer function is

$$\frac{10}{0.01s + 1}$$

11. The output response of a temperature transducer is given by the transfer function

$$\mathrm{TF} = \frac{1}{10s + 1}$$

The transducer is used in a wind tunnel and is subject to cyclic variation in temperature. Determine the maximum frequency (of temperature variation) that can be measured with this transducer if the measurement errors are limited to 10%.

12. Draw the frequency response of the following using MATLAB (or a similar program), and determine the values of maximum and minimum gain and phase angle that can exist over the entire frequency range.

a. $\dfrac{s^2 + s + 10}{s}$

b. $\dfrac{100}{(s + 2)(s + 6)}$

c. $\dfrac{s(0.01s + 1)}{(s + 1)(0.25s^2 + 0.2s + 1)}$

7

COMMON TRANSFER FUNCTIONS

Peak gain

$G_p = \frac{1}{2\zeta \cdot \sqrt{1-\zeta^2}}$ $\zeta = 3$

INTRODUCTION

The general nature of the time and frequency response of a system is completely described by its transfer function. The transfer function of a system does not provide any information about the physical nature of the system, and it is possible to have two entirely different systems to have identical transfer functions. Similar systems have similar functions with possibly different coefficient values and generate identical responses. A number of transfer functions are commonly found in control system analysis, and in this chapter some of the commonly occurring transfer functions are identified. Transient response (Chapter 5) and frequency response (Chapter 6) characteristics of these common transfer functions are discussed.

Consider two series *RC* circuits with each circuit using a different value of resistor and capacitor. Even though the two circuits may be numerically different, these are very similar in nature and are classified as *first-order lag* systems. The transfer function of an *RC* circuit is given as

$$\text{TF} = \frac{1}{(RCs + 1)}$$

With different values of resistors and capacitors, the transfer functions may differ in the coefficient values. However, when subjected to a unit step input, both circuits exhibit identical transient and frequency responses.

EXAMPLE 7.1

Two *RC* circuits are shown in Figure 7.1. Component values for each of these two circuits are different, and as a result the two transfer functions (V_{out}/V_{in}) are as shown in the figure.

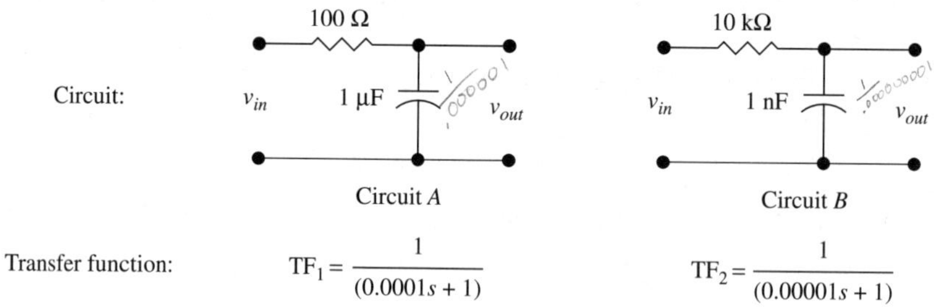

Figure 7.1

The standard form of the transfer function for a first-order lag-type system is given as

$$\text{transfer function} = \frac{1}{(\tau s + 1)}$$

where the value of τ is dependent upon the components in the system.

All systems and/or elements that can be modeled by a first-order lag-transfer function exhibit identical time and frequency responses (Figure 7.2). When subjected to a unit step input (1 V dc), both the preceding circuits will reach the 63.2% (0.632 V) of final output (1 V) in one time constant. Because the numerical value of the time constant is different for two systems, the time taken in seconds to reach this value is going to be different for the two circuits. But it is the similarity in behavior that is the dominant reason to classify the transfer functions into various groups, with each group displaying unique characteristic behavior.

The concept of the standard form of transfer functions is not limited to processes, which are physically of the same type. Processes from different disciplines (mechanical, electrical, thermal, etc.) can be grouped together. A servohydraulic control system used for a material testing may be similar to the transfer function of an oven-temperature control system. Because of similar transfer functions, both systems will exhibit similar response. It is possible to classify transfer functions representing various processes into several groups, with each group displaying unique operating characteristics.

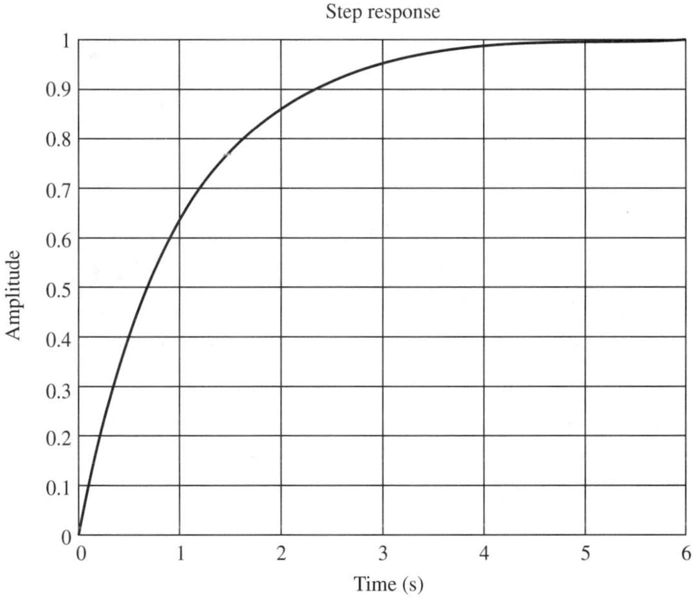

Figure 7.2 Unit step response: first-order lag system.

COMMON BLOCKS (TRANSFER FUNCTIONS)

The transfer function of a control system can be expressed as a polynomial of s. The polynomial may be of third order or higher, but it can possibly be factored into simpler first- and second-order terms.

Note: The term *system* is used to represent the unit under consideration, which may be a simple control element, a group of elements, or even a complete control system.

EXAMPLE 7.2

Express the following polynomial in simpler terms:

$$\mathrm{TF} = (s^3 + 2s^2 + 2s + 1)$$

Solution

It is a third-order transfer function and can be factored into two terms: a first-order term and a second-order term (Figure 7.3).

The block with the third-order polynomial is equivalent to two blocks in cascade containing second-order and first-order polynomials. The frequency response of the

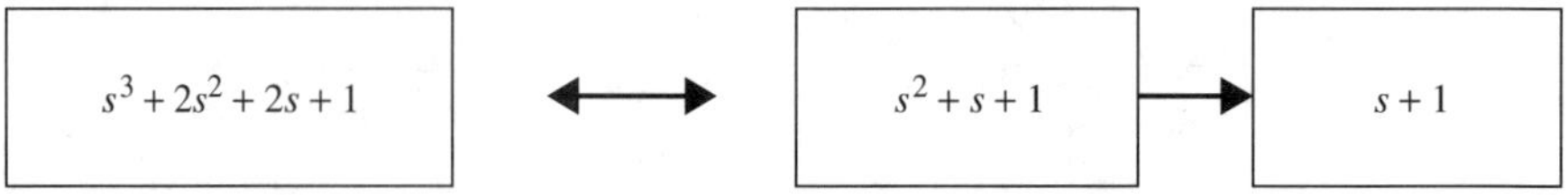

Figure 7.3 Factoring into lower-order terms in Example 7.2.

first- and second-order terms can be easily plotted, and overall response can be obtained through simple addition or subtraction.

A higher-order transfer function can usually be factored into one or more of the following simpler transfer functions. Some of these are also used to represent the complete transfer function of simpler systems, such as a spring mass system, RC circuit, and dc motor. Each of these transfer functions is referred to as a block (implying a system, factor, term, or a process):

1. Constant, or proportional, block
2. Integral block
3. Derivative block
4. First-order lag block
5. First-order lead block
6. Second-order lag block
7. Second-order lead block
8. Dead-time, or transportation, lag block

CONSTANT BLOCK

The constant block is the simplest of all and is the only transfer function that is totally independent of the operating frequency. Common examples are gear trains, potentiometers, and mechanical linkages. Most operational amplifiers operating at low to moderate gains can also be considered to provide constant transfer functions (Figure 7.4).

$$\mathrm{TF} = K$$

where K = system gain

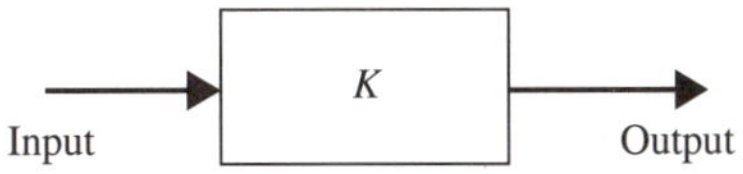

Figure 7.4 Constant block.

EXAMPLE 7.3

A 28-tooth gear wheel drives a 7-tooth pinion (smaller gear). Determine its transfer function.

Solution

$$\text{gear-train transfer function} = \frac{\text{output speed}}{\text{input speed}} = \frac{28}{7} = 4$$

See Figure 7.5.

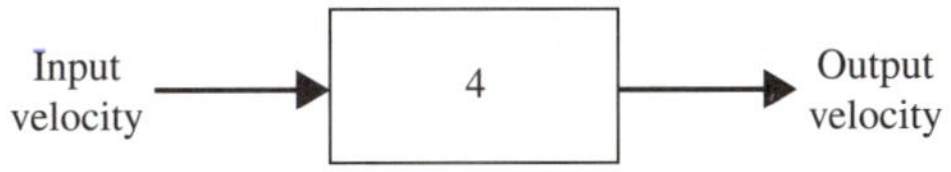

Figure 7.5 Diagram for Example 7.3.

The ratio between input and output rotational velocity always remains constant for all values of input velocity.

EXAMPLE 7.4

A potentiometer is connected across a 12-V dc source. It is adjusted to provide an output of 2.15 V.

a. Determine the transfer function.

b. The potentiometer can be readjusted to provide an output voltage ranging from 0 to 12 V. If this is the case, how can a potentiometer transfer function be considered a constant? See Figure 7.6.

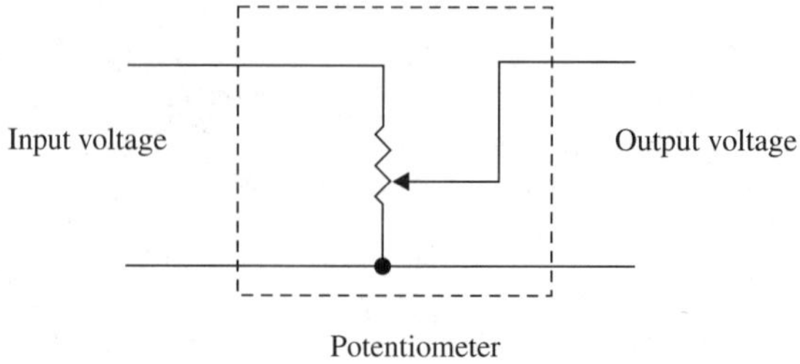

Figure 7.6 Potentiometer for Example 7.4.

Solution

a.

$$\text{potentiometer transfer function} = \frac{\text{output voltage}}{\text{input voltage}} = \frac{2.15}{12} = 0.179$$

See Figure 7.7.

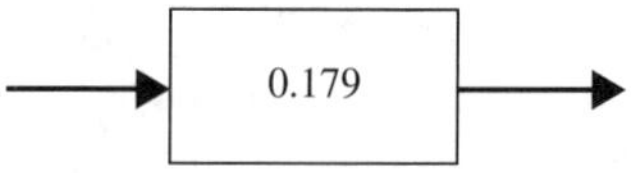

Figure 7.7 Block representing a potentiometer.

b. The potentiometer is an adjustable device, this being the very reason for its use in a circuit. Once a potentiometer is set to meet the circuit requirement, the relationship between output and input voltage is fixed and is given by a numerical constant.

EXAMPLE 7.5

Determine the (closed-loop) transfer function of the negative-feedback closed-loop control system in Figure 7.8 and decide if it can be represented as a proportional block (transfer function).

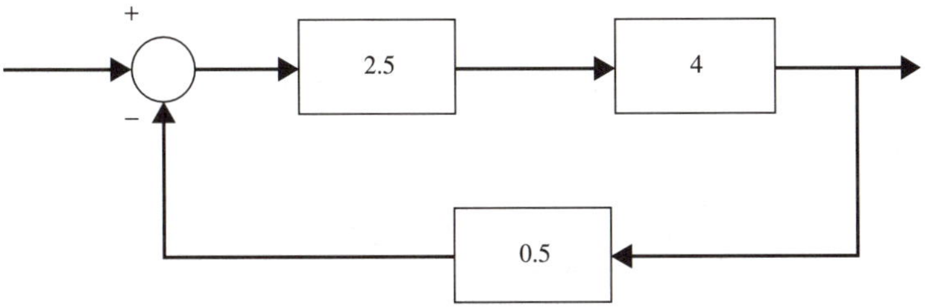

Figure 7.8 Block diagram for Example 7.5.

Solution

$$\text{closed-loop transfer function} = \frac{G}{1 + GH}$$
$$= \frac{2.5 \times 4}{1 + 2.5 \times 4 \times 0.5}$$
$$= 1.667$$

The transfer function is a numerical constant; hence the preceding closed-loop control system can be represented by a constant block (Figure 7.9).

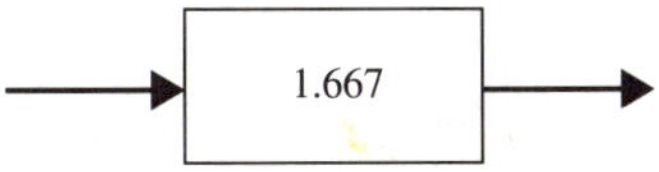

Figure 7.9 Overall TF for Example 7.5.

Time Response of a Constant TF

The transient response of a constant block, when subjected to a unit step input, is a straight line with constant amplitude equal to the transfer function (gain constant K). Figure 7.10 shows the step response of a proportional block with a transfer function of 0.5.

Frequency Response of a Constant Block

Because the transfer function of a constant block is independent of frequency, the magnitude plot is also a straight line with a phase difference of 0°.

$$\text{magnitude} = 20 \log(K)$$
$$\text{phase angle} = 0°$$

b/c resistive

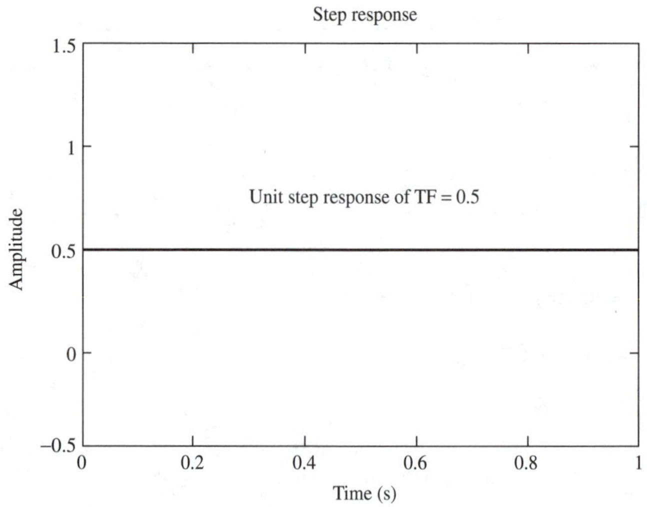

Figure 7.10 Unit step response: constant block.

The preceding Bode plot represents a constant block with a transfer function of 0.5 (Figure 7.11).

$$\text{magnitude} = 20\log(0.5)$$
$$= -6 \text{ dB}$$
$$\text{phase angle} = 0°$$

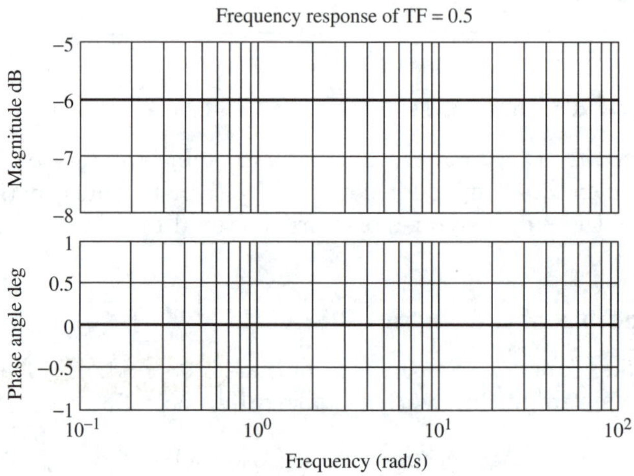

Figure 7.11 Frequency response: constant block.

EXAMPLE 7.6

A constant block has a magnitude of 20 dB at 100 rad/s. Determine each of the following:

a. Block transfer function
b. Magnitude at a frequency of 1000 rad/s
c. Phase angle at 1000 rad/s

Solution

a. The magnitude of a constant block is 20 log(K) dB. Therefore,
20 dB = 20 log(K) dB, or, $K = 10$.
The constant block has a transfer function of 10.

b. The magnitude of a constant block is the same at all frequencies. Therefore,

$$\text{magnitude at 1000 rad/s} = \text{magnitude at 100 rad/s}$$
$$= 20\,\text{dB}$$

c. There is no phase shift from input to output for a constant block. The phase difference is 0°.

EXAMPLE 7.7

A sinusoidal waveform, 10 sin(120t), is fed to a constant block with a transfer function of 0.675. Determine the following.

a. Signal gain (amplification)
b. Output amplitude
c. Phase difference between input and output

Solution

a. For a constant block, the signal gain is equal to the transfer function value.

$$\text{signal gain} = 0.675$$

b.

$$\text{output amplitude} = \text{input amplitude} \times \text{gain}$$
$$= 10\text{ V} \times 0.675$$
$$= 6.75\text{ V}$$

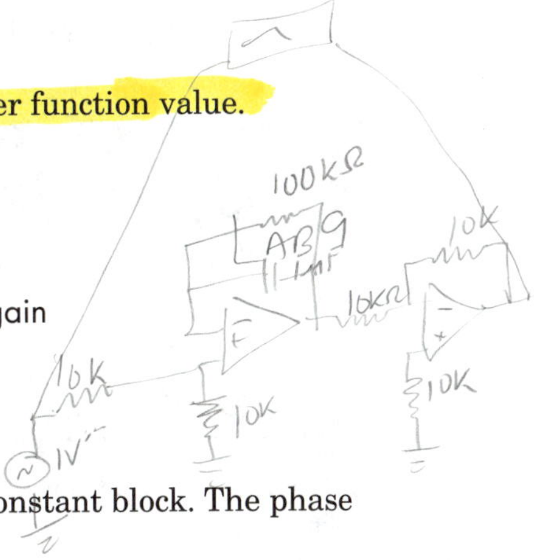

c. The output signal is in phase with the input signal for a constant block. The phase difference is 0°.

INTEGRAL BLOCK

In an integral system the output does not depend upon the current value, but rather the entire past history (time integral) of the input. Common examples are an integrating amplifier, a capacitor fed with a current source, a flow-totalizer, and a level control system with a constant-outflow device. Its transfer function is of the form

$$\text{TF} = \frac{1}{\tau s}$$

where τ is the integral time constant in seconds (Figure 7.12).

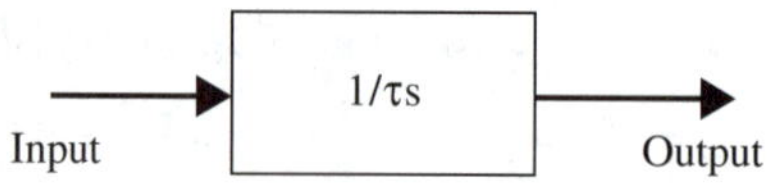

Figure 7.12 Integral block.

Time Response of an Integral Block

The unit step response of an integral block is a ramp (straight line with a constant, but nonzero, slope). The slope of the ramp is dependent upon the value of integral time constant τ and is given by

$$\text{ramp slope} = \frac{1}{\tau}$$

The output-signal amplitude, when subjected to a step input, is given as

$$\text{output amplitude} = A \times \text{ramp slope} \times T$$

where A = amplitude of step input
T = time elapsed since application of step input

EXAMPLE 7.8

The transfer function of an integral block is given by $\text{TF} = 1/10s$. Determine its step response.

Solution

The step response is shown in Figure 7.13.

$$
\begin{aligned}
\text{slope of ramp} &= \frac{\partial a \text{ (change in amplitude)}}{\partial t \text{ (change in time)}} \\
&= \frac{0.1}{1} \\
&= 0.1 \\
&= \frac{1}{10 \text{ (time constant)}}
\end{aligned}
$$

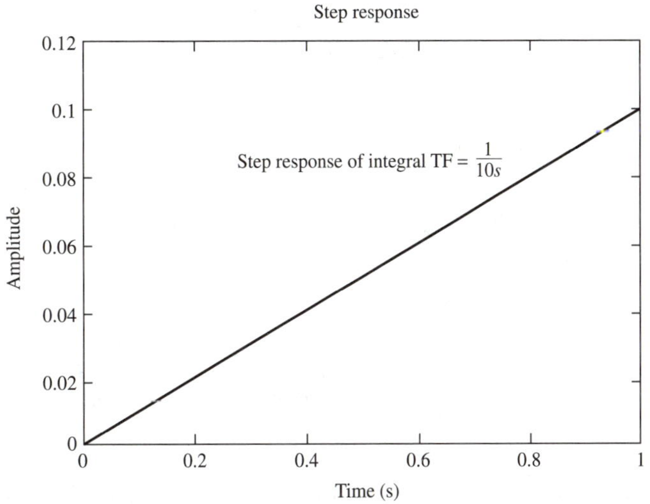

Figure 7.13 Unit step response for Example 7.8.

EXAMPLE 7.9

A system with a transfer function TF = $1/100s$ is subjected to a step input of 1 V dc. Determine the output signal amplitude after 10 s.

Solution

The output signal amplitude is $A \times$ ramp slope $\times\ T$. Here,

$$
\begin{aligned}
A &= 1\text{V} \\
\text{ramp slope} &= 0.01 \text{ V/V/s} \\
T &= 10 \text{ s}
\end{aligned}
$$

Therefore,

$$\text{output signal amplitude} = 1 \times 0.01 \times 10\ \text{V}$$
$$= 0.1\ \text{V}$$

Figure 7.14 shows the step response of this integral block for a time period of 0 to 100 s.

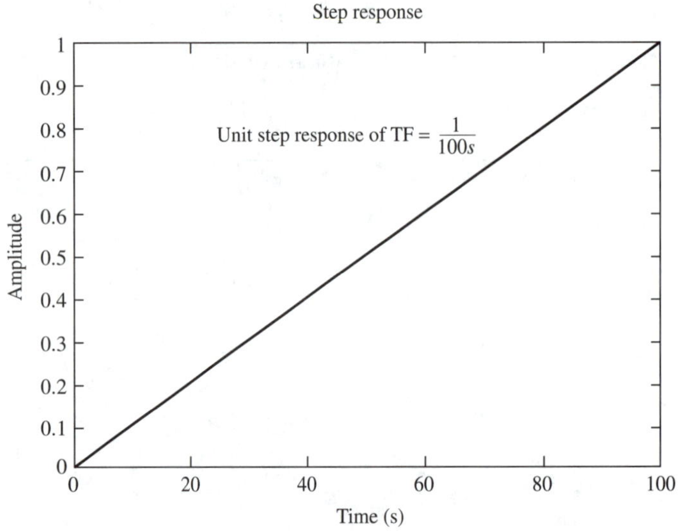

Figure 7.14 Step response for Example 7.9.

Frequency Response of an Integral Block

Signal gain for an integral block is inversely proportional to the signal frequency. It is infinite for zero frequency and drops at a constant slope to zero as the frequency approaches infinity. The gain rolls off at a rate of −20 dB/decade. The phase angle is constant at −90° for all frequencies. The integral time constant τ dictates the frequency at which the magnitude line crosses the 0-dB line and is given as

$$\text{0-dB crossover frequency} = \frac{1}{\tau}\ \text{rad/s}$$

EXAMPLE 7.10

Draw the frequency response of a block with transfer function TF = $1/10s$.

Solution

$$\begin{aligned}\text{0-dB crossover frequency} &= \frac{1}{\tau}\ \text{rad/s}\\ &= \frac{1}{10}\ \text{rad/s}\\ &= 0.1\ \text{rad/s}\end{aligned}$$

The frequency plot is shown in Figure 7.15.

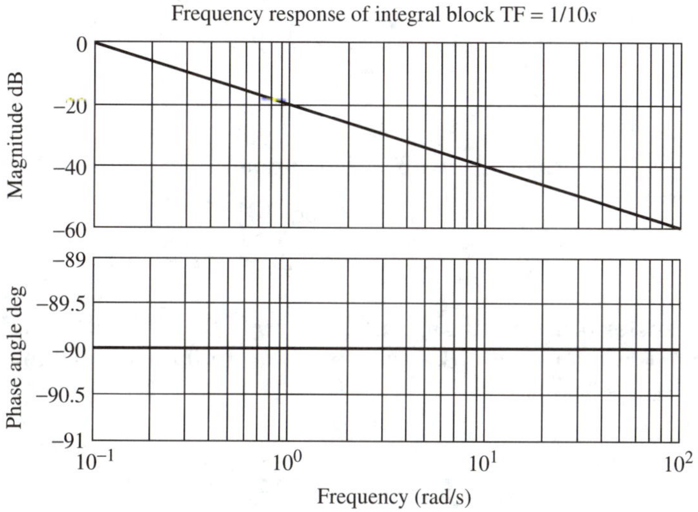

Figure 7.15 Frequency plot for Example 7.10.

It can be seen that the 0-dB crossover frequency is 0.1 rad/s and that the phase angle is a constant -90°. This implies that output signal is always 90° behind the input signal.

EXAMPLE 7.11

A sinusoidal signal of 10 V peak amplitude and a frequency of 100 rad/s is applied to an integral system. Determine the system transfer function if the peak amplitude of output signal is 1 V.

Solution

The standard form of the integral transfer function is

$$\text{TF} = \frac{1}{\tau s}$$

The transfer function is specified as soon as the integral time constant τ is determined. The time constant is related to break frequency, at which the magnitude (gain in dB) is 0 dB.

Here, both the input and output amplitudes are known. Thus, the signal gain and, correspondingly, the magnitude at this frequency (100 rad/s) can be determined.

$$\begin{aligned}
\text{gain} &= \frac{\text{peak output amplitude}}{\text{peak input amplitude}} \\
&= \frac{1\text{ V}}{10\text{ V}} \\
&= 0.1 \\
\text{magnitude in dB} &= 20\log(\text{gain}) \\
&= 20\log(0.1) \\
&= -20\text{ dB}
\end{aligned}$$

The integral transfer function has a magnitude roll-off of -20 dB/decade. Because the magnitude at 100 rad/s is -20 dB, it will be equal to 0 dB at one decade lower than current 100 rad/s.
Therefore,

$$\begin{aligned}
\text{crossover frequency} &= \frac{100\text{ rad/s}}{10} \\
&= 10\text{ rad/s} \\
\text{integral time constant } \tau &= \frac{1}{\text{crossover frequency}} \\
&= \frac{1}{10\text{ rad/s}} \\
&= 0.1\text{ s}
\end{aligned}$$

Therefore,

$$\begin{aligned}
\text{transfer function TF} &= \frac{1}{0.1\text{ s}} \\
&= \frac{10}{s}
\end{aligned}$$

The frequency response of this system is shown in Figure 7.16.

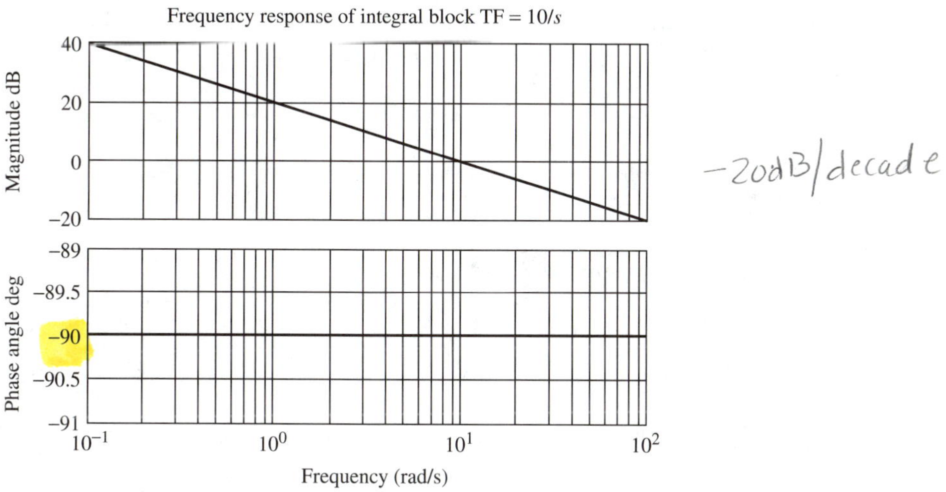

Figure 7.16 Frequency response for Example 7.11.

DERIVATIVE BLOCK

In a derivative system the output depends on neither the current value nor the past history. It depends simply on the rate at which the input signal is changing. Common examples are a derivative amplifier, the voltage across an inductive coil, and tachogenerator output voltage. Its transfer function is of the form

$$\text{TF} = \tau s$$

where τ = derivative time constant (s^{-1})

See Figure 7.17.

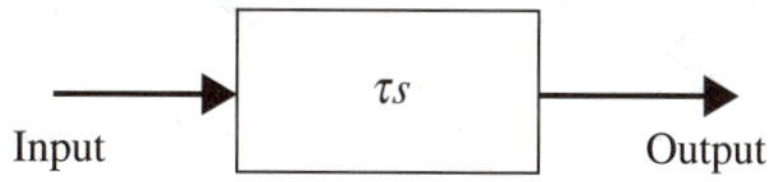

Figure 7.17 Derivative block.

Time Response of a Derivative Block

When a derivative block is subjected to a step input, a sharp impulse is generated at the instant input is applied. The impulse immediately collapses, and the output returns to zero. There is no further output from a derivative block. So, for all practical purposes it can be said that output response of a derivative block when subjected to a step input is zero (Figure 7.18).

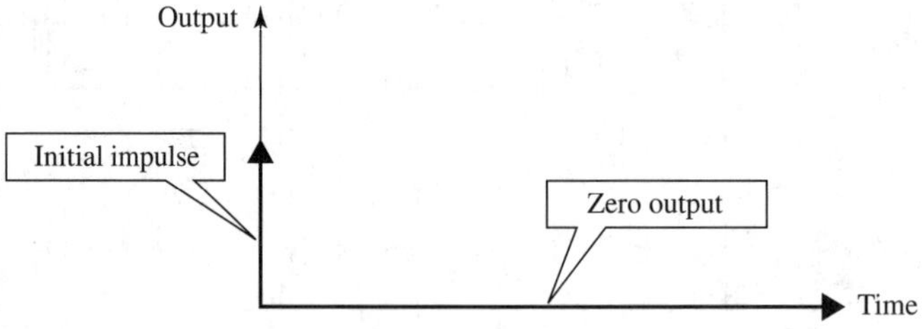

Figure 7.18 Unit step response: derivative block.

Frequency Response of a Derivative Block

For a derivative block, signal gain is directly proportional to the signal frequency. The signal gain (G) is zero for zero frequency and increases at a constant slope to infinity as frequency approaches infinity. The gain increases at a rate of +20 dB/decade. The phase angle is constant at +90° for all frequencies. The derivative time constant τ is inversely proportional to the crossover frequency, the frequency at which the magnitude curve crosses the 0-dB line.

$$\text{0-dB crossover frequency} = \frac{1}{\tau}\ \text{rad/s}$$

EXAMPLE 7.12

Draw the frequency response of a block with a transfer function TF = $10s$.

Solution

$$\begin{aligned}\text{0-dB crossover frequency} &= \frac{1}{\tau}\ \text{rad/s}\\ &= \frac{1}{10}\ \text{rad/s}\\ &= 0.1\ \text{rad/s}\end{aligned}$$

The frequency plot is shown in Figure 7.19.

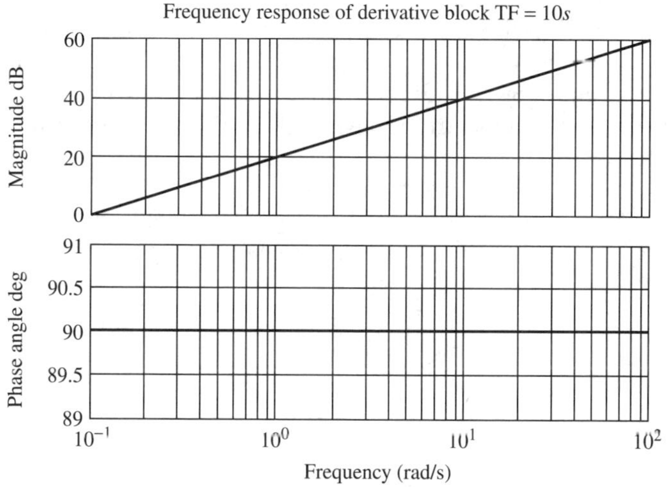

Figure 7.19 Frequency response for Example 7.12.

The 0-dB crossover occurs at a frequency of 0.1 rad/s. Phase angle is constant at +90° for all frequencies. In other words, for a derivative system the output signal always leads the input signal by +90°.

EXAMPLE 7.13

A signal generator applies a sinusoidal waveform to a derivative system. The signal frequency is varied over a range of 0.1 rad/s to 100 rad/s, and the data on the input and output peak amplitudes are recorded in Table 7.1. Determine the system transfer function.

Table 7.1 Frequency Data for Example 7.13

Frequency (rad/s)	***Input Amplitude (V)***	***Output Amplitude (V)***
0.1	0.25	0.0025
1	0.25	0.025
10	0.25	0.25
100	0.25	2.5

Solution

The general form of the transfer function for a derivative system is TF $= \tau s$.

The transfer function is completely defined once the value of derivative time constant τ is determined. This time constant τ is related to break frequency, which in turn is the frequency at which the magnitude (gain = 1) is 0 dB. So the problem boils down to determining the frequency in radian/seconds, at which the gain is equal to 1.

From the experimental data in Table 7.1, the gain at each of the four frequencies can be calculated.

$$\begin{aligned}\text{gain at } 0.1 \text{ rad/s} &= \frac{\text{peak amplitude of output signal}}{\text{peak amplitude of input signal}}\\ &= \frac{0.0025 \text{ V}}{0.25 \text{ V}}\\ &= 0.01\end{aligned}$$

Similarly,

$$\begin{aligned}\text{gain at } 1 \text{ rad/s} &= 0.1\\ \text{gain at } 10 \text{ rad/s} &= 1\\ \text{gain at } 100 \text{ rad/s} &= 10\end{aligned}$$

Because the crossover frequency is the frequency at which gain is unity, it is evident from the preceding calculations that this frequency is equal to 10 rad/s.
Therefore,

$$\begin{aligned}\text{derivative time constant } \tau &= \frac{1}{\text{crossover frequency}}\\ &= \frac{1}{10 \text{ rad/s}}\\ &= 0.1 \text{ s}\\ \text{system transfer function TF} &= 0.1\, s\end{aligned}$$

EXAMPLE 7.14

Determine the signal gain of the given derivative system at the following frequencies.

a. 1000 rad/s
b. 60 Hz

$$\text{system transfer function TF} = 10s$$

Solution

The transfer function (derivative time constant τ) indicates the crossover frequency at which the gain is unity.

$$\text{gain crossover frequency} = \frac{1}{\text{derivative time constant}}$$
$$= \frac{1}{10}$$
$$= 0.1 \text{ rad/s}$$

The gain of the derivative system increases at a rate of 20 dB/decade. Thus, knowing the gain crossover frequency, gain at any other frequency can be easily calculated.

a. The number of decades, n, between 1000 rad/s and 0.1 (the crossover frequency) is given by

$$n = \log\left(\frac{1000 \text{ rad/s}}{0.1 \text{ rad/s}}\right)$$
$$n = 4$$

Therefore, the gain at 4 decades after the crossover frequency is slope of

$$\text{gain frequency curve} \times \text{number of decades} = 20\,\text{dB/decade} \times 4\,\text{decades}$$
$$= 80\,\text{dB}$$

Amplification (simple gain) can be derived from the gain in dB:

$$\text{gain} = 10^{80/20}$$
$$= 10{,}000$$

The same result can be obtained by the following simplified approach:

$$\text{gain at 1000 rad/s} = 10^{(1000\text{ rad/s})/(0.1\text{ rad/s})}$$
$$= 10{,}000$$

b. The frequency information needs to be converted to radians per second:

$$\omega = 60 \times 2\pi \text{ rad/s}$$
$$= 376.991 \text{ rad/s}$$

Using the second approach,

$$\text{gain at 60 Hz} = 10^{(376.991\text{ rad/s})/(0.1\text{ rad/s})}$$
$$= 3769.91 \quad (71.5\,\text{dB})$$

The results can be verified from the frequency response plot for this system, as repeated in Figure 7.20.

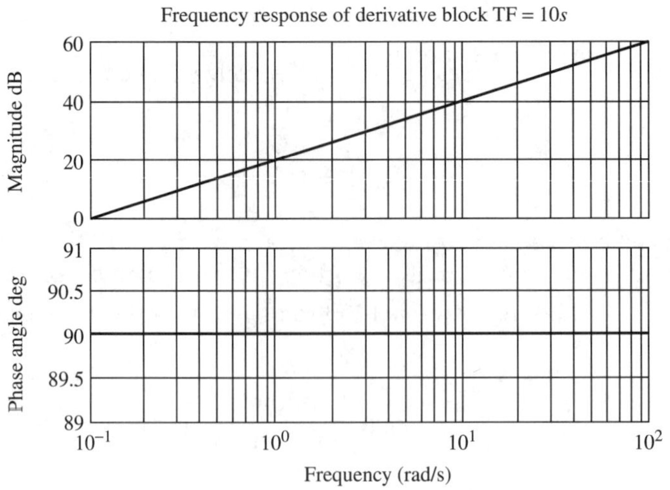

Figure 7.20 Frequency response for Example 7.14.

The Relationship Between Integral and Derivative Blocks

The characteristics of integral and derivative systems are complementary to each other. As a matter of fact, one can be considered as the ***mirror image*** of the other.

Phase Angle

- The integral system provides a 90° phase lag.
- The derivative system provides a 90° phase lead.

Magnitude

- The integral system provides gain, which constantly decreases with an increase in frequency at the rate of −20 dB/decade.
- The derivative system provides gain, which constantly increases with an increase in frequency at the rate of +20 dB/decade.

FIRST-ORDER LAG BLOCK

Systems that can be represented by this type of transfer function occur commonly in nature. Temperature-measuring devices, electric motors (simplified models), *RC* circuits, *RL* circuits, and tank-level control systems are some of the examples. Real-life operational amplifiers operating over a wide gain and frequency range also exhibit this characteristic. This type contains one energy-storage element and one dissipative element (Figure 7.21).

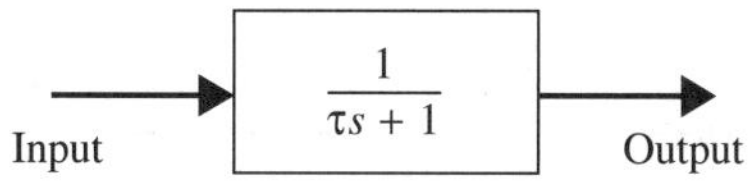

Figure 7.21 First-order lag block.

Real-Time Equation

General Form:

$$\tau \frac{dc}{dt} + c(t) = Kr(t)$$

where

τ = system time constant (s)
K = steady-state gain
$c(t)$ = output
$r(t)$ = input

Normalized Form:

$$\tau \frac{dc}{dt} + c(t) = r(t) \quad \text{(unity steady-state gain)}$$

Transfer Function

General Form:

$$\mathrm{TF} = \frac{C(s)}{R(s)} = \frac{K}{\tau s + 1}$$

Normalized Form:

$$\mathrm{TF} = \frac{C(s)}{R(s)} = \frac{1}{\tau s + 1}$$

Time Response to a Unit Step Input

When subjected to a unit step input, the output of a first-order lag process rises in an exponential manner. Output rises to 63.2% of the final output within one time constant and reaches a 99.3% value in 5 time constants. Figure 7.22 (reproduced from Figure 7.1) shows the output response of a normalized first-order lag process when subjected to a

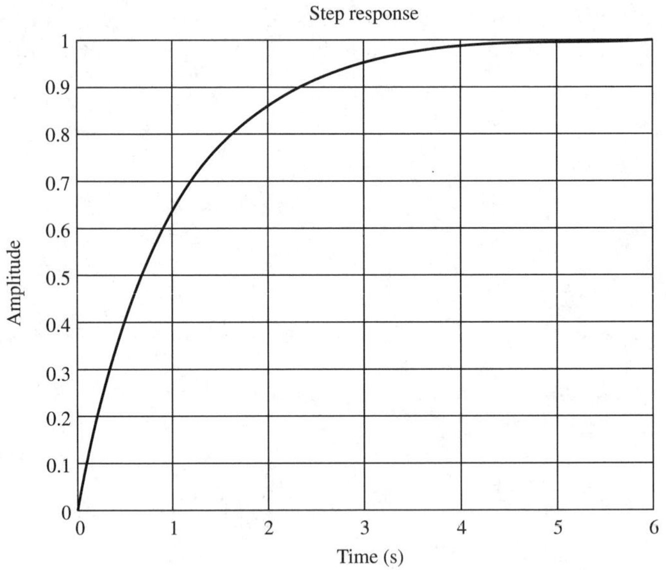

Figure 7.22 Unit step response: first-order lag block.

unit step input. The time constant in this case is 1. The final output for a normalized first-order lag system is same as the input, irrespective of the time constant.

Mathematically, the output step response at any time can be determined from the following expression:

$$\text{step response } c(t) = AK(1 - e^{-t/\tau})$$

where

$C(t)$ = output
A = amplitude of applied step input
K = steady-state system gain
τ = system time constant (s)

EXAMPLE 7.15

An input of 1.5 V is applied to each of the following first-order lag systems. Determine the time constant and final output in each case.

a. $\dfrac{1}{s+1}$

b. $\dfrac{1}{10s+1}$

c. $\dfrac{1}{0.2s+1}$

Solution

All these transfer functions are expressed in standard form. Thus, the time constant can directly be read from the transfer function in each case. It is equal to the coefficient of the *s* term.

Because the numerator coefficient in each case is 1, the steady-state gain is also equal to unity, and the final output amplitude will always be equal to the amplitude of the input signal.

a.

$$\begin{aligned}\text{time constant} &= 1\ \text{s}\\ \text{steady-state output} &= \text{applied input}\\ &= 1.5\ \text{V}\\ \text{output after one time constant} &= 63.2\% \times 1.5\ \text{V}\\ &= 0.9525\ \text{V}\end{aligned}$$

The output will reach 0.9525 V 1 s after the application of the step input of 1.5 V.

b.

$$\begin{aligned}\text{time constant} &= 10\ \text{s}\\ \text{steady-state output} &= \text{applied input}\\ &= 1.5\ \text{V}\end{aligned}$$

The output will reach 0.9525 V 10 s after the application of the step input.

c.

$$\begin{aligned}\text{time constant} &= 0.2\ \text{s}\\ \text{steady-state output} &= \text{applied input}\\ &= 1.5\ \text{V}\end{aligned}$$

The output will reach 0.9525 V 0.2 s after the application of the step input.

EXAMPLE 7.16

In a level control system, the fluid level is related to the inflow rate by the following transfer function:

$$\text{TF} = \frac{\text{level } H(s)}{\text{inflow } Q(s)} = \frac{0.75}{2500s + 1}$$

where

$$h = \text{level in m}$$
$$q = \text{inflow in m}^3/\text{s}$$

Initially the tank is empty. The inlet-flow control valve is suddenly opened to allow a flow rate of 0.5 m^3/s into the tank. Determine the following:

a. Steady-state gain
b. Time constant
c. Level after 10 min
d. Final steady-state level

Solution

The value of the steady-state gain and the time constant can be read directly from the transfer function.

a. Steady-state gain = 0.75
b. Time constant = 2500 s
c. Output response is given by the standard expression

$$\text{step response } c(t) = AK(1 - e^{-t/\tau})$$

Here,

$$A = 0.5$$
$$K = 0.75$$
$$\tau = 2500$$

Therefore,

$$\begin{aligned}\text{level} &= 0.5 \times 0.75(1 - e^{-t/2500}) \text{ m}\\ &= 0.375(1 - e^{-t/2500}) \text{ m}\end{aligned}$$

At time $t = 10$ min (600 s),

$$\begin{aligned}\text{level} &= 0.375(1 - e^{-600/2500}) \text{ m}\\ &= 0.08 \text{ m}\end{aligned}$$

d. The final level is the value reached at time $t \to \infty$.

$$\begin{aligned}\text{final level} &= 0.375(1 - e^{-\infty/2500})\\ &= 0.375 \text{ m}\end{aligned}$$

EXAMPLE 7.17

An unknown system was subjected to a unit step input and the output response was recorded on a chart recorder. It was observed that process took 12.5 s to reach 63.2% of the final (settled) value. From the experimental data, the steady-state gain was estimated to be 0.6. Assuming a first-order lag model for this process, determine the process transfer function.

Solution

The general form of the transfer function for a first-order lag process is given as

$$\mathrm{TF} = \frac{C(s)}{R(s)} = \frac{K}{\tau s + 1}$$

A first-order lag process takes 1 time constant to reach 63.2% of the final value. Thus, based upon experimental data the system time constant τ is 12.5 s. Because the steady-state gain is estimated to be 0.6, the process transfer function (assuming a first-order model) is

$$\frac{0.6}{12.5s + 1}$$

Frequency Response of a First-Order Lag Block

A first-order lag process contains one resistive (dissipative) and one capacitive (energy-storage) element. At very low frequency, the behavior is governed by the resistive element, and at higher frequency it is governed by the capacitive element. Low-frequency behavior resembles that of a proportional element, whereas the high-frequency behavior resembles that of an integral element. Figure 7.23 shows the Bode plot of a normalized (unity steady-state gain) first-order lag process with a transfer function of $\frac{1}{s+1}$.

Straight-Line Approximation If the frequency response is plotted over a much wider frequency range, a pattern begins to emerge. Figure 7.24 shows the Bode plot of the same transfer function, but covering a much wider frequency range. It is evident from the second plot that the frequency response of a first-order lag process can be approximated by straight-line segments. These segments are asymptotic to the curve. This assumption is more obvious for the magnitude plot than the phase-angle plot. The magnitude plot is approximated by two straight-line segments: a horizontal line asymptotic to the curve in the lower-frequency region and a straight line, sloping downward at higher frequency. The phase-angle plot is approximated by

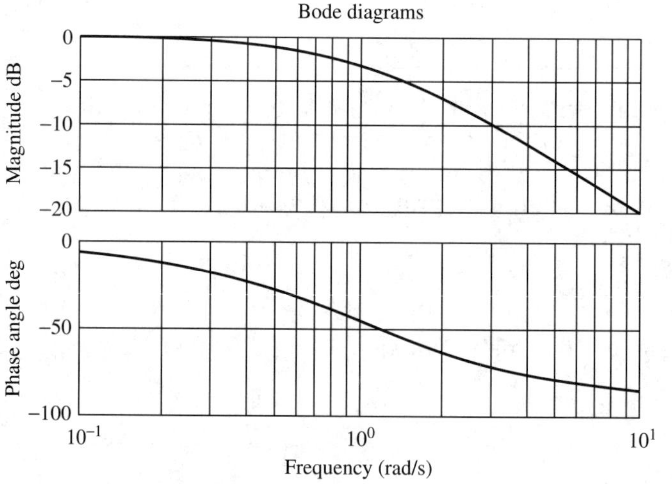

Figure 7.23 Frequency response: first-order lag block.

three straight-line segments, two horizontal and one sloping segment, which joins the horizontal ones.

These approximations allow quick sketching of frequency response plots and have been extensively used by control engineers. Lately, a number of powerful and yet economical software packages have become available. These software packages allow control personnel to quickly and accurately determine the frequency and time response of control components and systems.

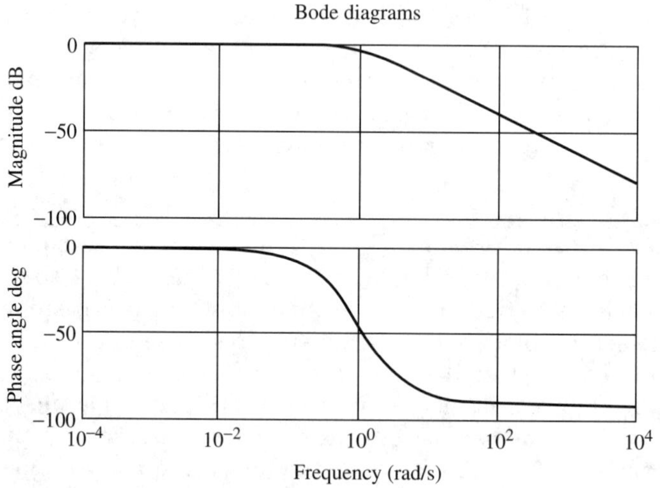

Figure 7.24 First-order response: extended frequency range.

The maximum error in the straight-line approximation and the actual curve occurs at the break frequency. At this frequency, the actual magnitude value is smaller than the one given by the straight-line approximation by 3 dB.

Corner Frequency: The frequency at which the two asymptotes meet is known as the corner frequency. It is also referred to as *break frequency*. It is related to the time constant τ, and is given as

$$\text{corner frequency} = \frac{1}{\tau} \text{ rad/s}$$

Magnitude Plot: The magnitude plot starts near zero frequency at a magnitude given by

$$M = 20 \log (\text{steady-state gain})$$

Magnitude then starts to decrease slowly; after break frequency, it starts to drop sharply at a rate of 20 dB/decade. When using asymptotic approximations, it can be assumed that magnitude remains constant at the preceding value until break frequency is reached. The magnitude after break frequency can be calculated by reducing the preceding value by −20 dB/decade of frequency.

Phase-Angle Plot: The phase angle near zero frequency is always equal to 0° and drops to −90° at very high frequency. At break frequency the phase angle is precisely −45°. The transition from 0° to −90° occurs mainly in the neighborhood of break frequency. These relationships hold true for all first-order lag processes, irrespective of time constant and/or steady-state gain.

When using straight-line approximations, it can be assumed that phase remains zero from zero to one decade before the break frequency. It then drops from 0° to −90° in a straight line. The drop starts one decade before the break frequency and reaches −90° one decade after the break frequency. After this frequency the phase angle is assumed to remain constant at −90°.

EXAMPLE 7.18

Determine the corner frequency for a first-order lag system with the given transfer function.

$$\text{TF} = \frac{C(s)}{R(s)} = \frac{1}{0.125s + 1}$$

Solution

The corner frequency for a first-order lag system is related to the time constant through the expression

$$\text{corner frequency} = \frac{1}{\tau} \text{ rad/s}$$

From the transfer function, the time constant for the given system is 0.125. Therefore,

$$\text{corner frequency} = \frac{1}{0.125}\ \text{rad/s}$$
$$= 8.0\ \text{rad/s}$$

EXAMPLE 7.19

Determine the frequency at which phase angle is $-45°$ for the following system.

$$\text{TF} = \frac{C(s)}{R(s)} = \frac{6.25}{50s + 10}$$

Solution

This system is a first-order lag system, and for such a system the phase angle reaches $-45°$ at the corner frequency. Before corner frequency can be evaluated, the transfer function has to be represented in standard form. Rearranging the terms,

$$\text{TF} = \frac{6.25}{10(5s + 1)}$$
$$\text{TF} = \frac{0.625}{5s + 1}$$

The time constant can now be read directly from the transfer function and is equal to 5. Therefore,

$$\text{corner frequency} = \frac{1}{5}\ \text{rad/s}$$
$$= 0.2\ \text{rad/s}$$

The phase angle is $-45°$ at a frequency of 0.2 rad/s.

EXAMPLE 7.20

The transfer function of a first-order lag system is as follows:

$$\text{TF} = \frac{2}{0.05s + 1}$$

Assuming straight-line approximation, determine the following:

a. Corner frequency in Hz
b. Magnitude at 0.1 rad/s

C1

FIRST ORDER LAG

c. Magnitude at 2000 rad/s
d. Phase angle at 2000 rad/s

Solution

a.

$$\text{corner frequency} = \frac{1}{0.05}\ \text{rad/s}$$
$$= 20\ \text{rad/s}$$

Normally, it is a good idea to sketch the Bode plot of this system and label the salient points, but it is not absolutely essential to do so.

b. A frequency of 0.1 rad/s is to the left of the corner frequency. The region to the left of the corner frequency is assumed to have a constant gain equal to the steady-state gain.
Thus,

$$\text{gain at } 0.1\ \text{rad/s} = \text{steady-state gain}$$
$$= 2$$
$$\text{magnitude} = 20 \log(2)$$
$$= 6\ \text{dB}$$

c. The number of decades between the current frequency (2000 rad/s) and the corner frequency (20 rad/s) is

$$\log\left(\frac{2000\ \text{rad/s}}{20\ \text{rad/s}}\right) = 2$$

After the corner frequency, the magnitude drops at a rate of −20 dB/decade.

$$\text{drop in magnitude in 2 decades} = -20 \times 2\ \text{dB}$$
$$= -40\ \text{dB}$$
$$\text{net magnitude at 2000 rad/s} = \text{magnitude at corner frequency} - \text{drop in magnitude}$$
$$= 6\ \text{dB} - 40\ \text{dB}$$
$$= -34\ \text{dB}$$

d. The phase angle is assumed to have reached to a constant −90° one decade after the corner frequency. Thus, at a frequency of 2000 rad/s, which is two decades after the corner frequency, the phase angle certainly should be at −90°:

$$\text{phase angle at 2000 rad/s} = -90^\circ$$

The analysis can be verified from Figure 7.25, which shows the frequency response plot of the given system.

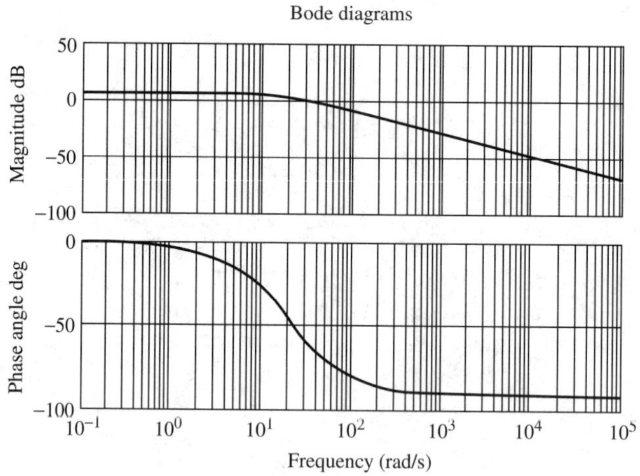

Figure 7.25 Bode plot for Example 7.20.

EXAMPLE 7.21

The output rotational velocity $n(t)$ of a permanent-magnet dc motor, when subjected to dc voltage $v(t)$, is modeled by the following transfer function.

$$\text{TF} = \frac{N(s)\ \text{rev/min}}{V(s)\ \text{V}} = \frac{50}{0.01s + 1}$$

It is subjected to an input of 10 V dc. Determine the output rotational velocity. Is there any phase difference between input and output signals?

Solution

From the transfer function the following can be observed:

$$\text{steady-state gain} = 50$$

$$\text{motor time constant} = 0.01\ \text{s}$$

Therefore,

$$\text{corner frequency} = \frac{1}{0.01}\ \text{rad/s}$$
$$= 100\ \text{rad/s}$$

The motor is subjected to dc voltage for which the signal frequency is zero. (A frequency of 0 rad/s can never be plotted on a log paper. Why?) This frequency point lies at an infinite distance to the left of the corner frequency.

The motor will accelerate to a final velocity, taking 50 ms (5 time constants) to reach 99.3% of final velocity.

$$\begin{aligned}\text{final rotational velocity} &= \text{input} \times \text{steady-state gain}\\ &= 10\,\text{V} \times \frac{50\ \text{rev/min}}{\text{V}}\\ &= 500\ \text{rev/min}\end{aligned}$$

The operation is at dc frequency, and the comparison of phase angle is meaningless at this frequency.

FIRST-ORDER LEAD BLOCK

A first-order lead system is complementary to a first-order lag system. Some electrical circuits and control subsystems are modeled by this transfer function. The physical model contains one energy-storage element (Figure 7.26).

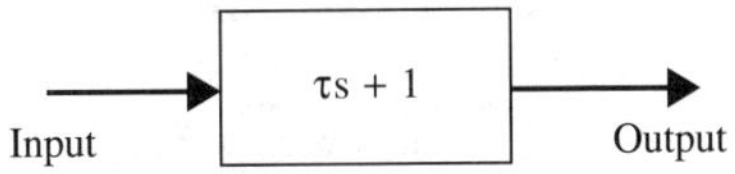

Figure 7.26 First-order lead block.

Real-Time Equation

General Form:

$$c(t) = K\left[\tau \frac{dr(t)}{dt} + r(t)\right]$$

where

$$\begin{aligned}\tau &= \text{system time constant (s)}\\ K &= \text{steady-state gain}\\ c(t) &= \text{output}\\ r(t) &= \text{input}\end{aligned}$$

Normalized Form:

$$c(t) = \tau \frac{dr(t)}{dt} + r(t) \quad \text{(unity steady-state gain)}$$

Transfer Function
General Form:

$$\text{TF} = \frac{C(s)}{R(s)} = K(\tau s + 1)$$

Normalized Form:

$$\text{TF} = \frac{C(s)}{R(s)} = \tau s + 1$$

Time Response to a Unit Step Input

The time response of a first-order lead system is an algebraic sum of the two components of the real-time equation. The first part responds to the derivative of applied input, whereas the second part is simply proportional to the applied input. When subjected to a unit step input, the result is an output, which is directly proportional to the applied input along with an impulse at the instant a step input was applied. The system time constant does not play any significant part in this type of response (Figure 7.27).

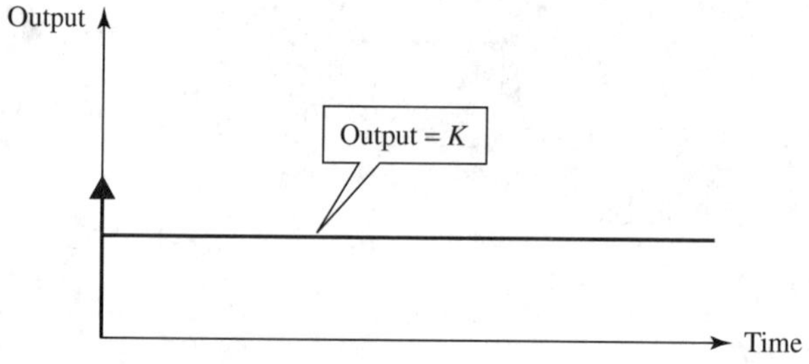

Figure 7.27 Step response: first-order lead system.

Frequency Response of a First-Order Lead Block

The frequency response of a first-order lead system is a mirror image of the response of a first-order lag system. At very low frequency, the gain is primarily equal to the steady-state gain K. At high frequency, the gain starts to increase in a logarithmic manner. The gain increases at a rate of +20 dB/decade.

Figure 7.28 shows the Bode plot of a normalized (unity steady-state gain) first-order lead process with a transfer function of $(s + 1)$.

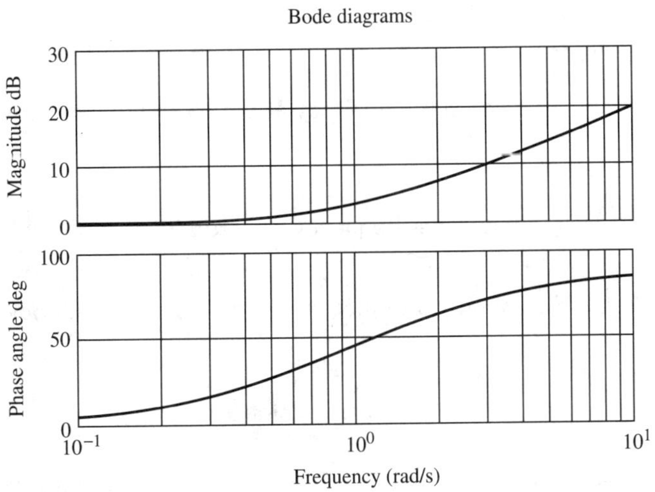

Figure 7.28 Frequency response: first-order lead.

Straight-line Approximation Similar to a first-order lag system, the frequency response of a first-order lead process can be approximated by straight-line approximations using asymptotes to the curve. Figure 7.29 shows the Bode plot of the same first-order lead block as in Figure 7.28 ($s + 1$), but covering a much wider frequency range.

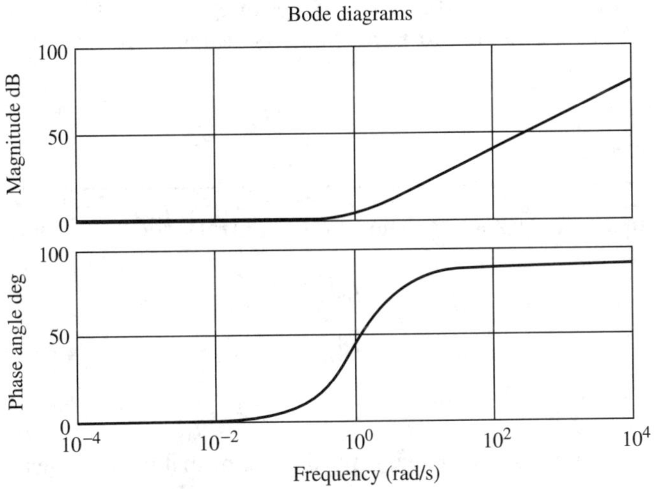

Figure 7.29 Extended frequency response: first-order lead.

The error in the straight-line approximation and the actual curve is largest at the break frequency and is equal to 3 dB.

Corner Frequency: The corner frequency at which the two asymptotes intersect is given as

$$\text{corner frequency} = \frac{1}{\tau}\ \text{rad/s}$$

Magnitude Plot: The magnitude plot starts near zero frequency at a magnitude given by

$$M = 20 \log (\text{steady-state gain } K)$$

Magnitude increases very slowly about a decade before the corner frequency. It then accelerates its growth and develops into a linear increase of +20 dB/decade. When using asymptotic approximations, it can be assumed that the magnitude remains constant between (near) zero frequency and the corner frequency.

Phase-Angle Plot: The phase angle near zero frequency is always equal to 0° and increases to +90° at very high frequency. At the corner frequency the phase angle is precisely +45°. The transition from 0° to +90° occurs mainly in the vicinity of the corner frequency. The conditions hold true for all first-order lead processes, irrespective of time constant and/or steady-state gain.

When using straight-line approximations, it can be assumed that phase remains zero from zero to one decade before the corner frequency. From one decade before the corner frequency to one decade after corner frequency, the phase angle changes linearly from 0° to +90°. After that, the phase angle is assumed to remain constant at +90°.

EXAMPLE 7.22

Determine the corner frequency for the following first-order lead system.

$$\text{TF} = \frac{C(s)}{R(s)} = 2s + 25$$

Solution

The transfer function needs to be represented in standard form. Rearranging the terms,

$$\text{TF} = 25(0.8s + 1)$$

Comparing it to the standard form TF $= K(\tau s + 1)$, the steady-state gain and time constant can be identified.

$$\begin{aligned} \text{steady-state gain} &= 25 \\ \text{time constant} &= 0.8\ \text{s} \end{aligned}$$

EXAMPLE 7.23

A first-order lead system is required to provide a unity dc gain and a gain of 25 at a frequency of 10 KHz. Determine the transfer function.

Solution

The standard transfer function is TF $= K(\tau s + 1)$. Because K represents the steady-state (dc) gain and a unity (0 dB) dc gain is required, the value of K can be established:

$$\begin{aligned} K &= 1 \\ \text{required gain at 10 KHz (62831.85 rad/s)} &= 25 \\ &= 20 \log (25) \\ &= 27.959\ \text{dB} \\ \text{increase in gain from dc} &= \text{gain at 10 KHz} - \text{dc gain} \\ &= 27.959\ \text{dB} - 0\ \text{dB} \\ &= 27.959\ \text{dB} \end{aligned}$$

Gain increases at the rate of +20 dB/decade, so the number of decades between 10 KHz and the corner frequency is

$$\frac{27.959}{20} = 1.3979\ \text{decade}$$

Therefore,

$$\begin{aligned} \text{corner frequency} &= \frac{10\ \text{KHz}}{10^{1.3979}} \\ &= 400\ \text{Hz} \end{aligned}$$

The same result could be obtained from the simple fact that gain was 25 at a frequency of 10 KHz; because gain increases in the same ratio as the frequency is multiplied, the corner frequency at which gain is expected to be equal to 1 should be 1/25 of 10 KHz, that is, 400 Hz.

SECOND-ORDER LAG BLOCK

A number of industrial processes and control systems can be described through the second-order lag model. This type of process has two separate energy-storage elements, which cannot be lumped together. A spring-mass-damper system, a temperature sensor installed in an insulating protective sheathing, a dc electric motor (more comprehensive model), and an *RLC* circuit are some examples (Figure 7.30).

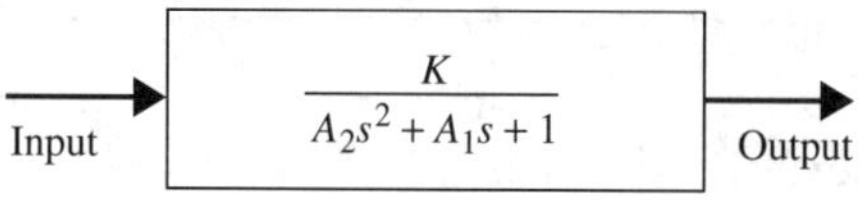

Figure 7.30 Second-order lag block.

Real-Time Equation
General Form:

$$A_2 \frac{d^2c}{dt^2} + A_1 \frac{dc}{dt} + c(t) = K r(t)$$

where

A_2, A_1 = system coefficients
K = steady-state gain
$c(t)$ = output
$r(t)$ = input

Normalized Form:

$$A_2 \frac{d^2c}{dt^2} + A_1 \frac{dc}{dt} + c(t) = r(t) \qquad \text{(unity steady-state gain)}$$

Transfer Function
General Form:

$$\text{TF} = \frac{C(s)}{R(s)} = \frac{K}{A_2 s^2 + A_1 s + 1}$$

An alternative form, which is more revealing, is

$$\text{TF} = \frac{C(s)}{R(s)} = \frac{K}{\frac{1}{\omega_n^2} s^2 + \frac{2\zeta}{\omega_n} s + 1}$$

where

ω_n = natural frequency of oscillation (rad/s)
ζ = damping ratio

Coefficients A_2 and A_1 are related to natural frequency and damping ratio:

$$\omega_n = \sqrt{\frac{1}{A_2}}$$

$$\zeta = \frac{A_1}{2\sqrt{A_2}}$$

$$A_2 = \frac{1}{\omega_n^2}$$

$$A_1 = \frac{2\zeta}{\omega_n}$$

Normalized Form:

$$\text{TF} = \frac{C(s)}{R(s)} = \frac{1}{A_2 s^2 + A_1 s + 1}$$

Time Response to a Unit Step Input

The step response of a second-order lag process depends on the value of the damping ratio, ζ. Time response can take one of the following three forms, based upon the value of ζ.

- Overdamped ($\zeta > 1$)
- Critically damped ($\zeta = 1$)
- Underdamped ($\zeta < 1$)

Overdamped Response When the damping ratio is greater than 1, the system response is considered overdamped. In this mode of operation, there are no oscillations in the output. Output reaches the final value in a somewhat exponential manner. Figure 7.31 shows the step response of a normalized (unity steady-state gain) second-order lag system with a damping ratio of 1.5.

$$\text{system TF} = \frac{1}{s^2 + 3s + 1}$$

Overdamped response is the slowest of the three damping modes. The rate of rise drops with an increase in damping ratio. Figure 7.32 shows the response of a similar system, but with a damping ratio of 2.5.

$$\text{system TF} = \frac{1}{s^2 + 5s + 1}$$

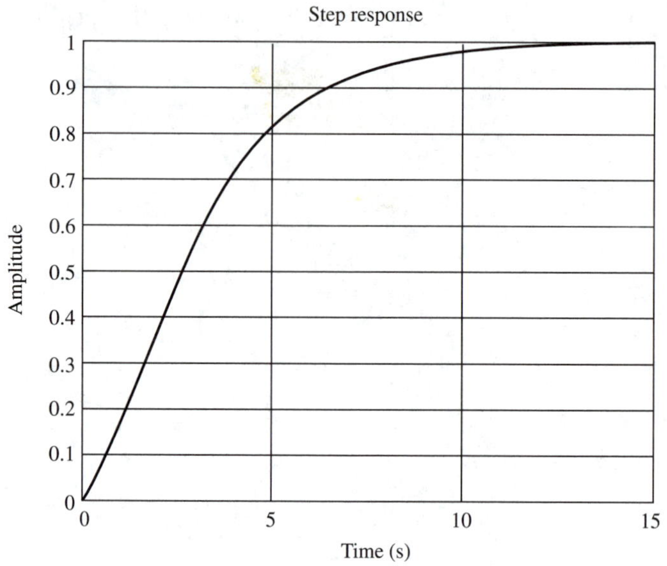

Figure 7.31 Overdamped step response: second-order lag.

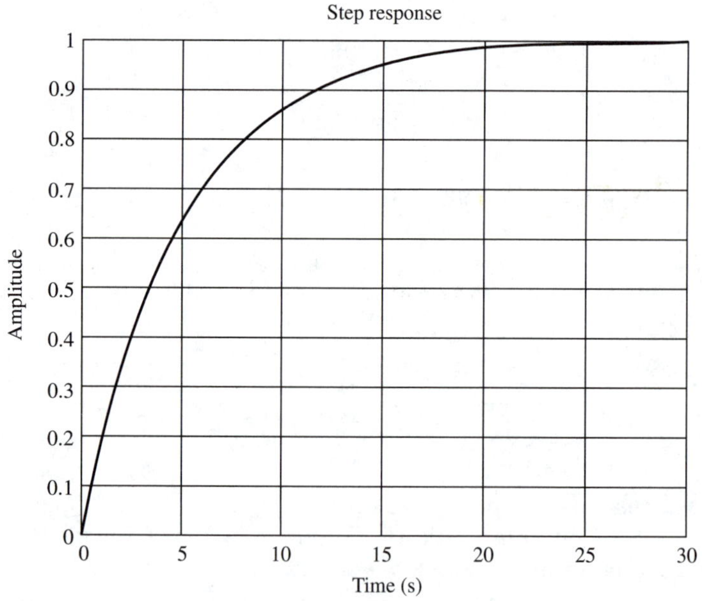

Figure 7.32 Second-order lag: damping ratio of 2.5.

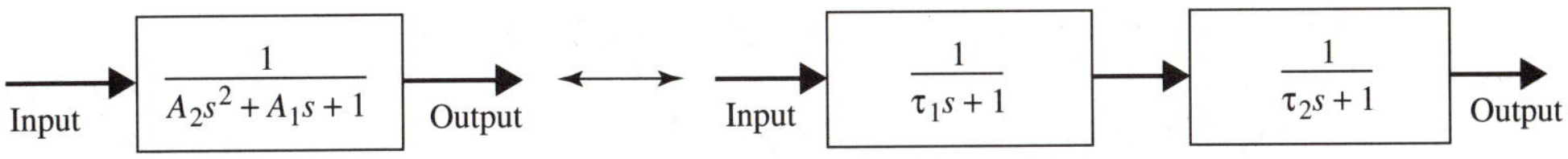

Figure 7.33 Two equivalent first-order lag blocks.

A second-order lag system with a damping ratio greater than 1 can be separated into two first-order lag processes (Figure 7.33).

Critically Damped Response With the damping ratio exactly equal to 1, system response is critically damped. There are no oscillations in the output, and the final value is reached in an exponential manner. A critically damped system can also be separated into two *identical* first-order lag systems. Figure 7.34 shows the step response of a normalized (unity steady-state gain) second-order lag system with a damping ratio of 1.0.

$$\text{system TF} = \frac{1}{s^2 + 2s + 1} = \frac{1}{(s+1)}\frac{1}{(s+1)}$$

Comparing this response to the overdamped response, it can be seen that a critically damped system provides a faster rise time. *A critically damped system provides the fastest rise time without any overshoot or oscillation*. In a number of applications, such as

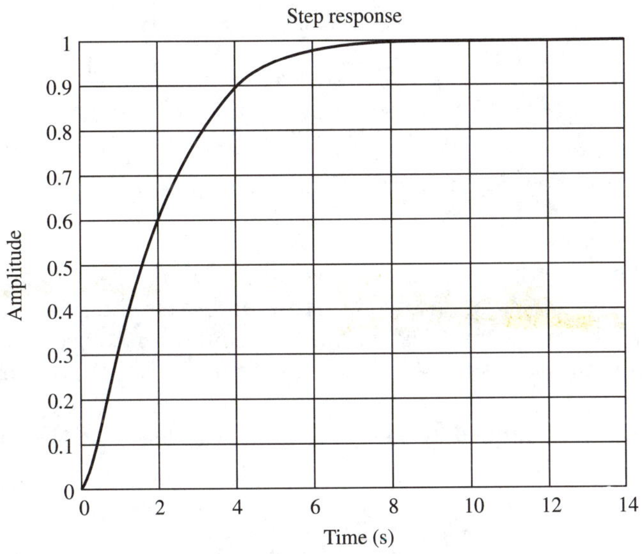

Figure 7.34 Second-order lag: critically damped step response.

machining, drilling, and automated seam welding, even the smallest overshoot is unacceptable. In these applications the system should ideally be tuned for critical damping.

Underdamped Response With a damping ratio less than 1, the system provides an underdamped step response. The output exhibits an oscillatory behavior. The amount of overshoot depends on the damping ratio; the lower the damping ratio, the greater is the overshoot. Figure 7.35 shows the step response of a normalized (unity steady-state gain) second-order lag system with a damping ratio of 0.75.

$$\text{system TF} = \frac{1}{s^2 + 1.5s + 1}$$

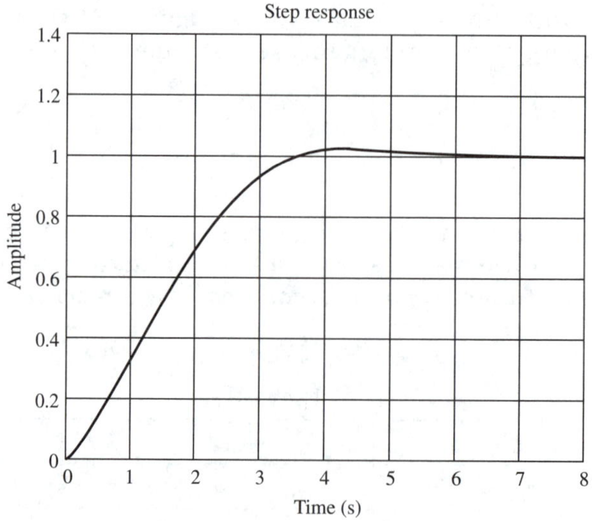

Figure 7.35 Second-order lag: underdamped step response.

On comparison with the response of overdamped and critically damped systems, it is apparent that an underdamped system has a faster rise time. But, it is accompanied by the output overshooting the final value and oscillating before it settles down. The amount of overshoot depends on the damping ratio; the lower the damping ratio, the greater is the overshoot.

Consider two underdamped second-order lag systems with the following transfer functions:

$$\text{TF}\,(\zeta = 0.5) = \frac{1}{s^2 + s + 1}$$

$$\text{TF}\,(\zeta = 0.3) = \frac{1}{s^2 + 0.6s + 1}$$

The first system has a damping ratio of 0.5, whereas the second system has a damping ratio of 0.3. The unit step time response for these two systems is shown in Figure 7.36. With a lower damping ratio of 0.3, the overshoot increases to 40% (1.4 for a unit output), although output rises faster than that for a damping ratio of 0.5. This type of behavior may be desirable in some control systems, where it is critical to reach the target in the shortest possible amount of time. Chemical mixing, temperature control, and motor-speed controls are some of the processes that exhibit this type of response.

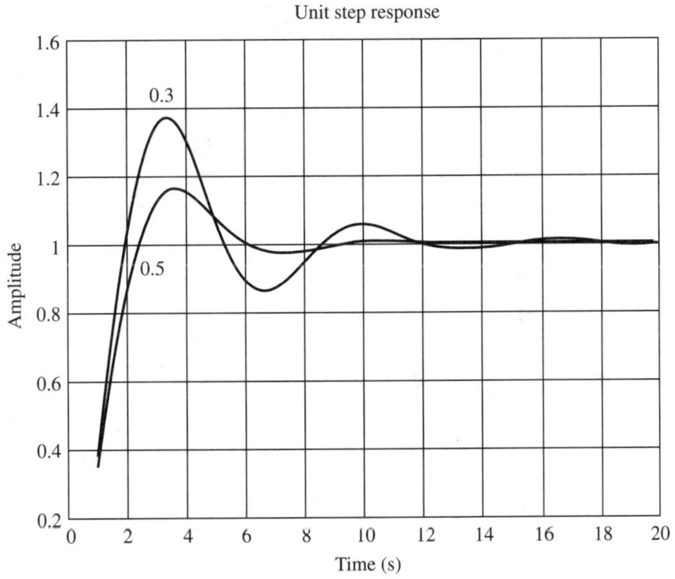

Figure 7.36 Second-order lag: underdamped step response.

EXAMPLE 7.24

Determine the type of response that will be obtained from the following second-order lag process when it is subjected to a unit step input.

$$\text{TF} = \frac{2.5}{3s^2 + 2s + 5}$$

Solution

The value of the damping ratio, ζ, is required to determine the type of step response that will be obtained from the system. The damping ratio is related to the coefficients of the transfer function once it is expressed in the proper form (the coefficient of s^0 term in the denominator should be unity). Rearranging the terms,

$$\text{TF} = \frac{2.5}{5(0.6s^2 + 0.4s + 1)}$$

$$\text{TF} = \frac{0.5}{0.6s^2 + 0.4s + 1}$$

Once the TF is in the proper form, the damping ratio can be calculated:

$$\begin{aligned} \text{damping ratio } \zeta &= \frac{A_1}{2\sqrt{A_2}} \\ &= \frac{0.4}{2\sqrt{0.6}} \\ &= 0.258 \end{aligned}$$

This value of the damping ratio is far below 1. Thus, the step response will be underdamped with a large overshoot and is shown in Figure 7.37.

From the rearranged form,

$$\text{steady-state gain} = 0.5$$

Therefore,

$$\text{final settled value} = 0.5 \times 1 = 0.5$$

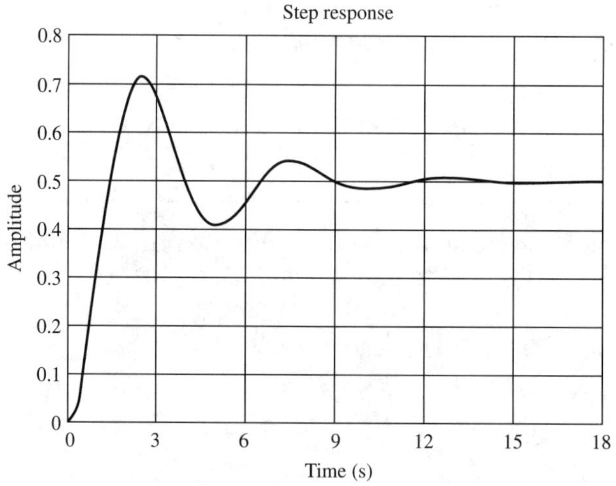

Figure 7.37 Underdamped step response.

Characteristics of Underdamped Response

A number of control systems are modeled as underdamped second-order lag systems. As a result, the transient response is of particular significance for the real-time operation of such a system. The following response characteristics (Figure 7.38) can be derived from the knowledge of the coefficient of the transfer function. The discussion is based on the normalized transfer function:

$$\mathrm{TF} = \frac{1}{A_2 s^2 + A_1 s + 1}$$

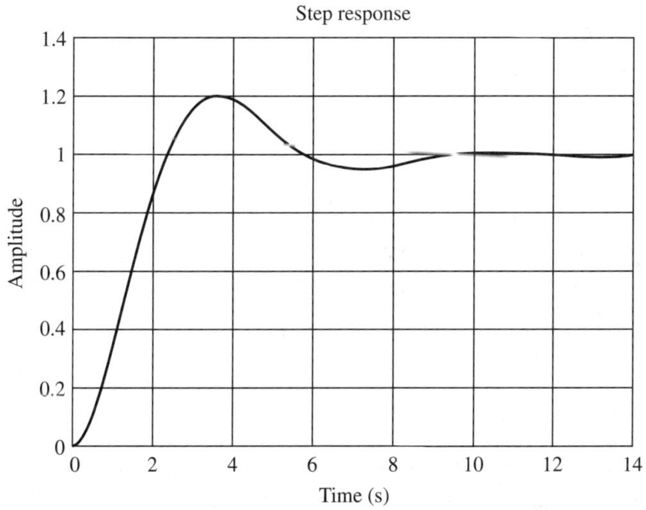

Figure 7.38 Second-order lag: transient response specifications.

Peak Value, M_P

Immediately after the application of the unit step input, the output starts to rise. It reaches the maximum value at time t_p later.

$$M_P = 1 + e^{-\pi\zeta/\sqrt{1-\zeta^2}}$$

Peak Time, t_P

Peak time is the amount of time taken to reach the peak amplitude. It is measured from the instant the step input was applied.

$$t_P = \frac{\pi}{\omega_n\sqrt{1-\zeta^2}}$$

***Percentage Overshoot,* P.O.**

Percentage overshoot is the difference between this peak value and the final steady-state value, expressed as the percentage of the final value.

$$\text{P.O.} = 100e^{-\pi\zeta/\sqrt{1-\zeta^2}}$$

Settling Time, T_s

Settling time is defined as the time taken for the system to settle down to within a certain percentage of the final value. It is the time taken to enter the given percentage band the last time before settling down. Percentage values of 2% and 5% are used, out of which 2% is fairly common. Settling time is given approximately as

$$2\%\text{ settling time } T_s = \frac{4}{\zeta\omega_n}$$

$$5\%\text{ settling time } T_s = \frac{3}{\zeta\omega_n}$$

Damped Natural Frequency, ω_d

The damped natural frequency is the frequency at which the system oscillates before settling down. It is smaller than the natural frequency of oscillation.

$$\omega_d = \omega_n\sqrt{1-\zeta^2}$$

Rise Time, t_r

Rise time is the time taken for the output amplitude to rise from 10% of final value to 90% of final value (the first time).

$$t_r = \frac{1}{\omega_n\sqrt{1-\zeta^2}}\tan^{-1}\left(\frac{\sqrt{1-\zeta^2}}{-\zeta}\right)$$

EXAMPLE 7.25

For the given system, determine the following step response characteristics.

a. Damped oscillation frequency
b. Rise time
c. Percentage overshoot
d. 2% settling time

$$\text{system TF} = \frac{1}{0.01s^2 + 0.1s + 1}$$

Solution

The natural frequency of the oscillation and the damping ratio should be calculated first.

$$\text{natural frequency of oscillation } \omega_n = \sqrt{\frac{1}{A_2}} = \sqrt{\frac{1}{0.01}} = 10 \text{ rad/s}$$

$$\text{damping ratio } \zeta = \frac{A_1}{2\sqrt{A_2}} = \frac{0.1}{2\sqrt{0.01}} = 0.5$$

a.

$$\text{damped oscillation frequency } \omega_d = \omega_n\sqrt{1-\zeta^2} = 10\sqrt{1-0.5^2} = 8.66 \text{ rad/s}$$

b.

$$\text{rise time } t_r = \frac{1}{\omega_n\sqrt{1-\zeta^2}}\tan^{-1}\left(\frac{\sqrt{1-\zeta^2}}{-\zeta}\right) = \frac{1}{10\sqrt{1-0.5^2}}\tan^{-1}\left(\frac{\sqrt{1-0.5^2}}{-0.5}\right) = 0.115\tan^{-1}(-1.732) = 0.115 \times 2.094 = 0.241 \text{ s}$$

c.

$$\text{percentage overshoot P.O.} = 100e^{-\pi\zeta/\sqrt{1-\zeta^2}} = 100e^{-0.5\pi/\sqrt{1-0.5^2}} = 16.3\%$$

d.

$$2\%\text{ settling time } T_s = \frac{4}{\zeta\omega_n} = \frac{4}{0.5 \times 10} = 0.8\text{ s}$$

The actual step response is shown in Figure 7.39.

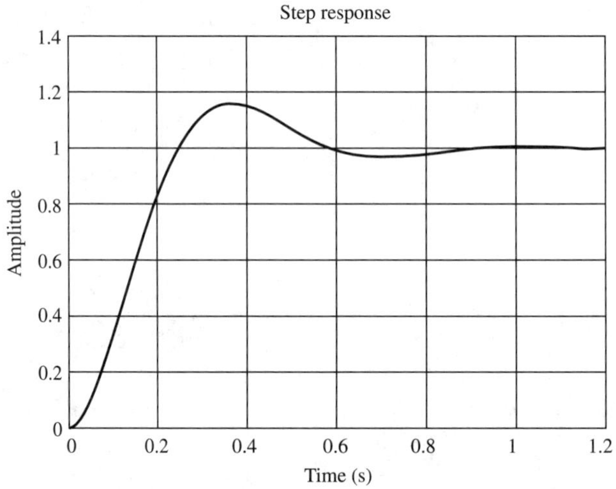

Figure 7.39 Step response for Example 7.25.

Frequency Response of a Second-Order Lag Block

Consider the following form of the transfer function for a second-order lag system.

$$\text{TF} = \frac{1}{\frac{1}{\omega_n^2}s^2 + \frac{2\zeta}{\omega_n}s + 1}$$

The frequency response of this system depends on the value of the damping ratio, ζ. These responses can be divided into three groups.

Damping Ratio Greater Than or Equal to 1 ($\zeta \geq 1$) If the damping coefficient is greater than or equal to 1 ($\zeta \geq 1$), the system can be represented as two first-order lag systems (in cascade).

Magnitude: Gain at very low frequency starts at 1 (magnitude of 0 dB) and gradually decreases until the natural frequency is reached. After that, the drop in gain accelerates and quickly reaches an asymptotic value of −40 dB/decade. It then continues its free fall. Using straight-line approximations, the following can be assumed:

- For $\omega = 0$ to $\omega = \omega_n$, gain is constant at 1 (magnitude = 0 dB).
- For $\omega > \omega_n$, gain drops at the rate of 1/100 for every tenfold increase in frequency (−40 dB/decade).

Phase Angle: Phase angle is the maximum at 0° at very low frequency. As the input signal frequency is increased, it decreases slowly until about $0.1\omega_n$. The drop in phase angle accelerates until a frequency of 10 ω_n is reached. The phase angle continues to drop slowly and eventually becomes equal to −180°. When the input signal frequency equals the natural frequency ω_n for the system, the phase angle is an exact −90°.

- For $\omega = 0$ to $\omega = 0.1\omega_n$, the phase angle is constant at 0°.
- For $0.1\omega_n < \omega < 10\omega_n$, the phase angle drops from 0° to −180°.
- For $\omega > 10\omega_n$, the phase angle is constant at −180°.
- At $\omega = \omega_n$, the phase angle is exactly −90°.

Figure 7.40 shows the frequency response of a second-order lag system with transfer function

$$\mathrm{TF} = \frac{1}{s^2 + 2s + 1}$$

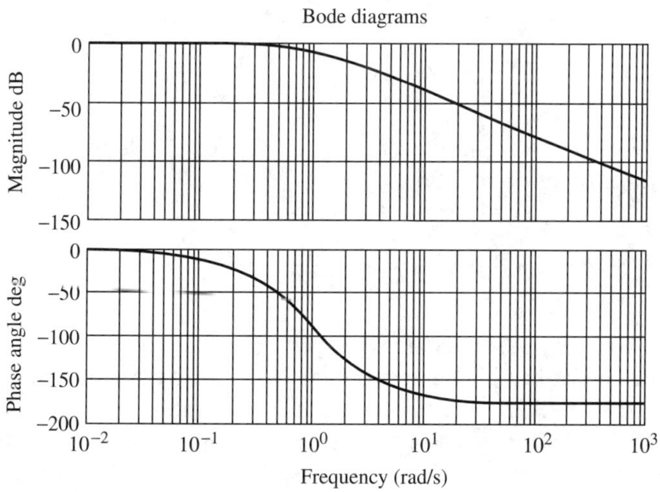

Figure 7.40 Second-order lag ($\zeta \geq 1$): frequency response.

It is clear that the magnitude plot can easily be constructed using straight-line approximation, with one horizontal line at 0 dB and the other sloping at the rate of −40 dB/decade. These two lines (asymptotes) intersect at the natural frequency, ω_n.

Damping Ratio Less Than 1 but Greater Than 0.707 (0.707 < ζ < 1) For a damping ratio less than 1, it is not possible to break the system into two first-order lag blocks. The frequency response of this system is otherwise equal to one with a damping ratio greater than 1.

Damping Ratio Less Than 0.707 (ζ < 0.707) For a damping ratio less than 0.707, a resonant behavior is obtained. Gain increases sharply in the vicinity of natural frequency. The phase angle also experiences rapid change from 0° to −180°. As the damping ratio is lowered, these actions become more pronounced. For a damping ratio of 0 (hypothetically speaking), the gain reaches infinity and phase angle changes from 0° to −180° in an absolute vertical fashion. Some of the characteristics of this type of frequency response are noted next.

Peak Gain, G_P: System gain peaks out at a certain frequency for a given value of damping ratio:

$$G_p = \frac{1}{2\zeta\sqrt{1-\zeta^2}}$$

Peak Frequency, ω_P: Peak frequency is the frequency at which peak gain occurs. It should be noted that it is not equal to natural frequency, except in case of a zero damping ratio. For other damping ratios, this frequency becomes smaller than natural frequency.

$$\omega_p = \omega_n\sqrt{1-2\zeta^2}$$

Figure 7.41 shows the frequency response plot of the following second-order lag system with a damping ratio of 0.05.

$$\mathrm{TF} = \frac{1}{s^2 + 0.1s + 1}$$

The system exhibits resonance as the operating frequency approaches the system natural frequency of 1 rad/s. At this frequency the gain rapidly increases to 10 and then starts to drop rapidly as the operating frequency continues to increase. At frequencies higher than natural frequency, the gain decreases at a rate of −40 dB/decade. Phase angle also changes rapidly near the natural frequency, but still is exactly −90° at the natural frequency (ω_n). Frequency response at other frequencies ($\omega \ll \omega_n$ and $\omega \gg \omega_n$)

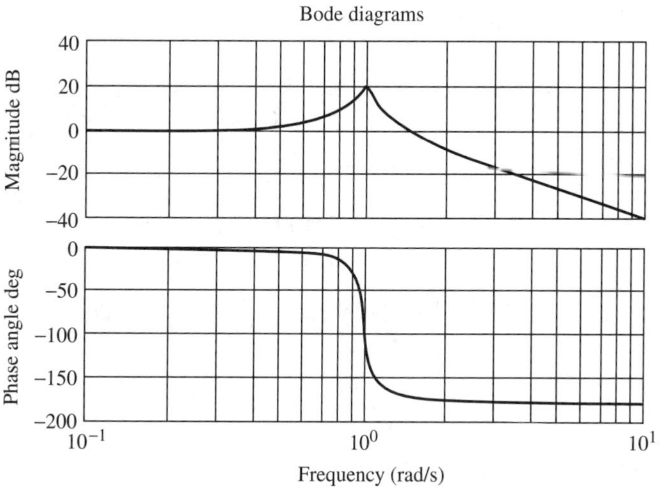

Figure 7.41 Frequency response: second-order lag ($\zeta < 1$)

is identical to that of any other second-order lag system. This is true irrespective of system damping ratio and/or natural frequency.

EXAMPLE 7.26

For the system $1/(s^2 + 0.1s + 1)$, determine the following:

a. Maximum gain

b. The frequency at which the maximum gain occurs.

c. Is this frequency the same as the natural frequency?

Solution

a. Maximum gain is given by

$$G_p = \frac{1}{2\zeta\sqrt{1-\zeta^2}}$$

$$\text{damping ratio } \zeta = \frac{A_1}{2\sqrt{A_2}} = \frac{0.1}{2\sqrt{1}} = 0.05$$

Substituting the value of damping ratio,

$$\text{maximum gain } G_p = \frac{1}{2 \times 0.05\sqrt{1 - 0.05^2}}$$
$$= 10.01$$

The frequency at which this gain occurs is given as $\omega_p = \omega_n\sqrt{1 - 2\zeta^2}$.

$$\text{natural frequency } \omega_n = \sqrt{\frac{1}{A_2}}$$
$$= \sqrt{\frac{1}{1}} = 1 \text{ rad/s}$$

Thus, peak frequency is

$$\omega_p = 1\sqrt{1 - 2 \times 0.05^2} \text{ rad/s}$$
$$= 0.9987 \text{ rad/s}$$

b. Peak frequency (0.9987 rad/s) is not the same as the natural frequency (1 rad/s). Because the damping ratio is greater than zero, this frequency should be less than the natural frequency, which is the case.

SECOND-ORDER LEAD BLOCK

A second-order lead transfer function cannot be used to describe a proper closed-loop control system. But it represents a number of electrical circuits, as well as other such subsystems (Figure 7.42).

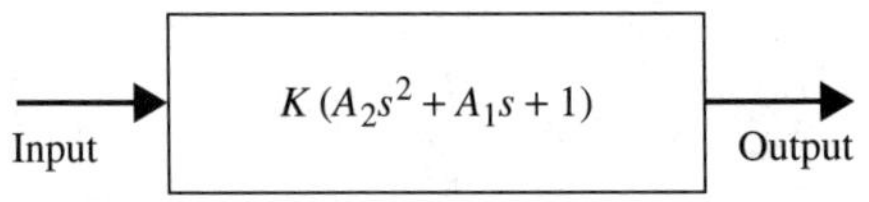

Figure 7.42 Second-order lead block transfer function.

Real-Time Equation
General Form:

$$c(t) = K\left(A_2\frac{d^2r}{dt^2} + A_1\frac{dr}{dt} + r(t)\right)$$

where
$$A_2, A_1 = \text{system coefficients}$$
$$K = \text{steady-state gain}$$
$$c(t) = \text{output}$$
$$r(t) = \text{input}$$

Normalized Form:

$$c(t) = A_2 \frac{d^2r}{dt^2} + A_1 \frac{dr}{dt} + r(t) \qquad \text{(unity steady-state gain)}$$

Transfer Function

General Form:

$$\text{TF} = \frac{C(s)}{R(s)} = K(A_2 s^2 + A_1 s + 1)$$

An alternative form, which is more revealing, is

$$\text{TF} = \frac{C(s)}{R(s)} = K\left(\frac{1}{\omega_n^2} s^2 + \frac{2\zeta}{\omega_n} s + 1\right)$$

where
$$\omega_n = \text{natural frequency of oscillation (rad/s)}$$
$$\zeta = \text{damping ratio}$$

The relation of coefficients A_2 and A_1 to natural frequency and damping ratio is

$$\omega_n = \sqrt{\frac{1}{A_2}}$$

$$\zeta = \frac{A_1}{2\sqrt{A_2}}$$

$$A_2 = \frac{1}{\omega_n^2}$$

$$A_1 = \frac{2\zeta}{\omega_n}$$

Normalized Form:

$$\text{TF} = \frac{C(s)}{R(s)} = A_2 s^2 + A_1 s + 1$$

Time Response to a Unit Step Input

The step response of a second-order lead process is essentially the sum of the individual responses of its three terms. The first term responds only to the second derivative (d/dt^2), which is the rate of change of the slope. For a step input, the output is zero except for a spike at time zero. The second term responds to the rate of change (d/dt). Its response to a step input is also a constant zero accompanied by spikes at the instant the step input is applied. The response of the last term is proportional to the step input with K (steady-state gain) as a constant of proportionality.

Frequency Response of a Second-Order Lead Block

The frequency response of the second-order lead block is a mirror image of the response of a second-order lag block. The response is again dependent upon the value of damping ratio ζ. (Refer to the discussion on the second-order lag system). If the damping coefficient is greater than or equal to 1 ($\zeta \geq 1$), the system can be represented as two first-order lead systems (in cascade).

Magnitude: Gain at very low frequencies starts at 1 (magnitude of 0 dB) and gradually increases until the natural frequency is reached. After that increase, gain accelerates and quickly reaches an asymptotic value of +40 dB/decade. Asymptotic assumptions are as follows:

- For $\omega = 0$ to $\omega = \omega_n$, gain is constant at 1 (magnitude = 0 dB).
- For $\omega > \omega_n$, gain increases at the rate of 100 for every tenfold increase in frequency (+40 dB/decade).

Phase Angle: Phase angle is the maximum at 0° at very low frequencies. The frequency increases slowly until about $0.1\omega_n$. The increase in phase angle accelerates until a frequency of $10\omega_n$ is reached. The phase angle continues to increase slowly and eventually becomes equal to +180°. At natural frequency ω_n, it is exactly equal to +90°. Asymptotic assumptions are as follows:

- For $\omega = 0$ to $\omega = 0.1\ \omega_n$, the phase angle is constant at 0°.
- For $0.1\ \omega_n < \omega > 10\ \omega_n$, the phase angle increases from 0° to +180°.
- For $\omega > 10\ \omega_n$, the phase angle is constant at +180°.
- At $\omega = \omega_n$, the phase angle is exactly +90°.

Figure 7.43 shows the frequency response of a second-order lead system with transfer function

$$\mathrm{TF} = s^2 + 3s + 1$$

From the observation of the plot, it is evident that the magnitude plot can be constructed using straight-line approximation: one straight line with a constant magnitude of 0 dB

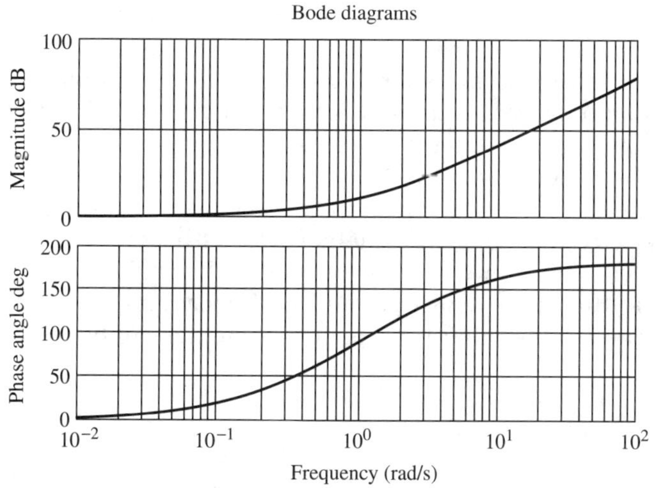

Figure 7.43 Second-order lead system: frequency response.

and the other straight line with a slope of +40 dB/decade. The lines intersect at the natural frequency.

DEAD TIME: TRANSPORTATION LAG BLOCK

Transportation lag is one of the most troublesome elements in a control system. It is frequently encountered as the delay in material movement (transport). A few examples are fluid flow, metal rolling, and solid/powder material-delivery systems. Figure 7.44 shows an example of transportation lag in a fluid-flow system.

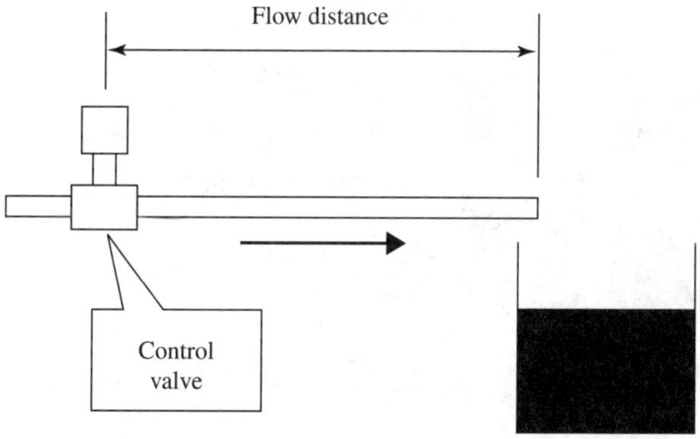

Figure 7.44 Transportation lag in a fluid-flow system.

The flow control valve is used to control the amount of fluid flowing into the tank. Once the valve spool is repositioned, it takes a finite amount of time before the flow rate into the tank changes. Similar situations exist in metal-rolling mills (Figure 7.45), boiler coal–feed belts, and conveyor systems in general. In a metal-rolling mill, the thickness transducer (measuring sheet-metal thickness) is located at some distance from the metal-rolling stand. When the metal rolls are adjusted, a time delay occurs before the transducer can sense a change in sheet-metal thickness. This time delay depends on the rotational velocity and diameter of metal rolls, the distance between the sensor, and the ratio of material thickness at the inlet and outlet. For a given production, equipment and material, the thickness reduction, and mill operating speed are generally held constant to optimum values, and the transportation lag can be assumed to be a fixed value of time T.

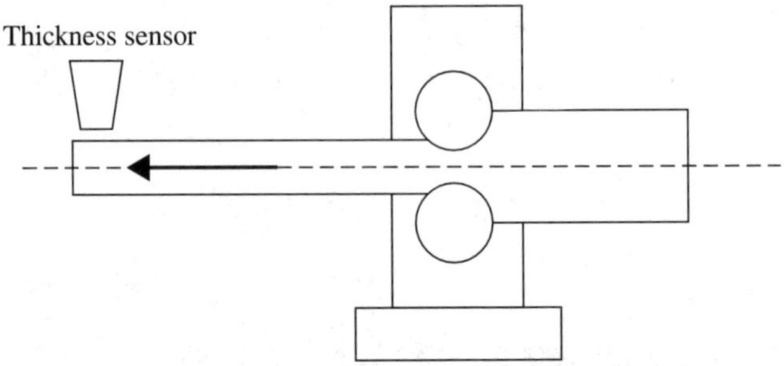

Figure 7.45 Transportation lag in metal-rolling mill.

Real-Time Equation

General Form:

$$c(t) = Kr(t - T)$$

where

$$\begin{aligned} K &= \text{steady-state gain} \\ c(t) &= \text{output} \\ r(t) &= \text{input} \\ T &= \text{dead time} \end{aligned}$$

Normalized Form:

$$c(t) = r(t - T) \qquad \text{(unity steady-state gain)}$$

Transfer Function

General Form:

$$\text{TF} = \frac{C(s)}{R(s)} = Ke^{-sT}$$

Normalized Form:

$$\text{TF} = \frac{C(s)}{R(s)} = e^{-sT}$$

$$\text{magnitude} = 0 \text{ dB}$$

$$\text{phase angle} = -\omega T \text{ (radians)}$$

$$= -57.3\,\omega T \text{ (degrees)}$$

Time Response to a Unit Step Input

The response of a dead-time process, when subjected to a unit step input, is also a step output, except that it is delayed by T time period (Figure 7.46).

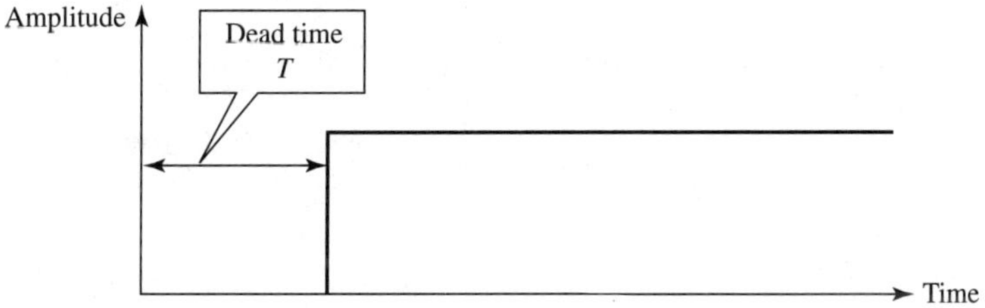

Figure 7.46 Step response: dead-time system.

Frequency Response of a Dead-Time Block

The gain of a dead-time block is always equal to unity. Hence the magnitude plot is a constant 0-dB line. Dead time causes the output signal to be delayed by a fixed time interval equal to the dead time (transportation lag) T. As a result, the output signal is always delayed compared to the input signal. This phase lag increases linearly with frequency. For low-frequency signals the phase lag is small, but with an increase in frequency, phase lag increases rapidly. Figure 7.47 shows a frequency response (phase only) plot of a dead-time process with a dead time of 0.15 s.

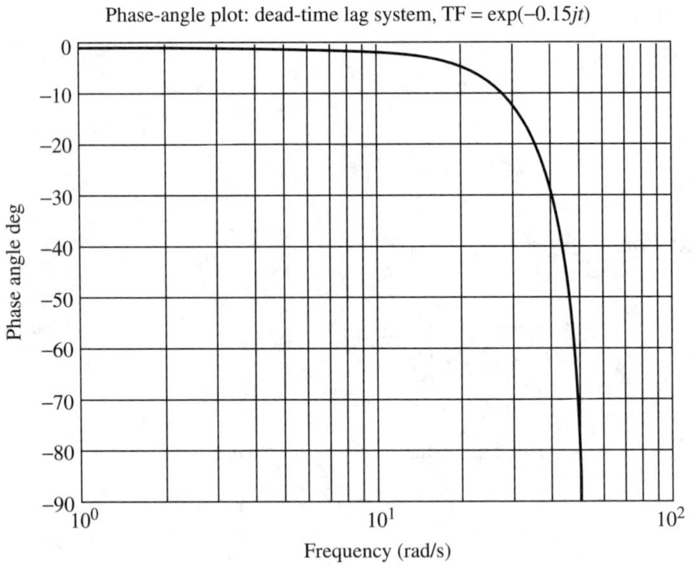

Figure 7.47 Frequency response: transportation lag.

PROBLEMS

1. Isolate the following transfer functions into a constant block (K) and a first-order lead block ($\tau s + 1$).
   a. $10s + 4$
   b. $s + 2$
   c. $2.5(3s + 10)$
   d. $8(s + 1.250)$
2. Isolate the following transfer functions into a constant block (K) and a first-order lag block ($\frac{1}{\tau s+1}$).
   a. $\dfrac{1}{s+2}$
   b. $\dfrac{2.5}{1.25s+5}$
   c. $\dfrac{3}{4(5s+2)}$
3. Isolate the following transfer functions into a constant block (K) and a pure integral block ($\frac{1}{s}$).
   a. $\dfrac{10}{s}$

b. $\frac{1}{2s}$

c. $\frac{2.5}{4s}$

4. Isolate the following transfer functions into a constant block (K) and a pure derivative block (s).

a. $2.5s$

b. $\frac{s}{8}$

c. $\frac{2s}{5}$

5. Isolate the following transfer functions into a constant block (K) and a second-order block ($As^2 + Bs + 1$).

a. $s^2 + 2.5s + 2$

b. $2s^2 + 5s + 10$

c. $4(2.5s^2 + s + 5)$

6. Isolate the following transfer functions into the simplest block form.

a. $1.2(8s + 12)$

b. $\frac{10s}{4s + 5}$

c. $\frac{6s + 2}{12.5s + 5}$

d. $\frac{3s(6s + 2)}{s^2 + 2.5s + 2}$

e. $\frac{s(3s + 12)}{12s(s^2 + 2.5s + 2)}$

7. Identify the break frequency for the system given by each transfer function.

a. $10s + 1$

b. $s + 2$

c. $\frac{1}{5s + 1}$

d. $\frac{1}{s + 5}$

e. $\frac{10}{s + 5}$

8. Using an asymptotic approximation of a Bode plot for the following transfer functions, determine the frequencies at which

- Magnitude reaches 20 dB
- Phase angle approaches 90°

a. $1.2s + 10$
b. $s + 2$
c. $10(s + 1)$

9. Using an asymptotic approximation of a Bode plot for the following transfer functions, determine the frequencies at which

- Magnitude reaches -20 dB
- Phase angle approaches $-90°$

a. $\dfrac{1}{s+4}$

b. $\dfrac{10}{s+4}$

c. $\dfrac{10}{4s+1}$

10. Isolate each transfer function into the simplest block form and draw a Bode plot using asymptotic approximation.

a. $(s + 1)$

b. $\dfrac{1}{10s+1}$

c. $\dfrac{(s+1)}{(10s+1)}$

d. $\dfrac{10(s+1)}{(10s+1)}$

e. $\dfrac{0.1(s+1)}{(10s+1)}$

f. $\dfrac{(s+1)}{s(10s+1)}$

g. $\dfrac{s(s+1)}{(10s+1)}$

h. $\dfrac{(100s+1)(s+1)}{(10s+1)}$

i. $\dfrac{5s(s+1)}{(10s+1)(s+10)}$

num =
den =
Bode(num, den)

11. Identify the type of response (underdamped, overdamped, or critically damped) that will be exhibited by the following transfer functions.

a. $\dfrac{1}{s^2+s+1}$

b. $\dfrac{1}{s^2 + 5s + 6}$

c. $\dfrac{1}{(s+2)(s+3)}$

d. $\dfrac{1}{s^2 + 2s + 1}$

12. For the given second-order lag system, determine the following characteristic parameters.

$$\text{TF} = \frac{1}{0.0016s^2 + 0.2s + 1}$$

a. Natural frequency of oscillations
b. Damped oscillation frequency
c. Rise time
d. Percentage overshoot
e. 2% settling time

8

STABILITY

INTRODUCTION

Stability is an important characteristic of a closed-loop control system. A control system must be stable in nature or else it is of no practical use. This chapter introduces the concept of stability and discusses the time response and frequency response method for stability analysis. The significance of characteristic equation and its roots in determining the stability is explained along with Routh criterion.

CONCEPT OF STABILITY

When a physical system is subjected to a bounded (limited) physical input, the system output should settle down to a certain value (no matter how large) and should not continue to increase indefinitely. This concept is best illustrated through an example. Figure 8.1 shows three different setups. The first contains a marble in a bowl (concave surface), the second one shows a marble on a flat surface, and the last one shows a marble balanced on top of an inverted bowl (convex surface). The standard test in this case is to give a gentle nudge to the marble in each of the three systems and observe the resultant marble movement, again considering two possibilities, zero friction (glass surface and perfect marble) and some friction (real surfaces and a rubberized marble).

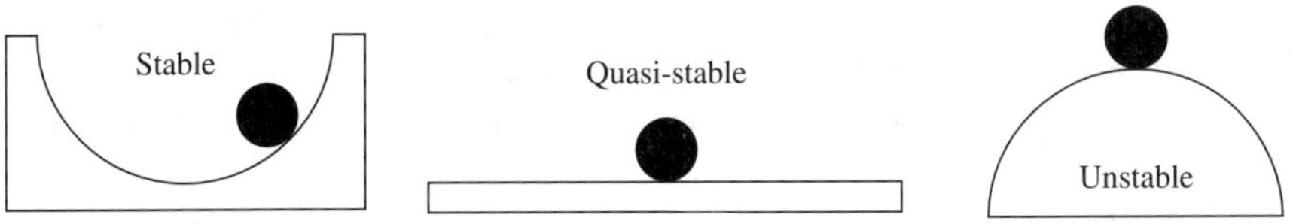

Figure 8.1 Stability.

Case I: Zero Friction

- *Stable system*: Once the marble is pushed gently, it will continue to rock back and forth without ever slowing down. A finite input (force) results in a finite movement.
- *Quasi-stable system*: Once a force is applied to the marble, it will continue to roll on the flat surface. Because there is no friction, the marble will continue to roll forever (provided a continuous and infinite amount of flat surface area is available). But it should be noted that the marble velocity is still a function of the applied force (second law of motion), and a finite force still results in a finite velocity.

  The system is still a stable system, but consider the possibility that the surface no longer remains horizontal or flat. It can happen when either the surface is tilted slightly or, being subjected to large temperature variation, the flat surface develops a slight convex curvature. Once disturbed from the position of rest, the marble will then continue to accelerate as it rolls down the convex (or tilted) surface. If this happens, the output is unbounded for a bounded input, and the system can be considered to be an unstable system.
- *Unstable system*: In this case even a gentle nudge will cause the marble to roll down the surface and accelerate. The amount of initial force establishes only the initial velocity, but irrespective of this, the marble continues to accelerate (and will fall off). A finite input force does not result in a finite (limited) velocity.

Case II: Some Friction Present

- *Stable system*: Once the marble is pushed gently, it will rock back and forth, but the loss of energy due to friction will cause the marble to slow down. Eventually, the marble will come to a standstill. A finite input (force) results in a finite movement.
- *Quasi-stable system*: In this case, the marble will again start to roll on the flat surface but will slow down and eventually come to a stop. It is still a stable system, but if the surface is tilted or develops a slight convex curvature, the marble will continue to roll, making the system an unstable one.
- *Unstable system*: In this case, even a gentle nudge will cause the marble to roll down the surface and accelerate. The marble acceleration rate will be hampered by friction, but it will continue to accelerate. A finite input force does not result in a finite (limited) velocity.

A stable system has a *zero impulse response*. Output of a stable control system, when excited by an impulse input, will eventually reach the zero state (back to where it started from). The following two criteria are often used to establish the stability of a system:

1. *Bounded-input, bounded-output (BIBO)*: When excited by a bounded (finite) input, system output response is also bounded (finite).
2. *Zero impulse response*: When subjected to an impulse input, system output response (given sufficient time) approaches zero.

EXAMPLE 8.1

The unit step response of a system with a closed-loop transfer function,

$$\frac{1}{(0.1s^3 + s^2 + 10s + 100)}$$

is shown in Figure 8.2.

It is evident that the system never settles down and continues to oscillate indefinitely. The oscillations do not continue to grow without bound and have settled within well-defined limits. But a change in the physical characteristics of the system due to aging or mechanical wear and tear will result in a change in transfer function, which

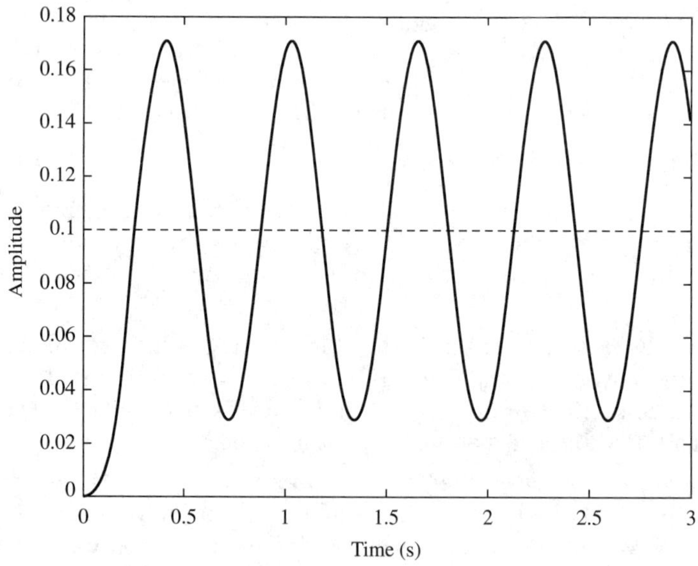

Figure 8.2 Step response.

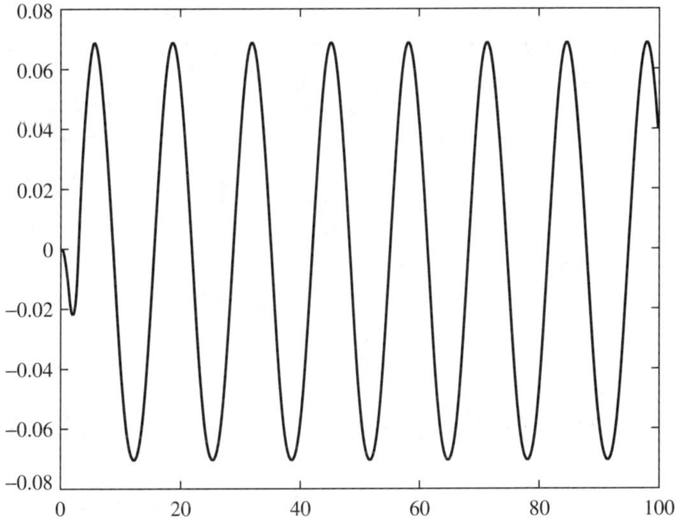

Figure 8.3 Impulse response.

may lead to an unstable operation. Thus, for all practical purposes it should be treated as an unstable system.

Similar observations can be made from the impulse response of the system, which is shown in Figure 8.3. Again, from the response it is evident that the system output never reaches the zero state.

EXAMPLE 8.2

When testing an automobile suspension, an auto mechanic generally applies a single and sudden downward force (impulse) on the corner of an automobile body (Figure 8.4).

Figure 8.4 Automobile suspension test.

This impulselike input excites the suspension system, resulting in small oscillations. In a properly working system, the oscillations quickly die out and the automobile body comes to rest (zero impulse response).

If the suspension system is not working properly, oscillations take much longer to die out. When such an automobile is driven on a gravel road or on a bumpy highway, it is quite possible that oscillations will continue to grow and eventually lead to total failure or breakdown of the suspension system.

In a real-life system such as an automobile suspension system, there are—fortunately—other factors such as the rubber tires, friction in joints, and other components connected to the automobile body that help prevent a disaster.

STABILITY AND THE OPEN-LOOP TRANSFER FUNCTION

The concept of stability can be better understood from analysis of a closed-loop transfer function. The transfer function of a closed-loop negative-feedback control system is given by

$$\frac{\text{output } C}{\text{input } R} = \frac{G}{1 + GH}$$

The control system output then is given by

$$\text{output } C = \frac{G}{1 + GH} \times \text{input } R$$

Considering the input R to be a finite quantity, the output C can become infinite only if the value of the CLTF is infinite. The CLTF, in turn, can take on an infinite value if either the numerator G becomes infinite or the denominator $(1 + GH)$ becomes zero. In a practical control system, forward gain G is a finite quantity, and even if G becomes very large, it still remains a finite quantity. This leaves only one possibility, that the denominator of CLTF can become zero. That is,

$$1 + GH = 0$$

$$GH = -1$$

The value of GH, the open-loop transfer function (gain) of the control system, depends on the value of forward gain G and feedback gain H. It is possible to have the two gain values set in such a way that their product may become equal to -1, and when this happens the output approaches infinity. Thus, any time the open-loop gain becomes equal to -1, the closed-loop control system becomes unstable. The following are some examples of unstable systems.

EXAMPLE 8.3

Determine the transfer function of the closed-loop system (Figure 8.5) and the system output when subjected to a unit input.

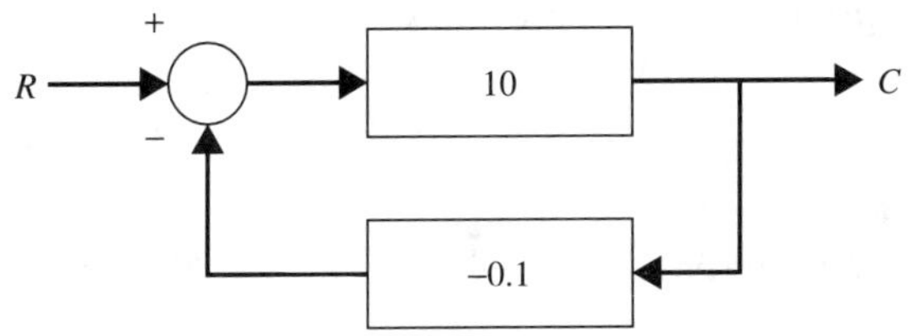

Figure 8.5 Block diagram for Example 8.3.

Solution

The closed-loop transfer function is

$$\text{CLTF} = \frac{G}{1+GH}$$
$$= \frac{10}{1+10(-0.1)}$$
$$= \frac{10}{1-1}$$
$$= \frac{10}{0} \to \infty$$
$$\text{system output } C = \text{input } R \cdot \text{CLTF}$$
$$= 1 \cdot \infty$$
$$\to \infty$$

EXAMPLE 8.4

Determine the transfer function of the closed-loop system (Figure 8.6) and the system output when subjected to a unit input.

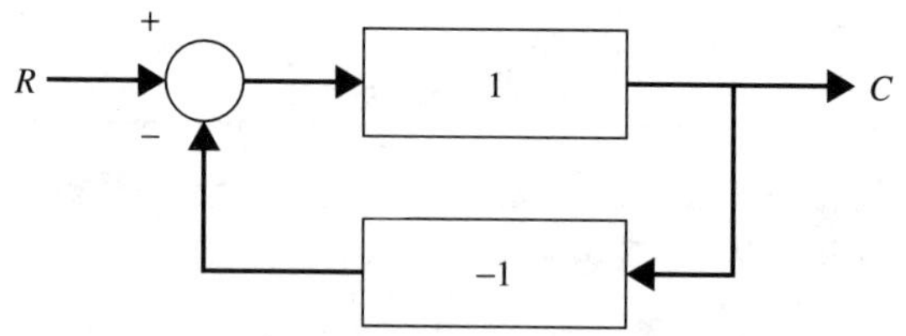

Figure 8.6 Block diagram for Example 8.4.

Solution

The closed-loop transfer function is

$$\begin{aligned} \text{CLTF} &= \frac{G}{1+GH} \\ &= \frac{1}{1+1(-1)} \\ &= \frac{10}{1-1} \\ &= \frac{10}{0} \to \infty \\ \text{system output } C &= \text{input } R \cdot \text{CLTF} \\ &= 1 \cdot \infty \\ &\to \infty \end{aligned}$$

EXAMPLE 8.5

Determine the transfer function of the unity feedback closed-loop system (Figure 8.7) and the system output when subjected to a unit input.

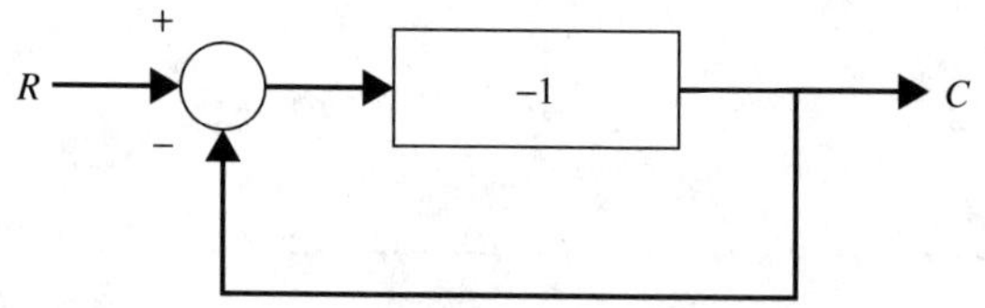

Figure 8.7 Block diagram for Example 8.5.

Solution

The closed-loop transfer function is

$$
\begin{aligned}
\text{CLTF} &= \frac{G}{1+GH} \\
&= \frac{-1}{1+(-1)1} \\
&= \frac{-1}{1-1} \\
&= \frac{10}{0} \rightarrow -\infty \\
\text{system output } C &= \text{input } R \cdot \text{CLTF} \\
&= 1(-\infty) \\
&\rightarrow -\infty
\end{aligned}
$$

In each of the previous three examples, forward and feedback gains are different, but the closed-loop transfer function still amounts to an infinite value, making the system unstable. Thus, it is not the individual value of forward gain G or feedback gain H, which is of most significance: it is the product of these two gains GH (also known as loop gain, or open-loop gain), which dictates the final outcome.

OPEN-LOOP GAIN AND OPERATING FREQUENCY

It is interesting to note that a negative sign does not necessarily mean that one of the two gain values, G or H, is negative in nature. A closed-loop control system is subjected not only to DC input values but also to sinusoidal signal varying over a wide range of frequencies.

Fourier analysis indicates that a step input signal (square wave) to a control system can be factored into an infinite number of sinusoidal signals, (Figure 8.8) varying from very low to very high frequency. A step (change) input is routinely applied to control systems. Even a sudden noise pulse, coupled to the control system through (say)

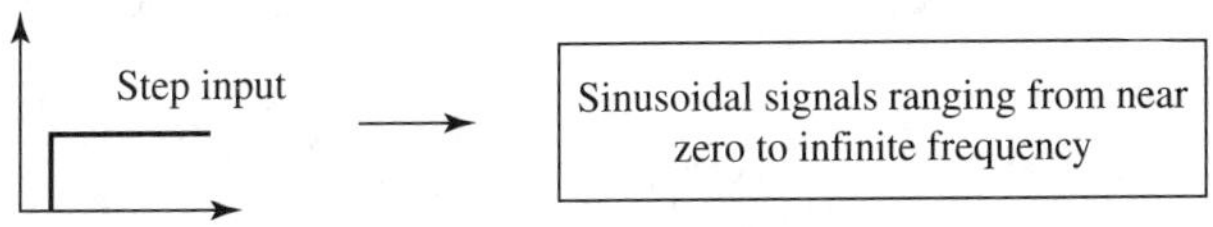

Figure 8.8 Step-signal equivalent.

electromagnetic interference, will effectively subject the control system to a sinusoidal signal covering a wide frequency range.

Note: For the sake of simplicity, in the following discussions it will be assumed that a closed-loop control system is a unity feedback type of system. This means that $H = 1$ and all the phase shift in the output is provided by the system forward gain, G.

As has been discussed in frequency response analysis, when a first- or higher-order control system is subjected to a sinusoidal input, there is a phase difference between control system output and applied input. This phase difference is not constant but changes with a change in frequency of input signal. Thus, in the case of sinusoidal signals, a sign inversion is equivalent to a phase delay of $-180°$. Real control systems are generally lag-type systems, where system output lags the input signal. These first- or higher-order lag systems exhibit an increase in phase shift with an increase in operating frequency. If the system is operating at a frequency where the phase difference between input and output reaches $-180°$, it will result in inversion of sign for open-loop gain (GH). Similarly, the gain of such a system is frequency-dependent, and if the loop gain becomes equal to 1, the overall gain (gain and phase difference) becomes equal to -1, and the condition of instability are met. An overall gain of -1 can be represented as a vector with an absolute magnitude of 1 and a phase angle of $-180°$ (Figure 8.9).

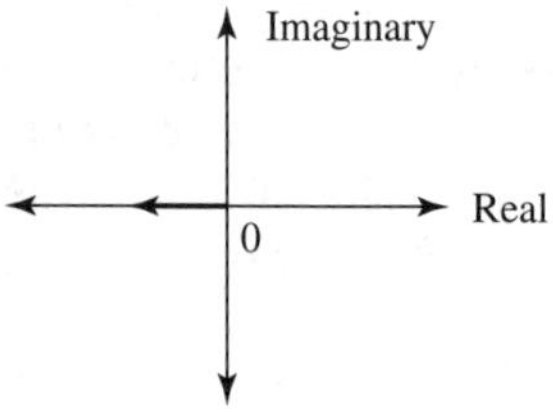

Figure 8.9 Open-loop gain vector $GH = -1$.

In a negative-feedback control system, this feedback signal goes through another inversion and becomes in phase with the input signal. As a result, the output increases in amplitude. If the loop gain is equal to 1, the output signal keeps on increasing without bound, rendering the system unstable.

EXAMPLE 8.6

Determine the open-loop gain of the system in Figure 8.10 at the following frequencies:

a. 1 rad/s
b. 10 rad/s
c. 100 rad/s

$$G = \frac{10}{s(s+1)(s+2)} \quad \text{and} \quad H = 0.5$$

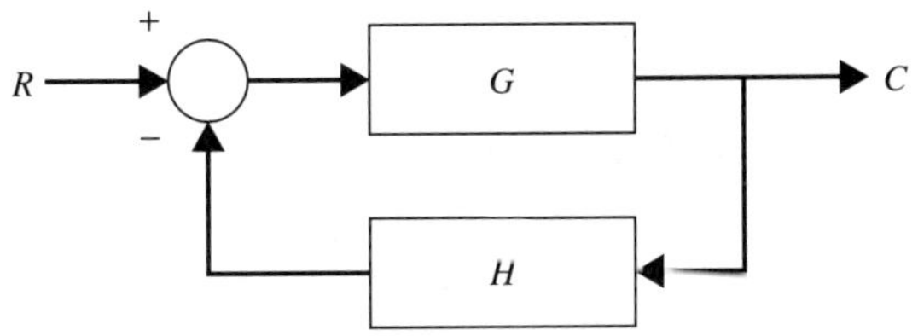

Figure 8.10 Block diagram for Example 8.6.

Solution

Open-loop gain and phase angle can be determined from the open-loop transfer function, by substituting $j\omega$ for every s term, where ω represents the frequency (rad/s) of interest. The open-loop transfer function is

$$GH = \frac{10}{s(s+1)(s+2)} \cdot 0.5$$
$$= \frac{5}{s(s+1)(s+2)}$$

Substituting $s = j\omega$,

$$GH(j\omega) = \frac{5}{j\omega(j\omega+1)(j\omega+2)}$$
$$= 5 \cdot \left(\frac{1}{\omega} \angle -90^\circ\right) \cdot \left(\frac{1}{\sqrt{1+\omega^2}} \angle -\tan^{-1}\omega\right) \cdot \left(\frac{1}{\sqrt{4+\omega^2}} \angle -\tan^{-1}\left(\frac{\omega}{2}\right)\right)$$

a.

$$GH = 5 \cdot \left(\frac{1}{1} \angle -90^\circ\right) \cdot \left(\frac{1}{\sqrt{1+1^2}} \angle -\tan^{-1}1\right) \cdot \left(\frac{1}{\sqrt{4+1^2}} \angle -\tan^{-1}\left(\frac{1}{2}\right)\right)$$
$$GH = 5 \cdot (1 \angle -90^\circ) \cdot (0.707 \angle -45^\circ) \cdot (0.447 \angle -26.56^\circ)$$
$$GH = 1.581 \angle -161.56^\circ$$

b.

$$GH = 5 \cdot \left(\frac{1}{10} \angle -90^\circ\right) \cdot \left(\frac{1}{\sqrt{1+10^2}} \angle -\tan^{-1}10\right) \cdot \left(\frac{1}{\sqrt{4+10^2}} \angle -\tan^{-1}\left(\frac{10}{2}\right)\right)$$
$$GH = 0.0049 \angle -252.98^\circ$$

c.

$$GH = 5 \cdot \left(\frac{1}{100} \angle -90^\circ\right) \cdot \left(\frac{1}{\sqrt{1+100^2}} \angle -\tan^{-1}100\right)$$
$$\cdot \left(\frac{1}{\sqrt{4+100^2}} \angle -\tan^{-1}\left(\frac{100}{2}\right)\right)$$
$$GH = 0.00005 \angle -268.28^\circ$$

The calculations show that the open-loop gain of the control system does not remain constant but decreases with an increase in frequency. On the other hand, the phase shift (delay) between output and input signal increases with frequency.

EXAMPLE 8.7

Determine if the system in Example 8.6 is stable or unstable.

Solution

The control system is unstable if the following two conditions are met.

1. At some frequency the phase angle crosses the -180° line.
2. The open-loop gain at this crossover frequency is greater than 1.

Calculations carried out previously for frequency values of 1, 10, and 100 rad/s clearly indicate that at higher frequencies, the phase angle does cross the -180° threshold. Having met the first condition, the system will be unstable if the open-loop gain is greater than 1 at the crossover frequency. On the other hand, if the gain at that frequency is less than 1, then the open-loop system is stable.

The problem is now reduced to

- Finding the frequency at which the phase angle becomes equal to -180°;
- Determining the open-loop gain at this frequency.

At a frequency of 1 rad/s, the phase angle is -161.50°, and it drops to -252.98° at a frequency of 10 rad/s. Thus, the phase angle must reach -180° somewhere in between 1 and 10 rad/s. In order to get a better insight, open-loop gain and phase angles need to be calculated at several points during this frequency range.

Performing calculations at a frequency of 2 rad/s, the following data are obtained:

$$\text{open-loop gain} = 0.395$$
$$\text{phase angle} = \angle -198.4^\circ$$

Thus it is evident that the phase angle crosses the -180° line between the frequencies of 1 and 2 rad/s. The data in Table 8.1 show open-loop gain, phase angle, and frequency data in this frequency range.

Table 8.1 Example 8.7

Frequency (rad/s)	Gain	Phase Angle (degrees)
1.0000	1.5811	−161.5651
1.1000	1.3396	−166.5371
1.2000	1.1437	−171.1582
1.3000	0.9831	−175.4553
1.4000	0.8503	−179.4543
1.5000	0.7396	−183.1798
1.6000	0.6467	−186.6544
1.7000	0.5681	−189.8990
1.8000	0.5014	−192.9326
1.9000	0.4443	−195.7727
2.0000	0.3953	−198.4349

Calculated data indicate that the phase angle reaches $-180°$ at a frequency just greater than 1.4 rad/s. At a frequency of 1.4 rad/s, the open-loop gain has already dropped to 0.8503 (<1) and continues to drop as frequency increases. Thus, at the frequency at which the phase angle crosses the $-180°$ threshold, the open-loop gain is going to be less than 1. The second criterion of instability (gain > 1 with phase angle $= -180°$) is thus not met, and the closed-loop control system is stable.

STABILITY FROM BODE PLOT

A system is considered stable if the open-loop gain is less than unity, when the phase angle (delay) has reached $-180°$. It is common to deal with magnitude (dB) rather than gain (amplification factor) in control system analysis, so using magnitude instead of gain, stability criterion can be restated as follows.

Stable Closed-Loop System

A closed-loop control system is **stable** if the open-loop gain is less than 0 dB at a frequency at which the phase angle reaches $-180°$.

A Bode plot provides information about magnitude and phase angle of transfer function, and from these data the stability of a system can be easily determined.

EXAMPLE 8.8

Using a Bode plot, determine if the following system is stable (Figure 8.11).

$$G_1 = \frac{1}{(s+1)}, \qquad G_2 = \frac{2}{(s+10)}, \quad \text{and} \quad H = \frac{1}{s}$$

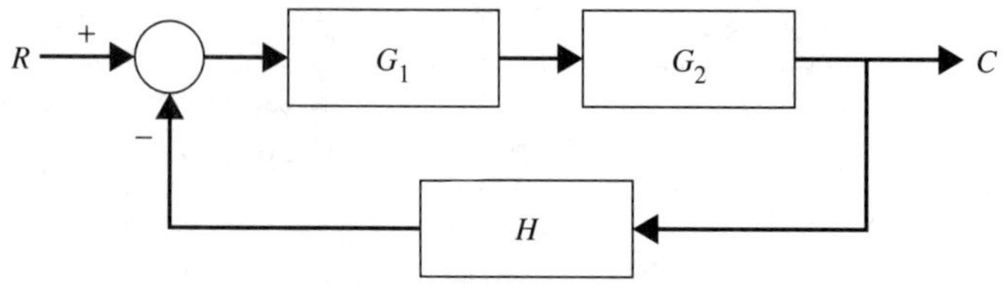

Figure 8.11 Block diagram for Example 8.8.

Solution

The open-loop transfer function is

$$GH = G_1 \cdot G_2 \cdot H$$

$$GH = \frac{1}{(s+1)} \cdot \frac{2}{(s+10)} \cdot \frac{1}{s}$$

$$GH = \frac{2}{s(s+1)(s+10)}$$

$$GH = \frac{2}{s^3 + 11s^2 + 10s}$$

The Bode plot of this open-loop transfer function is shown in Figure 8.12.

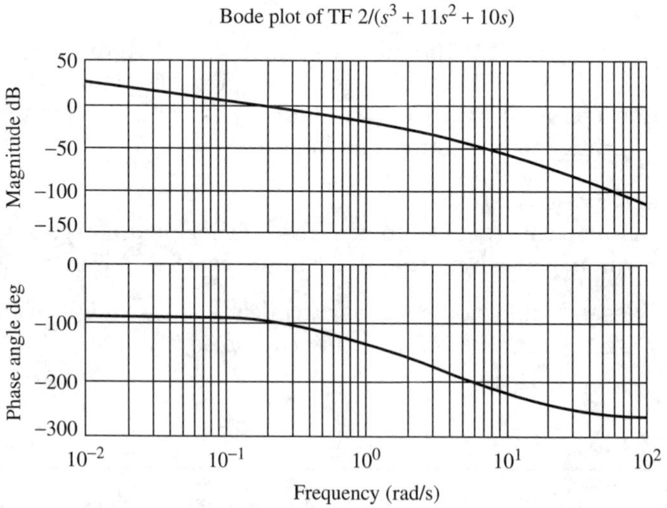

Figure 8.12 Bode plot for Example 8.8.

The plot indicates that the phase angle crosses -180° at a frequency of about 3 rad/s. At this frequency, the magnitude is less than -30 dB (gain of 0.03). Because it (magnitude) is less than 0 dB, the closed-loop control system is stable.

GAIN AND PHASE MARGINS

It has been stated that an open-loop gain of 1 with a phase angle of -180° represents the limiting condition for stability. In practice, these values of gain and phase angle are never used as the target values. An attempt is made to keep the open-loop gain well below 1 at the phase crossover frequency. This provides a margin of safety for variation in gain, and it can be changed slightly without running a risk of instability (gain $= 1$). Similarly, an attempt is made to maintain the phase angle above -180°, at the frequency at which open-loop gain drops to 1. These margins are necessitated by the following:

- System modeling may not represent the exact behavior of field components.
- With age, component characteristics may deviate and change the frequency response characteristics.
- Controller tuning (gain adjustment) can be carried out without the immediate concern that gain will reach unstable limits.

Gain and Phase Crossover Frequencies

The following frequencies are significant in establishing the stability of a control system.

Gain Crossover Frequency

Gain crossover frequency is the frequency at which the open-loop gain drops to 0 dB (gain of 1).

Phase Crossover Frequency

Phase crossover frequency is the frequency at which the open-loop phase angle drops to -180°.

Gain Margin (GM)

Gain Margin

Gain margin is the additional gain that can be added to open-loop gain before a closed-loop system becomes unstable in operation.

Some of the characteristics related to gain margin are:

- It is commonly expressed in dB.
- It is measured at the phase crossover frequency.
- It is the difference between the 0-dB line and the instantaneous open-loop magnitude (gain in dB).
- It is considered positive if, at phase crossover frequency, magnitude is less than 0 dB (gain < 1).
- It is considered negative if, at phase crossover frequency, magnitude is greater than 0 dB (gain > 1).
- The closed-loop control system is stable if GM is positive.

gain margin GM = 0 dB − magnitude at phase crossover frequency

EXAMPLE 8.9

Determine the gain margin of a control system from the data in Table 8.2.

Table 8.2 Example 8.9

Frequency (rad/s)	*Gain*	*Magnitude (dB)*	*Phase Angle (degrees)*
0.1	5	14	0
1	1.5	3.5	−90
10	0.25	−12	−180
100	0.1	−20	−235

Solution

Gain margin is measured at the phase crossover frequency. From the given data, the phase reaches -180° at a frequency of 10 rad/s.

$$\text{phase crossover frequency} = 10\,\text{rad/s}$$

$$\text{gain (magnitude) at phase crossover frequency} = -12\,\text{dB}$$

$$\begin{aligned}\text{gain margin} &= 0\,\text{dB} - \text{magnitude at phase crossover frequency}\\ &= 0\,\text{dB} - (-12\,\text{dB})\\ &= +12\,\text{dB}\end{aligned}$$

Because the gain margin is positive, the closed-loop system is stable. The gain at phase crossover frequency is 0.25. It can be increased to a maximum of 1 before the system becomes unstable.

Therefore, the additional gain that can be added to a system is

$$\frac{1 \quad \text{(limiting gain)}}{0.25 \quad \text{(current gain)}} = 4$$

$$= 20\log(4)$$

$$= 12\ \text{dB} \qquad \text{(gain margin)}$$

The gain margin indicates that an additional gain of 12 dB can be added to a system before it becomes unstable.

Phase Margin (PM)

Phase Margin

Phase margin is the additional drop in phase angle that can be added to a system before it becomes unstable.

Some of its characteristics are as follows:

- It is expressed in degrees.
- It is measured at the gain crossover frequency.
- It is the difference between the -180° line and the instantaneous open-loop phase angle.
- It is considered negative if the phase angle is greater than -180°.
- It is considered positive if the phase angle is less than -180°.
- A closed-loop control system is stable if the phase margin is positive.

phase margin $PM = 180^\circ +$ phase angle at gain crossover frequency

EXAMPLE 8.10

Determine the phase margin for a system from the data in Table 8.3.

Table 8.3 Example 8.10

Frequency (rad/s)	*Gain*	*Magnitude (dB)*	*Phase Angle (degrees)*
1	10	20	−50
10	2	6	−100
100	1	0	−155
1000	0.1	−20	−235

Solution

Phase margin is measured at the gain crossover frequency. From the given data, gain crossover (gain of 1) occurs at a frequency of 100 rad/s.

$$\begin{aligned}
\text{gain crossover frequency} &= 10\ \text{rad/s.}\\
\text{phase angle at gain crossover frequency} &= -155^\circ\\
\text{phase margin} &= 180^\circ + \text{phase angle at gain crossover frequency}\\
&= 180^\circ + (-155^\circ)\\
&= +25^\circ
\end{aligned}$$

Because phase margin is positive, the closed-loop system is stable. At the gain crossover frequency, phase angle is -155° and can be decreased by an additional 25° before a system becomes unstable.

Desired Gain and Phase Margins

From stability considerations, it is desirable to maintain a large gain and phase margin. A large value for gain margin means a lower system gain, which results in a sluggish system response. With low values of gain, the system will not be able to respond quickly to changes in set point as well as load. This requirement contradicts the requirement for stability, which dictates a lower gain. In order to satisfy both requirements, a compromise has to be made. The system gain is kept as high as possible while still maintaining a sufficient gain margin—that is, keeping the gain below 1 at phase crossover frequency. The following values of gain and phase margin are commonly used for setting (tuning) control system operations.

$$\begin{aligned}
\text{gain margin} &= 6 \text{ to } 12\ \text{dB}\\
\text{phase margin} &= 45^\circ
\end{aligned}$$

EXAMPLE 8.11

Figure 8.13 shows a closed-loop control system incorporating an amplifier with a gain of K. Use a Bode plot to determine the following.

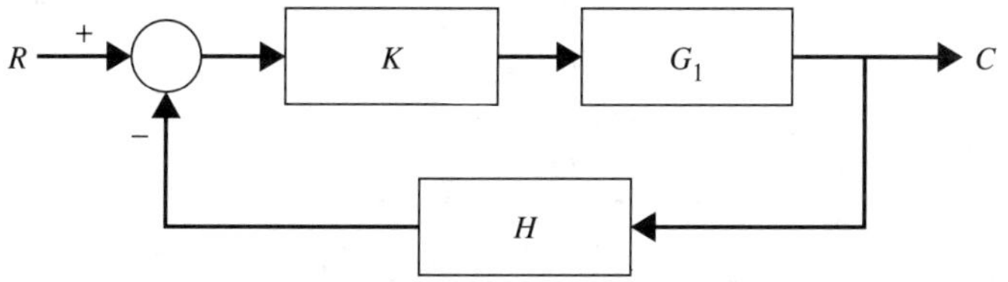

Figure 8.13 Block diagram for Example 8.11.

a. Is the system stable for a gain (K) of 1?
b. What is the maximum value of gain (K) before the system becomes unstable?

$$G_1 = \frac{2.5}{(s+1)(5s+12)} \qquad H = \frac{2}{(3s+4)}$$

Solution

The open-loop transfer function is

$$GH = K \cdot G_1 \cdot H$$
$$GH = K \cdot \frac{2.5}{(s+1)(5s+12)} \cdot \frac{2}{(3s+4)}$$
$$GH = \frac{5K}{(15s^3+71s^2+104s+48)}$$

The Bode plot can be drawn only after a numerical value is substituted for the undetermined amplifier gain K. Assuming an initial value of 1 for K, a Bode plot can be drawn. The maximum gain value can then be determined from this plot. Substituting the value of 1 for amplifier gain K, the open-loop transfer function becomes

$$GH = \frac{5}{(15s^3+71s^2+104s+48)}$$

From the Bode plot of this transfer function (Figure 8.14), it can be observed that the phase angle crosses the -180° line at a frequency of 3.2 rad/s. At this frequency the system magnitude is -42.9 dB (gain = 0.007). Because the magnitude is significantly below the 0-dB line, corresponding to a phase shift of -180°, the closed-loop system is stable for a gain of 1.

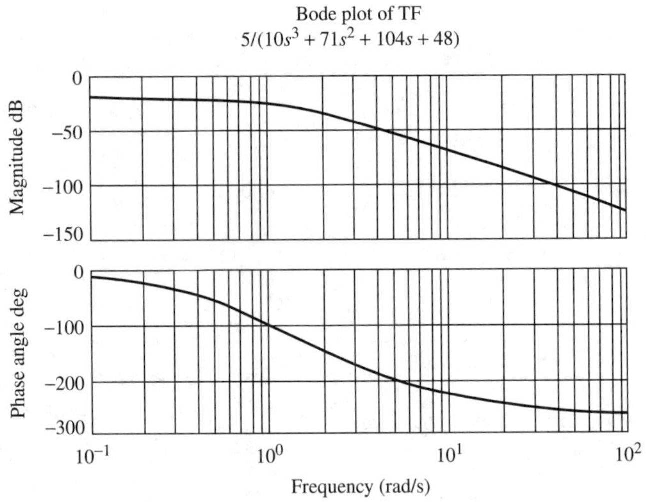

Figure 8.14 Bode plot for Example 8.11.

Maximum value of gain: Variation in gain affects only the magnitude plot, and does not affect the phase angle. The magnitude plot shifts vertically upward with an increase in gain and shifts downward with a decrease in gain. Amplifier gain can be increased until magnitude approaches 0 dB at a phase crossover frequency of 3.2 rad/s. With an assumed value of the gain equal to 1 ($K = 1$), the magnitude at this frequency is -42.9 dB. Thus, there is some margin and gain can be increased (theoretically) until

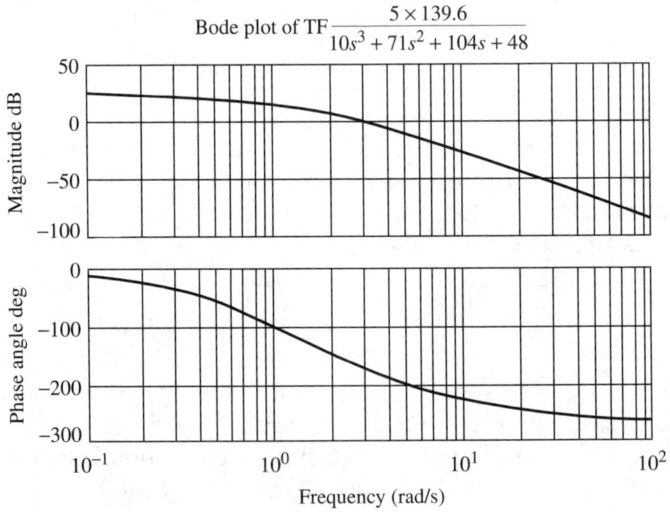

Figure 8.15 Example 8.11: Bode plot with maximum gain.

the magnitude approaches 0 dB. When gain is increased by this amount, the system can become unstable.

$$\text{maximum increase in magnitude} = 42.9 \text{ dB}$$

$$\begin{aligned}\text{maximum gain } K &= 20 \log^{-1}(42.9/20) \\ &= 139.6\end{aligned}$$

In practice, gain will never be increased by this amount but will be limited to a somewhat smaller value in order to retain a margin of safely. The Bode plot of the system with an amplifier gain of 139.6 (Figure 8.15) shows that at the gain crossover frequency, the phase angle has dropped to the limiting value of $-180°$.

EXAMPLE 8.12

Determine if the following unity feedback closed-loop system is unstable for a frequency range of 0.10 to 100 rad/s.

$$G(s) = \frac{25}{s(0.1s + 1)(0.02s + 1)}$$

Solution

Because it is a unity feedback system ($H = 1$), the open-loop transfer function is

$$\begin{aligned}GH &= G * ! \\ &= G \\ GH &= \frac{25}{s(0.1s + 1)(0.02s + 1)} \\ GH &= \frac{25}{0.002s^2 + 0.12s^2 + s}\end{aligned}$$

The MATLAB commands for plotting the frequency response are shown next:

MATLAB Commands

```
n = 25
d = [0.002, 0.12, 1, 0]
ω = logspace(-1, 2, 10)
[g,p,ω] = bode(n,d,ω)
m = 20*log10(g); Calculate magnitude in dB
[ω,g,m,p] ; Display numerical result
bode(n,d,ω) ; Plot the graph
```

The frequency response data in tabular form is shown in Table 8.4.

This system has a very high open-loop gain of 250 (48 dB) at a frequency of 0.1 rad/s. Gain drops rapidly to a very low value of 0.0013 at a frequency of 100 rad/s. The phase angle starts at $-90.7°$ and drops to $-266.6°$ at high frequency. High initial gain and a

Table 8.4 Frequency Response Data for Example 8.12

Frequency (rad/s)	*Gain*	*Magnitude (dB)*	*Phase Angle (degrees)*
0.1000	249.9870	47.9583	−90.6875
0.2154	116.0117	41.2900	−91.4811
0.4642	53.8006	34.6157	−93.1894
1.0000	24.8710	27.9138	−96.8564
2.1544	11.3332	21.0870	−104.6254
4.6416	4.8645	13.7409	−120.2025
10.0000	1.7334	4.7782	−146.3099
21.5443	0.4487	−6.9615	−178.4118
46.4159	0.0831	−21.6042	−210.7129
100.0000	0.0111	−39.0741	−237.7244

phase angle less than -180° at higher frequencies usually indicate a potential problem with system stability. In this case, the phase angle changes from -178.4° at a frequency of about 21.5 rad/s to -210.7° at 46.4 rad/s. The open-loop gains corresponding to these frequencies are −6.96 and −21.6, respectively. Assuming continuity (sometimes this needs further investigations), it can be stated that gain at a frequency where the phase angle reaches -180° is less than 1, and thus the system is stable. This is verified by the Bode plot in Figure 8.16.

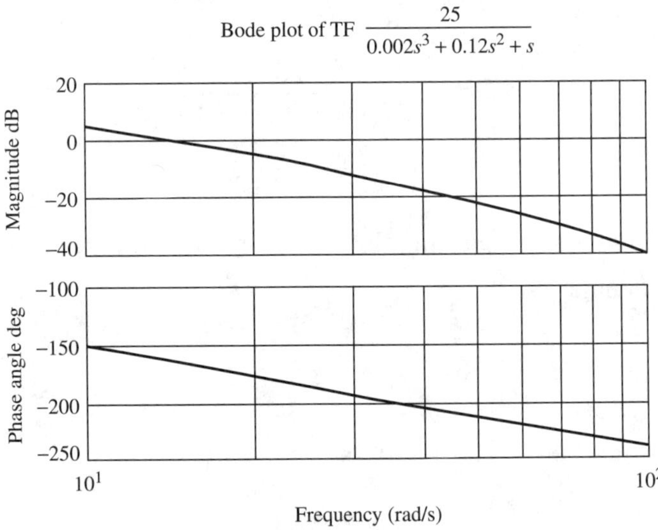

Figure 8.16 Bode plot for Example 8.12.

STABILITY FROM THE CHARACTERISTIC EQUATION

The roots of the characteristic equation of a closed-loop system provide significant information about the impulse response of a system. If a root lies on the left-hand side of the s-plane (*LHP*), the system output due to this root eventually dies out. On the other hand, if a root lies on the right-hand side of the s-plane (*RHP*), output continues to grow without bound. For a system with multiple roots, even if all the roots lie on the LHP except one, the output due to the roots in the LHP will eventually die, but output due to the root in the RHP will continue to rise without bound. As a result, the overall system output will continue to rise indefinitely, which may eventually lead to a catastrophic failure. Stability in terms of the roots of the characteristic equation can now be defined.

Stable System

A closed-loop control system is considered **stable** if all the roots of the characteristic equation lie in the LHP (**L**eft-**H**and side of the s-**P**lane).

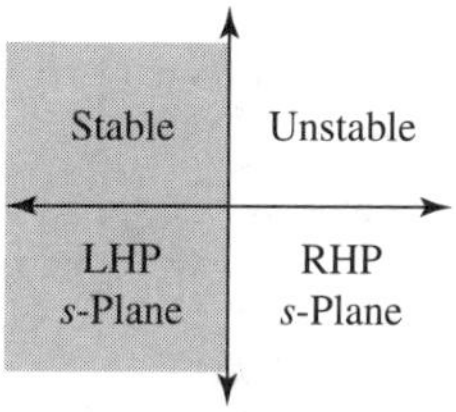

Unstable System

A closed-loop control system is considered **unstable** if any one (or more) root of the characteristic equation lie in the RHP (**R**ight-**H**and side of the s-**P**lane).

Marginally Stable System

A closed-loop control system is considered **marginally stable** (quasi-stable) if roots of the characteristic equation lie on the imaginary axis. For all practical purposes, a marginally stable control system can be treated as unstable.

EXAMPLE 8.13

Determine if the following system is stable:

$$G(s) = \frac{1.2}{s(s+2)} \quad \text{and} \quad H(s) = 1$$

Solution

Stability can be determined by finding the location of the roots of the characteristic equation of the closed-loop control system. The characteristic equation is

$$1 + GH(s) = 0$$

Substituting the values of $G(s)$ and $H(s)$ in the equation,

$$1 + \frac{1.2}{s(s+2)} \cdot 1 = 0$$

$$1 + \frac{1.2}{s^2 + 2s} = 0$$

$$\frac{s^2 + 2s + 1.2}{s^2 + 2s} = 0$$

$$s^2 + 2s + 1.2 = 0$$

Roots of this second-order equation can be determined from the quadratic formula:

$$s = \frac{-2 \pm \sqrt{2^2 - 4 \cdot 1 \cdot 1.2}}{2 \cdot 1}$$

$$s = -1 \pm j0.447$$

There are two roots of the characteristic equation, which appear as a complex conjugate pair. Both of these roots have a negative real component of −1. Thus, both these roots lie in the LHP and the closed-loop system is stable.

EXAMPLE 8.14

Determine the stability of the following system:

$$G(s) = \frac{1}{s^2 + 6s + 6} \quad \text{and} \quad H(s) = \frac{1}{s}$$

Solution

The characteristic equation is

$$1 + GH(s) = 0$$

Substituting the values of $G(s)$ and $H(s)$,

$$1+\frac{1}{s^2+6s+6}\cdot\frac{1}{s}=0$$

$$1+\frac{1}{s^3+6s^2+6s}=0$$

$$\frac{s^3+6s^2+6s+1}{s^3+6s^2+6s}=0$$

$$s^3+6s^2+6s+1=0$$

Factoring the polynomial,

$$(s+1)(s^2+5s+1)=0$$

The equation is satisfied if each factor is equal to zero.

1. From the first factor,

$$(s+1)=0$$
$$s=-1$$

2. From the second factor,

$$(s^2+5s+1)=0$$
$$s=\frac{-5\pm\sqrt{5^2-4\cdot1\cdot1}}{2\cdot1}$$
$$s=-0.209,\ -4.791$$

The characteristic equation has three roots ($s=-1$, $s=-0.209$, and $s=-4.791$), and all three roots lie in the LHP. The closed-loop control system is thus stable.

ROUTH CRITERION FOR STABILITY

As the order of the characteristic equation increases, it becomes increasingly difficult to solve for the roots of the characteristic equation. Even a third-order equation with complex roots may require a considerable amount of time for its solution. In order to determine the stability of a system, it is not necessary to obtain the roots of the characteristic equation. It is simply sufficient to determine if any one of the roots lies in the RHP; if so, the system is unstable.

E. J. Routh (1892) developed a simple test to determine if any root of the characteristic equation lies in the RHP. The procedure is as follows.

1. Determine the characteristic equation.
2. Simplify the characteristic equation so that there are no fractional terms. It is done by multiplying the whole equation by the denominator.
3. Arrange the resulting polynomial in ascending order, with the leftmost term being of the highest power.

$$a_0 s^n + a_1 s^{n-1} + \cdots + a_{n-1} s + a_n = 0$$

4. Ensure that every power term is present, even if the numerical coefficient of that term is zero.

EXAMPLE 8.15

Arrange the following characteristic equation for use with Routh stability criteria:

$$s^3 + 2s + 1 = 0$$

Solution

The characteristic equation $s^3 + 2s + 1 = 0$ is arranged as

$$s^3 + 0s^2 + 2s + 1 = 0$$

The s^2 term is missing in the original equation but is added in the final expression with a zero coefficient.

---

5. The last term should be a constant term, even if it is zero.
6. Create the Routh array.
7. Substitute zero for the terms that do not exist (or have zero coefficient).
8. Determine the number of sign changes in the entries for the first column. This number represents the number of roots that lie in the RHP.
9. A closed-loop system is stable if there is no sign change in the first column.

Routh Array

[Polynomial $a_0s^4 + a_1s^3 + a_2s^2 + a_3s + a_4 = 0$]				
s^4	a_0	a_2	a_4	a_6
s^3	a_1	a_3	a_5	a_7
s^2	b_1	b_2	b_3	
s^1	c_1	c_2		
s^0				

where

$$b_1 = \frac{a_1 a_2 - a_0 a_3}{a_1}$$

$$b_2 = \frac{a_1 a_4 - a_0 a_5}{a_1}$$

$$b_3 = \frac{a_1 a_6 - a_0 a_7}{a_1}$$

$$c_1 = \frac{b_1 a_3 - a_1 b_2}{b_1}$$

$$c_2 = \frac{b_1 a_5 - a_1 b_3}{b_1}$$

Special Case

10. If a term in the first column turns out to be zero, the subsequent terms can still be evaluated by replacing zero with ε, an infinitesimally small number, and then taking the limit of ε.
11. If an entire row of a Routh array is zero, an auxiliary polynomial array can be created.

EXAMPLE 8.16

Using the Routh criterion, determine if the following unity feedback system is stable:

$$G(s) = \frac{12}{(s+1)(s^2+s+1)}$$

Solution

The system characteristic equation is

$$1 + GH(s) = 0$$

$$1 + \frac{12}{(s+1)(s^2+s+1)} \cdot 1 = 0$$

$$s^3 + 2s^2 + 2s + 13 = 0$$

The polynomial for creating the Routh array is thus $s^3 + 2s^2 + 2s + 13$.

Routh Array

s^3	a_0	a_2
s^2	a_1	a_3
s^1	b_1	b_2
s^0	c_1	c_2

where

$$a_0 = 1$$

$$a_1 = 2$$

$$a_2 = 2$$

$$a_3 = 13$$

$$b_1 = \frac{a_1 a_2 - a_0 a_3}{a_1} = \frac{2 \cdot 2 - 1 \cdot 13}{2} = -11$$

$$b_2 = \frac{a_1 a_4 - a_0 a_5}{a_1} = \frac{2 \cdot 0 - 1 \cdot 0}{2} = 0$$

(The a_4 and a_5 terms do not exist and are considered to be zero.)

$$c_1 = \frac{b_1 a_3 - a_1 b_2}{b_1} = \frac{-11 \cdot 13 - 2 \cdot 0}{-11} = 13$$

All the other terms are zero.

s^3	1	2
s^2	2	13
s^1	−11	0
s^0	13	0

There are two sign changes in the first column. The first sign change occurs in the third row, and the second sign change occurs in the fourth row. This indicates that there are two roots that lie in the RHP and the closed-loop system is unstable.

Check: The result can be verified by determining the roots of the characteristic equation. The characteristic equation is

$$s^3 + 2s^2 + 2s + 13 = 0$$

The roots of the characteristic equation are

$$s = -2.8761$$
$$s = 0.4381 + j2.0804$$
$$s = 0.4381 - j2.0804$$

The last two roots appear as a complex conjugate pair and lie in the right-hand plane.

EXAMPLE 8.17

Identify whether the following system is stable and also determine the GM and PM.

$$\text{forward gain } G(s) = \frac{5}{s(0.045s+1)}$$
$$\text{feedback gain } H(s) = \frac{2}{(0.1s+1)}$$

Solution

Stability, GM, and PM can all be determined from the frequency response of the open-loop transfer function:

$$GH(s) = \frac{5}{s(0.045s+1)} \times \frac{2}{(0.1s+1)}$$
$$GH(s) = \frac{10}{s(0.045s+1)(0.1s+1)}$$
$$GH(s) = \frac{10}{0.0045s^3 + 0.145s^2 + s}$$

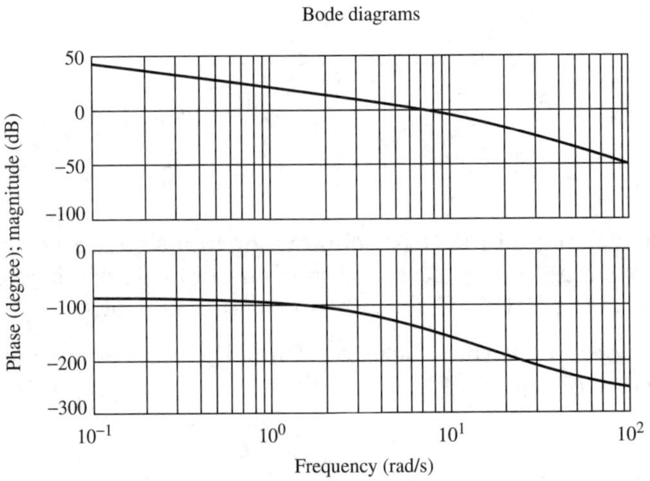

Figure 8.17 Bode plot for Example 8.17.

The following can be observed from the Bode plot:

$$\text{gain crossover frequency} = 7.5 \text{ rad/s}$$
$$\text{phase crossover frequency} = 14.9 \text{ rad/s}$$
$$\text{gain margin} = 10.2 \text{ dB}$$
$$\text{phase margin} = 34.2^\circ$$

Because both the gain margin and phase margin are positive, the closed-loop system is stable.

EXAMPLE 8.18

Using a Bode plot, determine the maximum value of gain K for which the following system (Figure 8.18) is stable, given that

$$G(s) = \frac{K}{s^3 + 10s^2 + s}$$

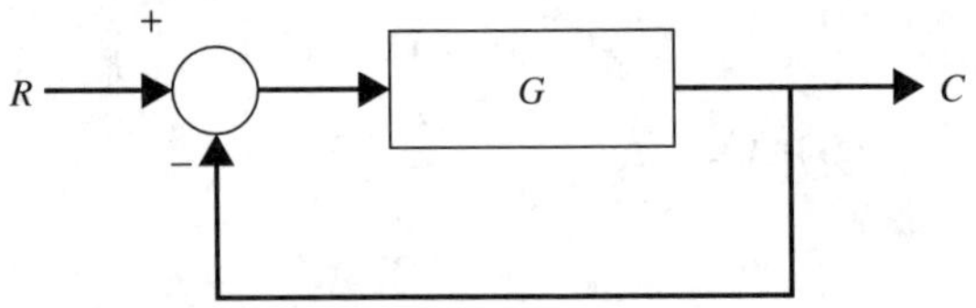

Figure 8.18 Block diagram for Example 8.18.

Solution

The open-loop transfer function of the given system is

$$GH(s) = \frac{K}{s^3 + 10s^2 + s}$$

The frequency response curve cannot be plotted for this open-loop transfer function, because the value of the gain coefficient K is not known. In such a case the following procedure can be used:

- Assume the value of the gain coefficient K to be unity ($K = 1$).
- Obtain the open-loop frequency response for this value.
- Determine the gain and phase margin.
- If the phase margin is negative, even a gain of 1 is excessive and needs to be reduced. The reduction in gain is indicated by the negative gain margin. On the other hand, if the phase margin is positive, the gain can be increased. The amount through which gain can be increased is indicated by the positive gain margin.

Assuming a gain of 1 ($K = 1$), the open-loop transfer function is

$$GH(s) = \frac{1}{s^3 + 10s^2 + s}$$

The plot (Figure 8.19) indicates a gain margin of 20 dB and a phase margin of 16.5°. Thus, the system is stable and it is possible to increase the gain. Again, gain can be

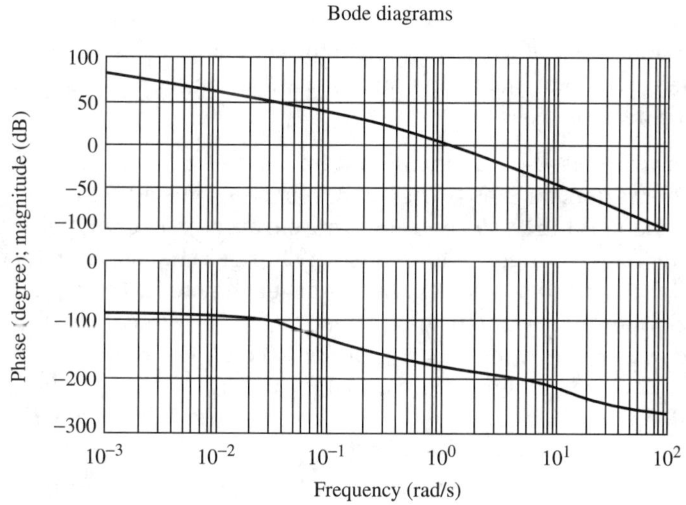

Figure 8.19 Example 8.18: Bode plot ($K = 1$).

(*theoretically*) increased to an extent that will render a zero gain margin. Thus,

$$\begin{aligned}\text{permissible increase in gain} &= 20\ \text{dB}\\ &= 10^{(20/20)}\\ &= 10\\ \text{new (maximum) gain} &= \text{current gain} \times \text{increase in gain}\\ &= 1 \times 10\\ &= 10\end{aligned}$$

The maximum gain K that can be used is 10.

Note: This is only a theoretical value. The gain of 10 brings the system on the verge of instability (zero gain margin). In practice this will not be acceptable, and a lower ($K < 10$) gain value will be used.

ROOT LOCUS ANALYSIS

Root locus analysis was first proposed by W. R. Evans. It allows the determination of the path of the closed-loop poles of a control system from the knowledge of its open-loop transfer function for various values of one system parameter. The root loci are plotted on the s-plane and clearly show the movement of pole locations as the variable changes.

Root Locus

The **root locus** is the path of the poles of a closed-loop control system as one of the system parameters; usually the system gain factor K is varied over a certain range.

Pole locations of a closed-loop control system dictate some very significant characteristics of system operation, such as transient response, damping ratio, damped frequency, and the control loop stability. When a system parameter is changed, the control loop characteristic equation changes, which in turn changes the location of poles and zeros. As this parameter is varied over a certain range, each pole traces a path in the s-plane. These paths provide an insight into the system performance, related to the system parameter under consideration. The root locus plot helps visualize the transient response and can be used as an effective tool for control system design.

The closed-loop transfer function for the given system (Figure 8.20) is

$$\text{CLTF} = \frac{KG(s)}{1 + KG(s)H(s)}$$

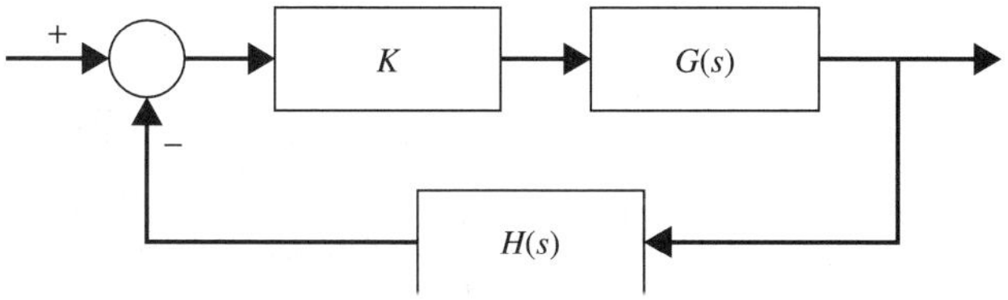

Figure 8.20 Closed-loop system for a root locus.

The characteristic equation is

$$1 + KG(s)H(s) = 0$$

Rearranging gives

$$KG(s)H(s) = -1$$

or

$$G(s)H(s) = -\frac{1}{K}$$

This expression simply states that for every value of K, a set of roots of $G(s)H(s)$, the open-loop function, will be found. The pole locations and the angles of all vectors drawn to the poles from the origin are influenced by the value of system parameter K. A number of rules were developed, which aided in the root locus plot. W. R. Evans even came up with a mechanical aid for plotting the root locus. Root locus plots required considerable effort on the part of the control analyst. Today, MATLAB and similar computer programs have made it extremely easy to develop a root locus (and time and frequency response) for practically any type of control system.

Root locus techniques are applied to control system design, where the objective is to place the poles at a certain location in the s-plane and achieve a desired performance. These techniques can also be used to study the influence of certain system parameters on performance and stability.

EXAMPLE 8.19

Sketch the root locus of a closed-loop control system with the given transfer function as the variable K is varied over a range of 0 to ∞.

$$\text{CLTF}(s) = \frac{K}{s(s + 2K)}$$

Solution

The characteristic equation is

$$s(s + 2K) = 0$$
$$s^2 + 2Ks = 0$$

Because it is a second-order equation, there will be two poles (roots of the characteristic equation) and hence two loci (paths).One root is always equal to zero, and the other root depends upon the value of K. The roots are

$$s = 0 \quad \text{and} \quad s = -2K$$

The location of the second pole as K is varied as shown in Table 8.5.

Table 8.5 Location of Second Pole for Example 8.19

	K	Pole Location
1	0	0
2	1	0.5
3	10	5
4	100	50
5	$\rightarrow \infty$	$\rightarrow \infty$

When represented graphically, it traces the locus shown in Figure 8.21.

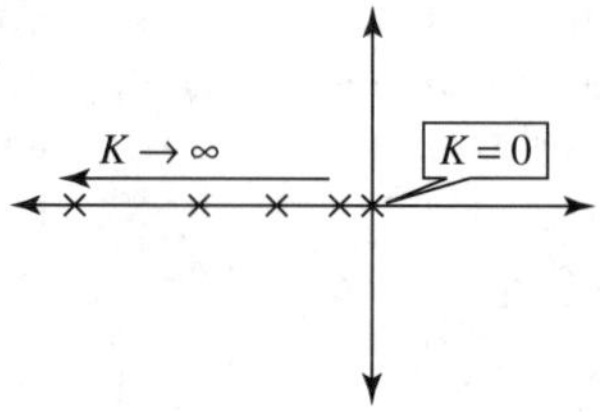

Figure 8.21 Root locus for Example 8.19.

The complexity of root locus analysis increases in exponential order with an increase in the order of the system transfer function. The mechanical aid developed by Evans and the set of rules for root locus plotting seemed to ease the process to some extent. MATLAB makes the process fairly simple. The following are some examples of using MATLAB for root locus analysis.

Root Locus Through MATLAB

Assuming a system transfer function of the type

$$1 + K\frac{\text{Num}(s)}{\text{Den}(s)} = 0$$

MATLAB can calculate the roots of the characteristic equation as the system variable K is varied through a given range.

```
rlocus(num,den,K)
```

If the value of K is not specified, MATLAB plots the root locus for the value of K ranging from 0 to ∞. The syntax for obtaining this root locus is

```
rlocus(num,den)
```

EXAMPLE 8.20

Plot the root locus of the closed-loop system with the following loop (open) transfer function:

$$\text{OLTF} = \frac{Ks}{s^3 + 4s^2 + 5s + 2}$$

Solution

Because the transfer function incorporates a third-order polynomial, MATLAB will be used to plot the root locus.

Entering the data: MATLAB requires the open-loop transfer function and not the closed-loop transfer function to be entered. There is no need to assign any value to the gain factor K when specifying the numerator. MATLAB automatically uses a range of 0 to ∞ for K.

Specify the numerator of the OLTF:

```
num = [1 0]
```

Specify the denominator of the OLTF:

```
den = [1 4 5 2]
```

Plotting the root locus: The following command plots the root locus:

```
rlocus(num,den)
```

The resulting plot is shown in Figure 8.22.

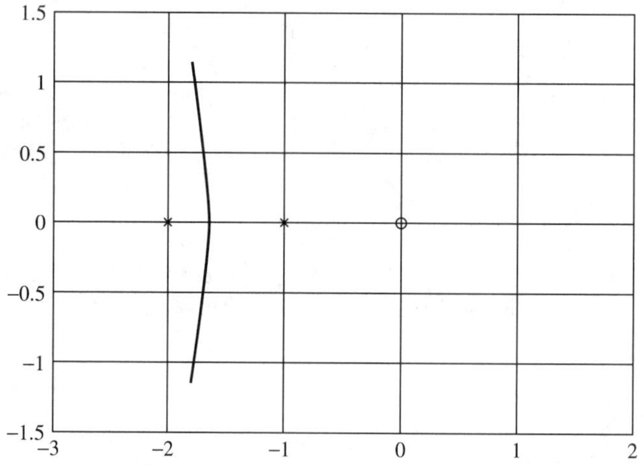

Figure 8.22 Root locus for Example 8.20.

EXAMPLE 8.21

Determine if the system with the given OLTF remains stable if the proportional gain K_P is varied over a range of 1 to 10.

$$\text{OLTF} = \frac{0.35K_P}{s^3 + s^2 + 2s}$$

Solution

The system remains stable if all the roots of the characteristic equation remain in the LHS of the s-plane or, in other words, the root locus does cross into the RHS of the s-plane. Enter the data in MATLAB:

Specify the range of the gain factor:

```
K = 1:10
```

Specify the numerator:

```
num = 0.35
```

Specify the denominator:

```
den = [1 1 2 0]
```

The root locus can now be plotted:

```
rlocus(num,den,k)
```

Figure 8.23 shows the root locus plot. It is evident from the plot that for certain values of proportional gain, the closed-loop poles move into the RHS of the s-plane and render the system unstable.

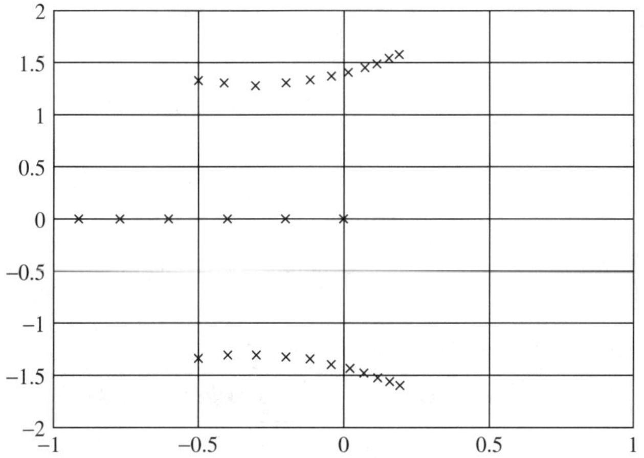

Figure 8.23 Root locus: for Example 8.21, $K_P = 1:10$.

The plot is repeated in Figure 8.24 for reduced values of the proportional gain coefficient K_P ranging from 1 to 5. For these values of K_P, there is no crossing over of paths to the RHS of the s-plane, and the system remains stable. Further analysis and modification to controller parameters may be needed for satisfactory performance.

In this case it will be helpful to determine the exact pole locations for the different values of the gain coefficient K_P. It can be determined as follows. First, specify the

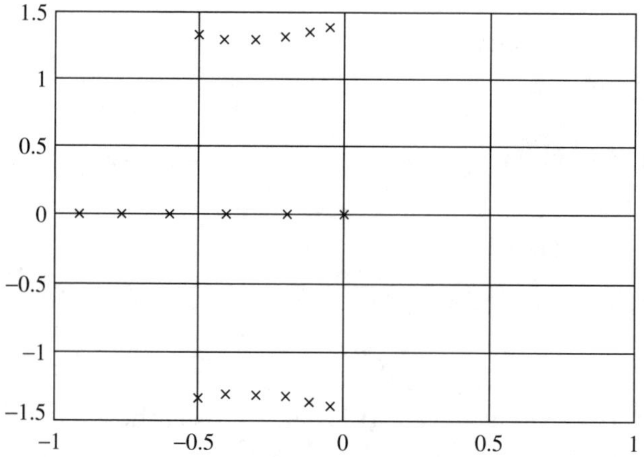

Figure 8.24 Root locus: for Example 8.21, $K_P = 1:5$.

range of the gain coefficient:

```
K = 1:5
```

The numerical value of the pole locations corresponding to gains of 1 to 5 can be obtained using the command

```
[p,K] = rlocus(num,den)
```

which yields the results in Table 8.6.

Table 8.6 Pole Locations for Example 8.21

Gain K_P	***Complex Pole 1a***	***Complex Pole 1b***	***Real Pole***
1	−0.4395 −1.6433*i*	−0.4395 −1.6433*i*	−0.1210
2	−0.3756 −1.6346*i*	−0.3756 −1.6346*i*	−0.2488
3	−0.3101 −1.6335*i*	−0.3101 −1.6335*i*	−0.3798
4	−0.2455 −1.6401*i*	−0.2455 −1.6401*i*	−0.5091
5	−0.1838 −1.6534*i*	−0.1838 −1.6534*i*	−0.6323

Sometimes the desired system parameter K is embedded in the expressions, and it is not possible to isolate it in the following form, as required by MATLAB.

$$1 + KGH(s) = 0$$

where

$$GH(s) = \text{OLTF}.$$

In such a case some algebraic manipulation is required so that the characteristic equation can be represented in the required form.

EXAMPLE 8.22

A PI controller with a transfer function $G_1(s)$ is controlling a process with the transfer function $G_2(s)$ (Figure 8.25). Determine if the unity feedback closed-loop control system is stable for all values of proportional gain, K_P.

$$\text{PI controller TF:}\quad G_1(s) = K_P + \frac{2}{s} \qquad \text{process TF:}\quad G_2(s) = \frac{1}{s^2 + s + 1}$$

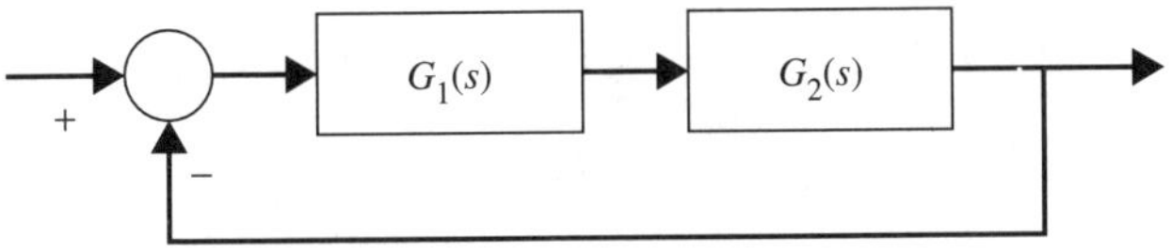

Figure 8.25 Block diagram for Example 8.22.

Solution

The system open-loop transfer function is

$$GH(s) = \left(K_P + \frac{2}{s}\right)\left(\frac{1}{s^2 + s + 1}\right) \cdot 1$$

or

$$GH(s) = \left(\frac{K_P s + 2}{s}\right)\left(\frac{1}{s^2 + s + 1}\right)$$

or

$$GH(s) = \frac{K_P s + 2}{s^3 + s^2 + s}$$

The system closed-loop transfer function is

$$\mathrm{CLTF} = \frac{\dfrac{K_P s + 2}{s^3 + s^2 + s}}{1 + \dfrac{K_P s + 2}{s^3 + s^2 + s}}$$

or

$$\mathrm{CLTF} = \frac{K_P s + 2}{s^3 + s^2 + s + K_P s + 2}$$

The system characteristic equation is

$$s^3 + s^2 + s + K_P s + 2 = 0$$

Rearranging and isolating the term containing the desired parameter, K_P, gives

$$(s^3 + s^2 + s + 2) + K_P s = 0$$

Dividing the entire equation by the terms not containing K_P,

$$1 + \frac{K_P s}{(s^3 + s^2 + s + 2)} = 0$$

or

$$1 + K_P \frac{s}{(s^3 + s^2 + s + 2)} = 0$$

This expression is now in the same format as the one required for the MATLAB program. The root locus can now be plotted for the proportional coefficient ranging from 0 to ∞.

Specify the numerator:

```
num = [1 0]
```

Specify the denominator:

```
den = [1 1 2 0]
```

Plot the root locus:

```
rlocus(num, den)
```

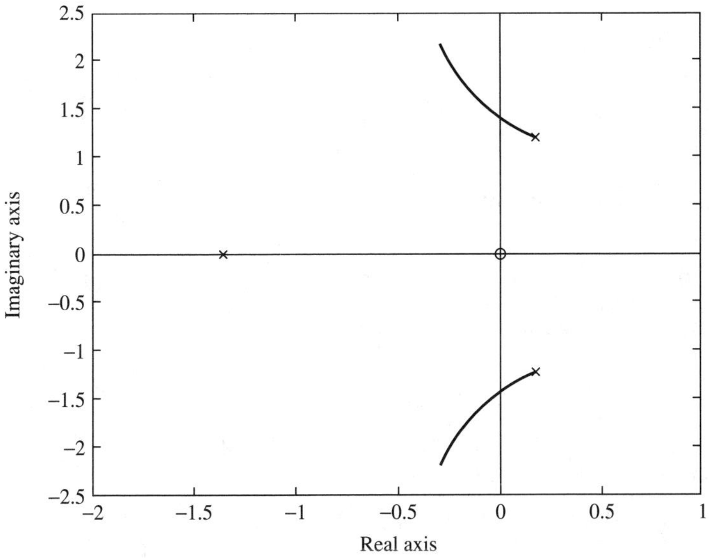

Figure 8.26 Example 8.22: root locus ($K_P = 0 : \infty$).

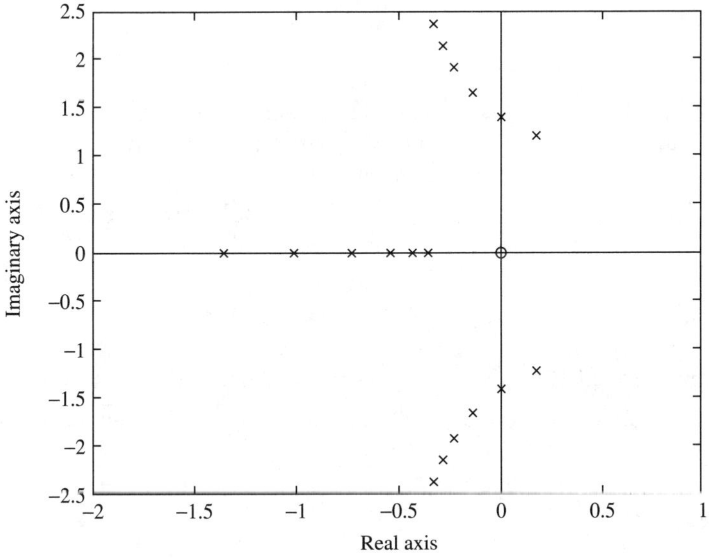

Figure 8.27 Example 8.22: root locus ($K_P = 1 : 5$).

The resultant root locus is given in Figure 8.26. It is an interesting graph, indicating that for very small values of the proportional coefficient, two poles lie in the right half of the s-plane and the closed-loop system is unstable. In order to determine the values of K_P for which the system is stable, the root locus can be plotted for a set of predefined values.

Specify the range of K_P: 1 to 5.

```
k = [1:5]
```

Plot the root locus for these values of K_P:

```
rlocus(num, den. k)
```

It is evident from the graph (Figure 8.27) that for a stable system, the proportional gain has to be greater than 1.

The root locus is an excellent tool for visualizing the impact of any system parameter on closed-loop stability. It is also useful in control system design when targeting specific values of damping ratios.

PROBLEMS

1. Explain the concept of stability.
2. A negative-feedback closed-loop control system is subjected to a ramp input. It is observed that, in response, the control system output continues to rise. Can it be stated that the control system is unstable?
3. A unity gain closed-loop system is given by the following transfer function:

$$\text{CLTF} = \frac{G}{1 + G}$$

   Identify the value of the forward gain G that will render this system unstable.
4. A closed-loop control system is given by the following transfer function:

$$\text{CLTF} = \frac{K}{s + 10}$$

   a. Assuming that K is equal to 1, is there any frequency for which the system becomes unstable?
   b. Can this system be made unstable by increasing the value of K?

5. Identify if the following closed-loop systems are stable.

a. $\dfrac{10}{(s+1)(s+2)}$

b. $\dfrac{1.25}{s(s+1)}$

c. $\dfrac{1}{(s+1)(s+2)(s+3)}$

d. $\dfrac{10(s+2)}{s(s+0.1)(s+10)}$

e. $\dfrac{100}{s(s+0.1)(s+1)^2}$

6. Identify if the following system is stable.

$$K = 10, \qquad G_1(s) = \frac{1}{s(s+2)}, \qquad G_2(s) = \frac{1}{(s+4.8)}$$

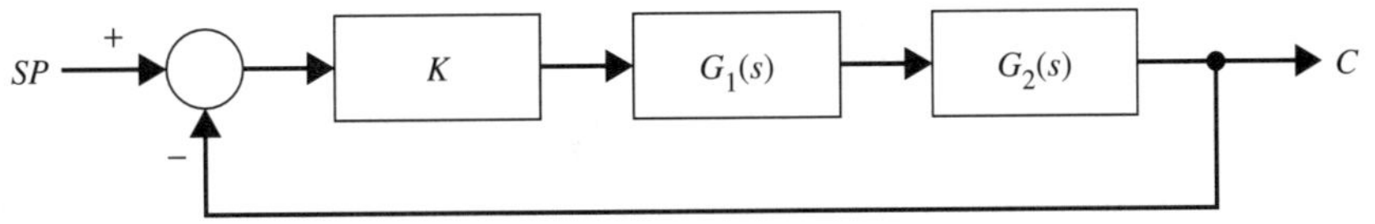

7. In Problem 6, what is the maximum value of gain K that can be used before the system becomes unstable?

8. Determine the location of the roots of the characteristic equation for the following unity feedback systems and identify

1. System response when subjected to a step input

2. System stability

a. $G(s) = \dfrac{2.5}{(s+1)(s+100)}$

b. $G(s) = \dfrac{124s}{(s+1)(s^2+5s+10)}$ (Use a program such as MATLAB.)

c. $G(s) = \dfrac{45(s+1)}{s(s+4)(s^2+s+100)}$ (Use a program such as MATLAB.)

9. Repeat Problem 8 using the Routh criterion, and determine the stability of the systems.

10. Plot the frequency response of the given systems and determine the gain and phase margins.

a. $\dfrac{4}{s(s+1)(s+2)}$

b. $\dfrac{2.5}{s(s+2)^2}$

c. $\dfrac{0.2s}{(s+1)(s^2+1.2s+4)}$

11. Using the Routh criterion, determine the number of roots in the RHP for systems with the given CLTF.

a. $\dfrac{3s+25}{s(s+4)(s+12)}$

b. $\dfrac{10}{(s^2+0.5s+1)(s+1)^2}$

9

CONTROLLERS: THE ON-OFF CONTROLLER

INTRODUCTION

Controllers are the brains behind control system operation. These have been delegated tasks previously carried out under human control. Theoretically speaking, it is possible to have any system (including socioeconomic) run under automatic control, as long as it can be explicitly defined. Automatic control of physical systems is a much simpler matter, and many different types of controllers are available to carry out these tasks. In this chapter the types of controllers and some of the terms used for controller operation are discussed. The various types of on-off controllers and the implementation and operation of electronic on-off controllers are described.

ROLE OF A CONTROLLER

A controller serves as the front end in a closed-loop control system. It accepts the error signal as input and generates an output in order to drive the plant to a desired state. A controller may have very simple mechanical, pneumatic, or electrical construction or may consist of a complex real-time computer system capable of executing some very smart algorithms.

Although a summation block is considered a separate identity and is shown separately in a block diagram, it should essentially be thought of as an integral part of the controller (Figure 9.1). It is especially true of a digital controller, which uses

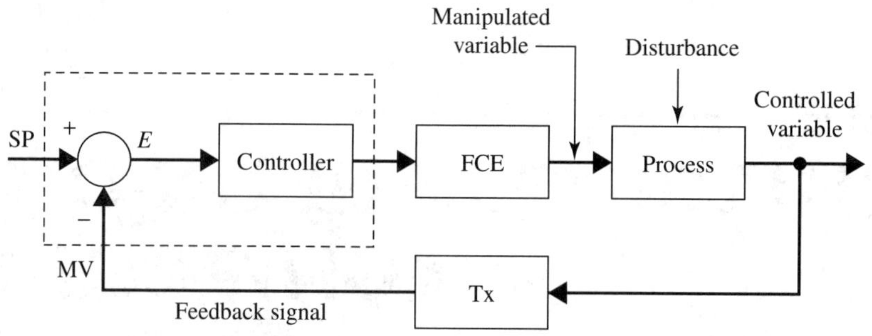

Figure 9.1 Controller in a closed-loop control system.

analog-to-digital converters to get process variable (feedback) and set-point information.

Note: A number of textbooks refer to a controller as a *compensator*. A controller (or compensator) is essentially used to modify (compensate) the process control loop characteristics, hence the name compensator. In this book the term *controller* is used to describe all the different forms of controllers (compensators).

TYPES OF CONTROLLERS

Controllers are classified according to their modes of operation. In spite of some overlap, all controllers can be classified as follows:

- On-off controller
- Analog controller
  - Proportional controller (P)
  - Integral controller (I)
  - Derivative controller (D)
  - Proportional integral and derivative controller (PID)
  - Lag-lead controller
- Digital controller
- Fuzzy controller

An on-off controller is not a continuous controller and can fall both in the category of analog controller and digital controller. As a matter of fact, an on-off controller can be considered to be a continuous proportional controller operating at a very high gain. A PID controller is a full three-term (mode) controller and contains the characteristics of all proportional (P), integral (I), and derivative (D) controllers. It is the most versatile of all analog controllers and can be configured to operate as simple proportional, proportional + integral (PI), or proportional + derivative (PD) controller. A controller is generally not configured to operate in pure integral or derivative mode but is combined with proportional mode of control.

Digital controllers were initially developed as the computer implementation of analog controllers, but a modern digital controller is lot more versatile than its analog counterpart. A digital controller can be programmed to operate in a multitude of control modes. It not only can duplicate the operation of a full PID controller, but it is also capable of executing some very novel control algorithms.

A modern fuzzy controller is essentially a digital controller that is implementing a fuzzy algorithm. Fuzzy control has made strong gains in recent years. Its popularity has grown steadily with the availability of low-cost microcontrollers. Its applications range from the control of household appliances, such as a microwave oven or a vacuum cleaner, to building elevator and industrial process control. In Japan, fuzzy controllers have even been applied to the real-time control of high-speed subway train systems. In subsequent chapters these controllers are discussed in detail.

USE OF PERCENT AS INPUT AND OUTPUT

A controller is generally not limited to a specific process and can be applied for control of many different types of processes. This is especially true of a commercial controller unit, where a model may be used for temperature control in one application and dc motor–velocity control in another. For the temperature control application, the measurements may be in degree C or degree F, whereas for motor speed control, rpm may be the preferred unit of measurement. It is obviously very difficult for controller manufacturers to anticipate the measurement unit in advance of the sale. A more meaningful alternative in such cases is to display the input and output values as percentages rather than actual units, such as degree C, volts, rpm, etc. When percentages are used, 0% corresponds to minimum controller output and 100% corresponds to the maximum output. Controller output at a given instant can be expressed in percent through the following expression.

$$\text{controller output in } \% = \frac{\text{current output} - \text{minimum output}}{\text{maximum output} - \text{minimum output}} \times 100$$

EXAMPLE 9.1

Output of an industrial controller ranges from 0 to 10 V. Express a controller output of 4.5 V as a percentage value.

Solution

$$\text{controller output \%} = \frac{\text{current voltage output} - \text{minimum voltage output}}{\text{maximum voltage output} - \text{minimum voltage output}} \times 100$$

$$= \frac{4.5 - 0}{10 - 0} \times 100$$

$$= 45\%$$

EXAMPLE 9.2

Express a current output of 8 mA as a percentage of controller output. The controller has an output range of 4 to 20 mA.

Solution

$$\text{controller output \%} = \frac{\text{current voltage output} - \text{minimum voltage output}}{\text{maximum voltage output} - \text{minimum voltage output}} \times 100$$

$$= \frac{8\text{ mA} - 4\text{ mA}}{20\text{ mA} - 4\text{ mA}} \times 100$$

$$= 25\%$$

EXAMPLE 9.3

The output of a pneumatic controller ranges from 3 to 15 psi. Determine the controller output in psi if the output indicator displays a value of 60%.

Solution

$$\text{controller output \%} = \frac{\text{current psi output} - \text{minimum psi output}}{\text{maximum psi output} - \text{minimum psi output}} \times 100$$

$$60\% = \frac{\text{current psi output} - 3\text{ psi}}{15\text{ psi} - 3\text{ psi}} \times 100$$

$$0.60 \times 12 = \text{output in psi} - 3\text{ psi}$$

$$\text{controller output in psi} = 10.2\text{ psi}$$

Measured Value as a Percentage Value

Even though measurement units differ from process to process, a measured value (MV) can again be conveniently expressed as a percentage value. This can be advantageous in a case where the minimum point in the operating range differs significantly from the zero point.

$$\text{measured value in \%} = \frac{\text{MV} - \text{minimum MV}}{\text{maximum MV} - \text{minimum MV}} \times 100$$

EXAMPLE 9.4

In a temperature control system, operating temperature ranges from 0° to 100°C. Express the temperature of 50°C as a percentage of the operating range.

Solution

$$\text{temperature \%} = \frac{\text{measured temperature} - \text{minimum temperature}}{\text{maximum temperature} - \text{minimum temperature}} \times 100$$

$$= \frac{50^\circ\text{C} - 0^\circ\text{C}}{100^\circ\text{C} - 0^\circ\text{C}} \times 100$$

$$= 50\%$$

EXAMPLE 9.5

In a temperature control system, operating temperature ranges from 40°C to 80°C. Express the temperature of 50°C as a percentage of the operating range.

Solution

$$\text{temperature \%} = \frac{\text{measured temperature} - \text{minimum temperature}}{\text{maximum temperature} - \text{minimum temperature}} \times 100$$

$$= \frac{50^\circ\text{C} - 40^\circ\text{C}}{80^\circ\text{C} - 40^\circ\text{C}} \times 100$$

$$= 25\%$$

Set Point as a Percentage Value

The set-point value (SP) can also be conveniently expressed as a percentage of the operating range. This can again be more meaningful in situations where the minimum operating point is anything other than zero.

$$\text{set point \%} = \frac{\text{SP} - \text{minimum operating value}}{\text{operating range}} \times 100$$

EXAMPLE 9.6

A dc motor in a rolling mill is required to operate in a range of 1000 to 1200 rpm. Express a set-point speed of 1050 rpm as a percentage of the operating range.

Solution

$$\text{set point \%} = \frac{\text{SP speed} - \text{minimum motor speed}}{\text{operating range}} \times 100$$

$$= \frac{1050 \text{ rpm} - 1000 \text{ rpm}}{1200 \text{ rpm} - 1000 \text{ rpm}} \times 100$$

$$= 25\%$$

Error as a Percentage Value

In a closed-loop control system, where both the SP and MV are expressed as percentage values, it is only natural to express error as a percentage value. It can be determined either from SP × MV, expressed as percentages, or it can be calculated from their unit values.

$$\text{error \%} = \text{\% SP} - \text{\% MV}$$

$$= \frac{\text{SP value} - \text{minimum value}}{\text{operating range}} \times 100 - \frac{\text{MV value} - \text{minimum value}}{\text{operating range}} \times 100$$

$$= \frac{\text{SP value} - \text{MV value}}{\text{operating range}} \times 100$$

EXAMPLE 9.7

Operating parameters of a motor speed control system are as follows:

$$\text{minimum operating speed} = 100 \text{ rpm}$$
$$\text{maximum operating speed} = 1500 \text{ rpm}$$
$$\text{set-point speed} = 1000 \text{ rpm}$$
$$\text{actual motor speed} = 980 \text{ rpm}$$

Determine the following:

a. % SP
b. % MV
c. Error
d. % error

Solution

a.

$$\% \text{ SP} = \frac{\text{SP speed} - \text{minimum motor speed}}{\text{operating range}} \times 100$$
$$= \frac{1000 \text{ rpm} - 100 \text{ rpm}}{1500 \text{ rpm} - 100 \text{ rpm}} \times 100$$
$$= 64.286\%$$

b.

$$\% \text{ MV} = \frac{\text{current speed} - \text{minimum motor speed}}{\text{operating range}} \times 100$$
$$= \frac{980 \text{ rpm} - 100 \text{ rpm}}{1500 \text{ rpm} - 100 \text{ rpm}} \times 100$$
$$= 62.857\%$$

c.

$$\text{error} = \text{SP} - \text{MV}$$
$$= 1000 \text{ rpm} - 980 \text{ rpm}$$
$$= 20 \text{ rpm}$$

d. % error can be calculated from actual error and operating range:

$$\% \text{ error} = \frac{\text{error}}{\text{operating range}} \times 100$$

$$= \frac{20 \text{ rpm}}{1500 \text{ rpm} - 100 \text{ rpm}} \times 100$$

$$= 1.429\%$$

% error can also be calculated from % SP and % MV values.

$$\% \text{ error} = \% \text{ SP} - \% \text{ MV}$$

$$= 64.286\% - 62.857\%$$

$$= 1.429\%$$

ON-OFF CONTROL

Although an on-off type of control cannot qualify as a continuous control, it is used in numerous residential, commercial, and industrial applications. It is commonly used in control of home appliances: hot plate and oven-temperature control, refrigerator-temperature control, and dishwasher-water-level control. It is by far the method of choice for environmental temperature control of all residential and commercial buildings. Some may even argue that a digital controller is nothing but an extremely fast and fancy on-off controller.

Types of On-Off Control

On-off controllers can be operated in three different ways.

- (Strict) on-off control
- Two-position control
- Floating control

(Strict) On-off control In this method of control, controller output changes at the instant the error changes from positive to negative, or vice versa. Whenever the controlled variable is below the set point (positive error) the controller output is at 100%, and if the controlled variable is higher than the set point (negative error) the controller output drops to 0%.

A strict on-off method of control is not a very practical one, because in the majority of applications, controller output would constantly be switching between full-scale (100%)

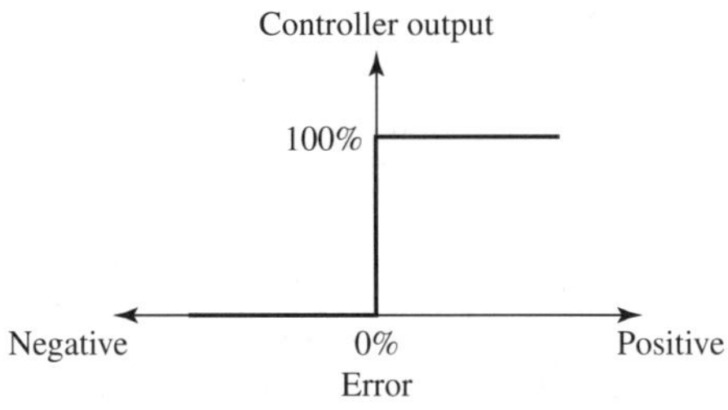

Figure 9.2 On-off control.

and zero (0%) outputs (Figure 9.2). Thus, a strict on-off method of control is seldom used. A modified form of on-off control, two-position control, is more commonly used.

Processes with Small Time Constants: On-off control cannot be used for a process with a small time constant, such as a DC motor with a light load. If an attempt is made to control the motor speed through a switching relay, the relay will be constantly chattering (switching on and off) and will not last very long.

Processes with Large Time Constants: On-off control can be used in processes where the process inertia is very large compared to the energy-supply unit, such as when a large building is heated with a relatively small or moderately sized furnace. On-off control can also be used in situations where measuring instruments have built-in hysteresis.

Electronic Implementation (Analog)

A strict on-off controller can be implemented with a comparator (Figure 9.3). The process variable (MV) is applied to the inverting terminal of the comparator, and the set

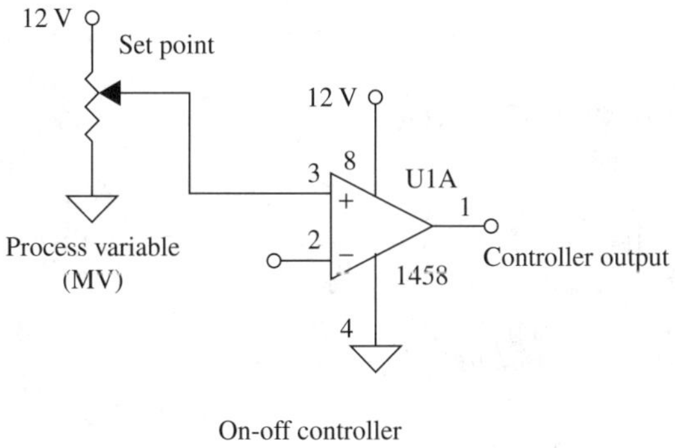

Figure 9.3 Electronic strict on-off controller.

point (SP) is applied to the noninverting terminal. If the process variable is below the SP value, controller output will swing to the maximum positive output ($+V_{sat}$). On the other hand, if the process variable exceeds the SP value, controller output will drop to zero ($-V_{sat}$).

Application: On-Off Temperature Control Figure 9.4 shows a circuit diagram for a temperature control system operating as an on-off control system. This can be utilized for controlling the temperature of a small oven. Such ovens are used in temperature-controlled Xtal oscillators, precision-pressure transducers, voltage references, etc.

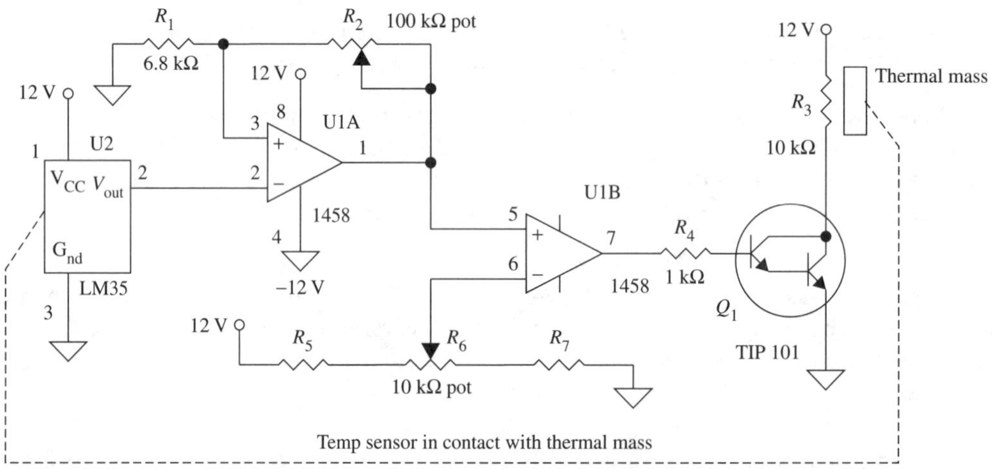

Figure 9.4 Schematic diagram of an on-off temperature controller.

LM35 is a temperature sensor, which provides an output of 10 mV/°C. Op-amp U1A amplifies the transducer signal by a gain of 10 to provide a larger output of 100 mV/°C. Potentiometer R_2 can be used either to adjust the gain or to accommodate transducers with different outputs. Op-amp U1B acts as the on-off controller. It is set up as a comparator, and its output swings between 0 V and 10.5 V ($+V_{sat}$). Darlington transistor TIP 101 switches the large current load (1.2 A in this case) through the heater resistor, R_3. Both the heater resistor and transistor provide the thermal energy to raise the temperature of the small oven. The temperature sensor is kept in physical contact with the oven but not directly in contact with the heating elements.

The temperature control system performs satisfactorily with constant switching at an alarmingly high rate. It is possible because of the following factors:

Temperature-sensor time constant

Power-transistor switching time

Thermal time constant of the oven

Potentiometer R_6 is used to mark the temperature set point. Resistors R_5 and R_7 establish the temperature control range.

$$\text{low limit} = \frac{R_7}{R_5 + R_6 + R_7} \times V_{cc}$$

$$\text{high limit} = \frac{R_6 + R_7}{R_5 + R_6 + R_7} \times V_{cc}$$

EXAMPLE 9.8

Set up the circuit of Figure 9.4 for a temperature control range of 40°C to 60°C.

Solution

$$\text{amplified temperature sensor output} = 100\ \text{mV}/^\circ\text{C}$$

$$\text{output corresponding to } 40^\circ\text{C} = 40 \times 100\,\text{mV}$$

$$= 4.000\ \text{V}$$

$$\text{sensor output at } 60^\circ\text{C} = 6.000\ \text{V}$$

The potentiometer limits should correspond to these two sensor outputs. Therefore,

$$\frac{R_7}{R_5 + R_6 + R_7} \times V_{cc} = 4.000\ \text{V}$$

$$\frac{R_6 + R_7}{R_5 + R_6 + R_7} \times V_{cc} = 6.000\ \text{V}$$

Using a 10-kΩ potentiometer for R_6 and a 12-V supply gives

$$\frac{R_7}{R_5 + 10{,}000 + R_7} \times 12\,\text{V} = 4.000\ \text{V}$$

$$\frac{10{,}000 + R_7}{R_5 + 10{,}000 + R_7} \times 12\,\text{V} = 6.000\ \text{V}$$

Solving the two equations, the values obtained are $R_5 = 30\,\text{k}\Omega$ and $R_7 = 20\,\text{k}\Omega$.

Application: Stepper Motor Chopper Control A stepper motor chopper control circuit allows a faster stepping rate than is possible with an RL drive. It also cuts down on the amount of heat dissipated with an RL drive.

Background: In a stepper motor, individual coils are energized, causing the rotor to move one step at a time. A step movement is initiated at the start of the step-movement command and is considered complete when the current in the relevant coil has reached the designed value. Because stepper motor coils are primarily inductive in nature, current through these cannot be changed instantaneously. Various schemes are employed to aid in fast buildup of current in these coils.

A chopper circuit is designed to control the amount of current in a stepper motor coil. It works on the principle that when a voltage much larger than the rated coil voltage is applied across a coil, the current rise in the coil is accelerated. As soon as current reaches the threshold value, the supply is disconnected. This causes the current to collapse. When current has dropped below the threshold value, it is triggered again and the process continues.

Figure 9.5 shows the schematic diagram for a stepper motor chopper control circuit. The circuit allows a coil current to be set in the range of 0–2 A. Motor current is sensed by voltage drop across a sense resistor through which the current flows. The op-amp comparator UL acts as an on-off controller. It compares the coil current with the

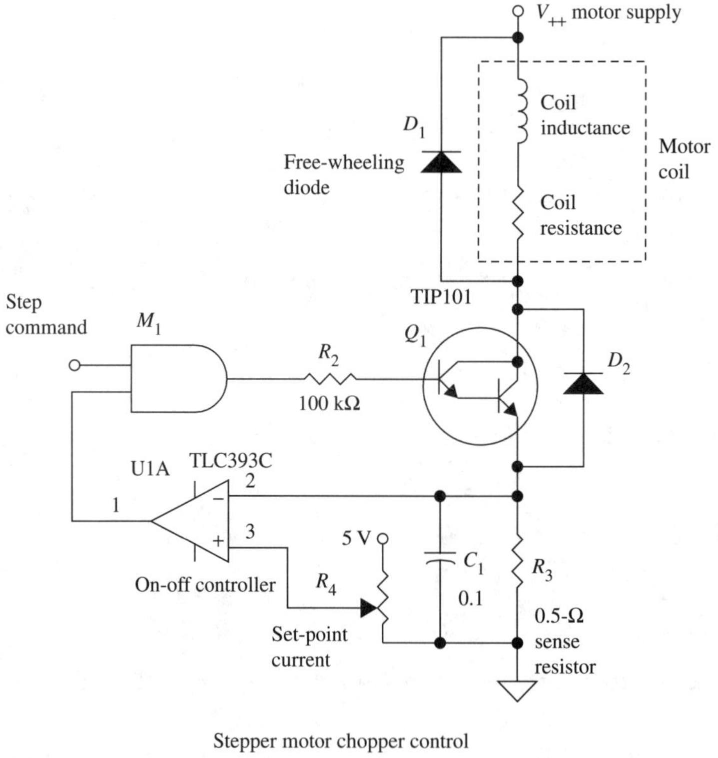

Figure 9.5 Stepper motor on-off control.

set-point value and generates a high (5-V) or low (0-V) output. Controller output is high when the voltage drop across the sense resistor is below the SP setting and goes low when the sense voltage drop exceeds the SP setting. The current set-point value can be adjusted by means of the set-point potentiometer.

Controller output is ANDED with the step command signal. The transistor is switched on when the AND gate output is high, which occurs when both the inputs to AND gate are high. When the motor is off, there is no current flowing through the coil, and the voltage drop across the sense resistor is zero. Because the SP potentiometer is set for the desired current limit, it has a positive output. The controller senses these two inputs, and as a result its output goes high. This output remains high until the current is switched in the coil and has risen to the set value.

As soon as a step command is received, the output of the AND gate goes high. This energizes the transistor and full supply voltage is applied to the motor coil. Because of the substantially larger supply voltage, current rapidly starts to rise in the coil. Soon a point is reached when the current through the coil (and through the sense resistor) increases above the set-point value. This causes the controller to switch its output to low. It forces the AND gate output to go low, which turns off the transistor. Stored inductive energy causes coil current to recirculate through the free-wheeling diode. Coil resistance dissates some of this energy into heat and current in the coil starts to decay.

When the voltage drop through sense resistor goes below the set-point value, controller output switches back to high and turns on the transistor. This process continues at a moderately high frequency (10–100 kHz). A number of factors dictate the overall switching rate:

- Transistor switching time
- Comparator switching time
- AND-gate switching time
- Stepper motor coil inductance and resistance
- Sense resistor and capacitor

Each of the these actions must take a finite amount of time; as a result, the switching is not instantaneous but is limited to a reasonable rate. In the preceding design, a low-side (between ground and load) transistor is used. A better design will use a high-pass (between V_{++} and load) transistor. This will allow the actual coil current to be sensed all the time, making the control more reliable. Unfortunately, a high-pass transistor circuit demands a more careful design and a higher component count.

TWO-POSITION CONTROL

Two-position control is an improvement over the basic on-off control. The controller output still swings between 0% and 100%, but an additional hysteresis element (neutral zone) is now incorporated (Figure 9.6).

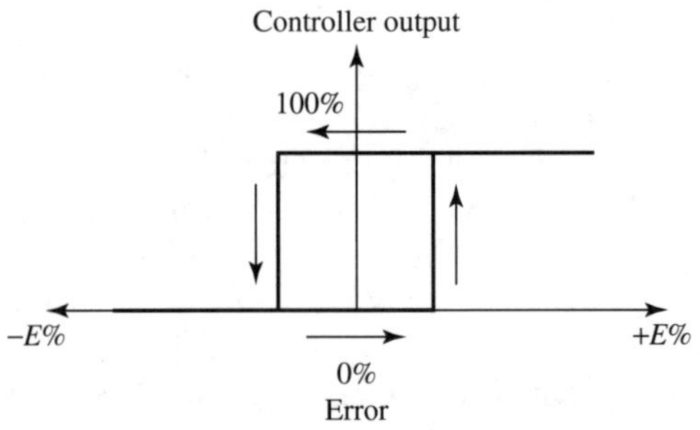

Figure 9.6 Two-position control.

Neutral Zone

Neutral zone

The **neutral zone** is the region of error over which the controller does not change its output.

The neutral zone is also known by the terms *dead band* and *error band*. Generally, the neutral zone is symmetrical about 0% error, but there is no rigid rule (Figure 9.7).

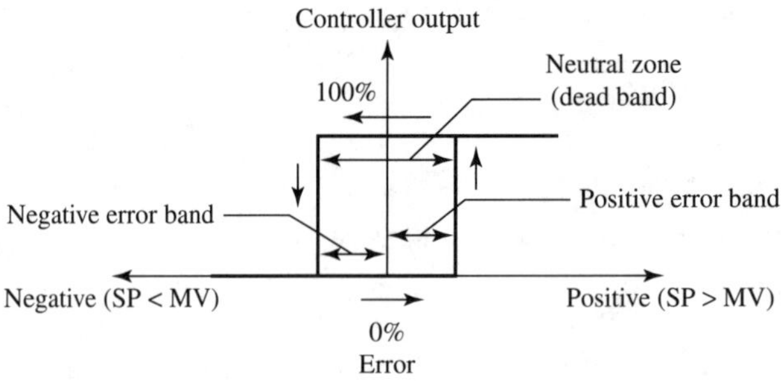

Figure 9.7 Neutral zone.

EXAMPLE 9.9

Draw a diagram showing a neutral zone of 4% for a two-position controller and determine the value of the following:

a. Dead band

b. Error band

Solution

a. Because the term neutral zone is synonymous with dead band, the controller has a dead band of 4%.

b. Assuming a neutral zone to be symmetrical about the 0 (% error) point, positive and negative error bands will be equal.

$$\text{positive error band} = \frac{1}{2} \times \text{dead zone}$$
$$= \frac{1}{2} \times 4\%$$
$$= 2\%$$

Similarly, negative error band = −2%.

Figure 9.8 shows this information graphically.

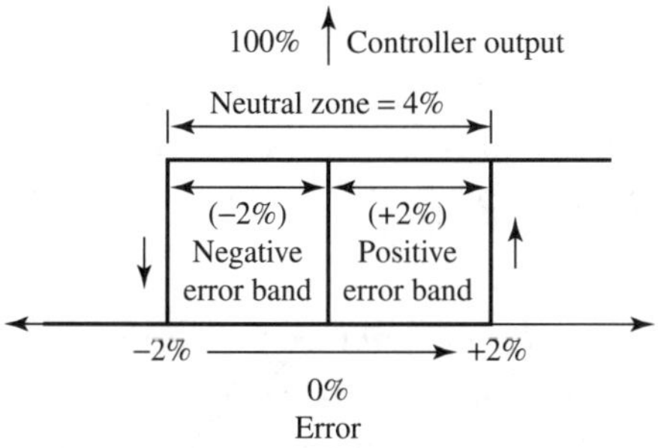

Figure 9.8 Dead band for Example 9.9.

EXAMPLE 9.10

Determine the error band for a two-position temperature control system from the following data:

$$\text{set-point temperature} = 480^\circ\text{C}$$
$$\text{allowable variation in temperature (about SP)} = \pm 5^\circ\text{C}$$
$$\text{operating temperature range} = 450^\circ - 550^\circ\text{C}$$

Solution

$$\text{positive error band} = \frac{\text{maximum permissible positive error}}{\text{temperature range}} \times 100$$

$$= \frac{5^\circ\text{C}}{(550 - 450)^\circ\text{C}} \times 100$$

$$= 5\%$$

$$\text{negative error band} = \frac{\text{maximum permissible negative error}}{\text{temperature range}} \times 100$$

$$= \frac{-5^\circ\text{C}}{(550 - 450)^\circ\text{C}} \times 100$$

$$= -5\%$$

$$\text{neutral zone} = \text{dead band} = (5 + 5)\%$$

$$= 10\%$$

Controller Action

The controller action of a two-position controller is very similar to that of a pure on-off controller. Controller operation can be best illustrated with an example of a home-heating system.

Home-Heating System A gas-fired furnace is used to heat a home. The controller works on a two-position control principle. The operating specifications of the controller are as follows:

$$\text{operating temperature range} = 15^\circ\text{–}30^\circ\text{C}$$

$$\text{dead band} = 2^\circ\text{C } (\pm 1^\circ\text{C})$$

Assume that the heating system is switched off for the summer and is turned on at the start of the fall season. The thermostat is set to the desired temperature. Assume the following set of values:

$$\text{room temperature} = 15^\circ\text{C}$$

$$\text{set-point temperature} = 18^\circ\text{C}$$

$$\text{error at the instant the heating system is turned on} = \text{SP} - \text{MV}$$

$$= 18^\circ\text{C} - 15^\circ\text{C}$$

$$= 3^\circ\text{C}$$

This error is greater than the positive error band of 1°C.

Error outside the neutral zone (positive)—Start: Because error is outside the neutral zone and is positive, the controller turns on and the furnace starts producing heat. This causes the room temperature (MV) to rise. As temperature (MV) rises, the error E (E = SP − MV) becomes smaller and smaller.

Error in between neutral zone and zero (positive error band)—Rising Temperature: With room temperature rising to 17°C, the error reaches the positive outer limit of the neutral zone, but there is no change in controller output. It has been fully (100%) on and continues to stay on. With more heat being put in the system, the error continues to be reduced. The error value enters the neutral zone and moves toward 0%, reaching it some time later. Even though the room temperature has reached the desired value (MV = 18°C), the controller still does not shut off. Heat continues to be added, and the room temperature continues to increase, driving the error into the negative zone.

Error in between zero and neutral zone (negative error band)—Rising Temperature: With more heat error entering the system, the temperature continues to rise and the error moves toward the negative outer limit of the neutral zone. Controller output still does not change. It is only when the temperature has risen to 19°C (the end of the neutral zone) that the controller shuts off, stopping further addition of heat.

Error outside neutral zone (negative)—Rising Temperature: Although the furnace is no longer generating heat, some heat energy is still trapped in air ducts and the heat exchanger. The temperature sensor also exhibits a first-order lag behavior, which is typical of most temperature sensors. This introduces a small delay in temperature sensing. As a result, the room temperature will actually exceed the upper limit of 19°C, before the furnace is switched off. Both these factors cause room temperature to rise beyond the upper limit by a small amount. Because the error is outside the neutral zone, the controller will stay switched off. With the controller off and the furnace not supplying any heat, temperature starts to drop, and error will gradually move from negative territory to positive territory.

Error in between neutral zone and zero (negative error band)—Falling Temperature: As the error enters the neutral zone, the controller stays off. Room temperature drops to the SP value of 18°C, but the controller output does not change (stays off).

Error in between zero and the neutral zone (positive error band)—Falling Temperature: Room temperature continues to drop and reaches a value of 17°C. As soon as this threshold temperature (positive outer limit of neutral zone) is reached, the controller changes its state and turns on (100%). This starts the furnace, adding heat to the system and causing the room temperature to rise again.

The whole cycle is repeated indefinitely. The action of a two-position controller is summarized in Table 9.1. Two-position control can be satisfactorily applied for control of plants with large capacity (inertia). Because of the large capacity, process error cannot change quickly and the controller does not have to switch rapidly between the on and off states.

Table 9.1 Two-Position Control: Controller Action

	Outside Neutral Zone (Positive)	***Neutral Zone***	***Outside Neutral Zone (Negative)***
Error	$E \geq$ Positive error limit	$E <$ Positive error limit $E >$ Negative error limit	$E \leq$ Negative error limit
Controller action	On	No change in state On if previously on Off if previously off	Off

Applications

This type of control is commonly used in the following applications:

- *Home-heating system:* A home-heating system works on a two-position control basis. The thermostat (controller) used in a typical system has an error band ranging from 0.5° to 1°C (deadband of 1° to 2°)
- *Air-conditioning system:* It works essentially on the same principle as the heating system.
- *Refrigerator-temperature control system:* The temperature controller used in a refrigerator works on the same principle. Actual temperature varies between two set limits. If this were not the case, then the refrigeration cycle would start immediately on opening the refrigerator door.
- *Tank-fluid level control:* A number of process/storage tanks work on the principle of maintaining fluid between two set limits. A pump starts filling up the tank when the low limit is reached and shuts off when the level reaches the upper limit. The pump does not restart until the level has fallen back to the low limit.

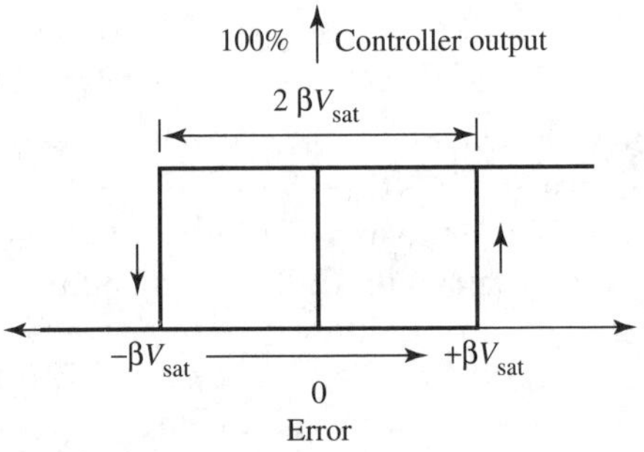

Figure 9.9 Op-amp implementation of the neutral zone.

Electronic Implementation

Figure 9.10 shows an electronic implementation of a two-position controller. IC1 (U1A) forms the error amplifier (summation block). Its output is given as

$$\text{error voltage } E = \text{set-point voltage} - \text{process variable voltage}$$

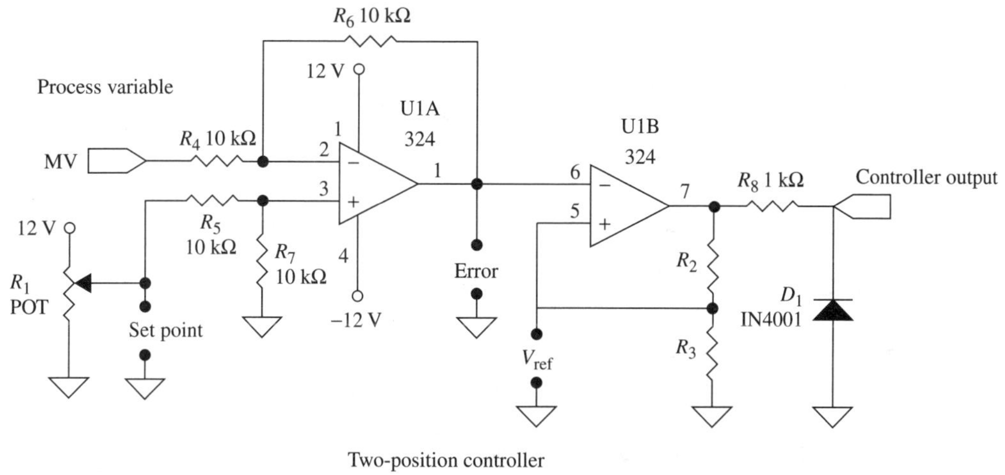

Figure 9.10 Schematic diagram: two-position controller.

IC2 (U1B) acts as the two-position controller. Voltage V_r at the noninverting terminal is governed by the saturation voltage, V_{sat} of IC2, and the two resistors R_2 and R_3. It varies between two limits (error bands) and can be determined from the following expression:

$$V_r = \pm\frac{R_3}{R_2 + R_3} \times V_{sat}$$
$$= \pm\beta V_{sat}$$

V_{ref} is positive when the IC2 output is at V_{sat+} and is negative when the IC2 output is at the V_{sat-} value.

$$0\% \text{ IC2 output} = V_{sat-}$$
$$= -10.5 \text{ V}$$
$$100\% \text{ IC2 output} = V_{sat+}$$
$$= +10.5 \text{ V}$$

The controller output switch's state is based upon its two inputs, error E and reference voltage V_{ref}. Let us assume that controller output is initially high. This will result in V_{ref} being a negative value. In order for the controller to switch state (to 0%), the error voltage has to become more positive than V_{ref}. Once this happens, controller output will go to the V_{sat-} value and V_{ref} will attain a negative value. The error will then have to become more negative than V_{ref} before the controller is turned on again.

The total dead band (neutral zone) for the circuit is

$$\text{dead band} = 2\beta V_{sat}$$

Resistor R_8 and diode D_1 limit the controller output to a voltage in between zero (−0.7 V) and V_{sat+} (approximately 10.5 V).

EXAMPLE 9.11

Refer to the schematic diagram shown in Figure 9.10. Assuming the following values for the circuit components, determine the error limits and dead band.

$$R_1 = 10\,\text{k}\Omega,\ R_2 = 100\,\text{k}\Omega,\ R_3 = 2.2\,\text{k}\Omega,\ R_4 - R_7 = 10\,\text{k}\Omega,\ V_{sat} = \pm 10\ \text{V}$$

Solution

Error bands are solely dependent upon R_2 and R_3 and saturation voltage.

$$\begin{aligned}\text{upper error band} &= +\beta V_{sat}\\ &= \frac{R_3}{R_2 + R_3} \times V_{sat}\\ &= \frac{2200}{100{,}000 + 2200} \times 10\,\text{V}\\ &= +0.215\,\text{V}\end{aligned}$$

Similarly,

$$\begin{aligned}\text{lower error band} &= -\beta V_{sat}\\ &= -0.215\,\text{V}\\ \text{dead band} &= 2\beta V_{sat}\\ &= 0.43\,\text{V}\end{aligned}$$

EXAMPLE 9.12

Using the data from Example 9.11, determine the controller output state for the following conditions:

a. SP = 5 V, MV − 4.85 V
b. SP = 2.5 V, MV = 2.6 V
c. SP = 5 V, MV = 4.5 V
d. SP = 5 V, MV = 5.4 V

Solution

From the previous data

$$\text{error band} = \pm\beta V_{sat} = \pm 0.215\,\text{V}$$

a.

$$\text{error} = SP - MV = 5\,\text{V} - 4.85\,\text{V} = 0.15\,\text{V} < 0.2.15\ \text{V} \quad \text{(upper error limit)}$$

This error value is within the dead-band region, and thus the controller will not switch state. If it was previously on, it will stay on, and if was previously off, then it will stay off. Because the information about the previous state of the controller is not known, the current state of the controller cannot be identified.

b.

$$\text{error} = SP - MV = 2.5\,\text{V} - 2.6\,\text{V} = -0.1\,\text{V} > -0.2.15\,\text{V} \quad \text{(lower error limit)}$$

This error value is again within the dead-band region, and due to lack of knowledge of the previous state, the current state of the controller cannot be identified

c.

$$\text{error} = SP - MV = 5\,\text{V} - 4.5\,\text{V} = 0.5\,\text{V} > +0.2.15\,\text{V} \quad \text{(upper error limit)}$$

Because this error value exceeds the upper threshold, the controller has to be on (100%).

d.

$$\begin{aligned} \text{error} &= \text{SP} - \text{MV} \\ &= 5\,\text{V} - 5.4\,\text{V} \\ &= -0.4\,\text{V} < -0.2.15\,\text{V} \qquad \text{(lower error limit)} \end{aligned}$$

Because this error value exceeds (is more negative than) the lower threshold, the controller has to be off (0%).

Application: Two-Position Temperature Control Figure 9.11 shows a schematic diagram of a two-position temperature control system. This is essentially a modified form of Figure 9.4, where the simple on-off controller has been replaced with a two-position controller.

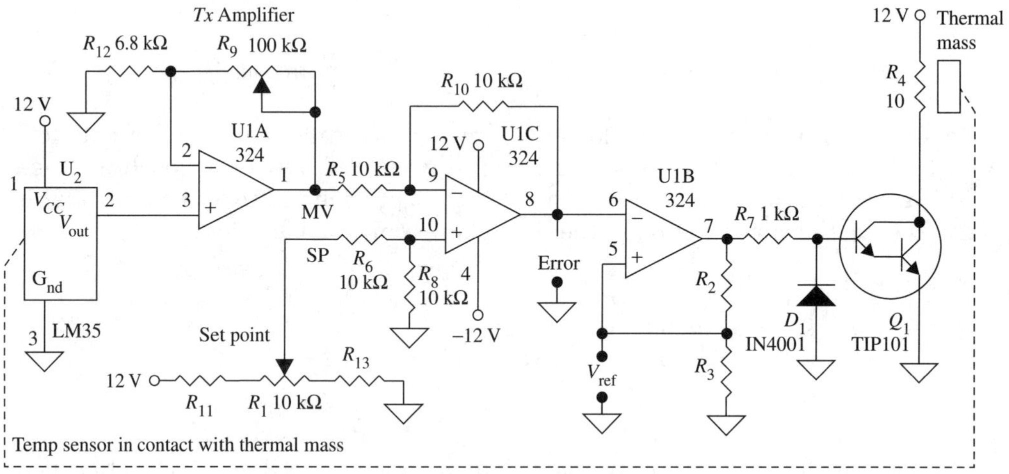

Figure 9.11 Two-position temperature control system.

Circuit Description: Op-amp U1A amplifies the temperature transducer signal to yield an output of 100 mV/°C. U1C serves as a summation amplifier. It accepts inputs from the set-point potentiometer and transducer amplifier and generates the error signal. Op-amp U1B is the heart of the unit, functioning as the two-position controller. Controller output switches between the V_{sat+} and V_{sat-} voltages. Diode D_1 and resistor R_7 limit this swing between 0 and V_{sat+} and feed it to a power transistor. The power transistor acts as a switching device and is on whenever controller output is high. When the transistor is switched on, current flows through the heater resistor.

Dead band: Resistors R_2 and R_3 establish the dead band.

$$\text{dead band} = \text{positive error band} + \text{negative error band}$$

$$\text{positive error band} = \frac{R_3}{R_2 + R_3} \times +V_{sat}$$

$$\text{negative error band} = \frac{R_3}{R_2 + R_3} \times -V_{sat}$$

Operating Temperature Range: Operating range is established by the amplified temperature transducer output from U1C and the set-point resistors

$$\text{minimum temperature } (^\circ\text{C}) = \frac{\dfrac{R_{13}}{R_{11} + R_1 + R_{13}} \times V_{cc}}{\text{amplified temperature } Tx \text{ output}/^\circ\text{C}}$$

$$\text{maximum temperature } (^\circ\text{C}) = \frac{\dfrac{R_1 + R_{13}}{R_{11} + R_1 + R_{13}} \times V_{cc}}{\text{amplified temperature } Tx \text{ output}/^\circ\text{C}}$$

EXAMPLE 9.13

Set up the circuit in Figure 9.11 for an operating temperature range of 50°C to 80°C.

Solution

Assume the following:

- Amplified temperature transducer output at U1C = 100 mV/°C
- V_{cc} of 12 V dc
- $R_1 = 10\,\text{k}\Omega$

Substituting these values in the expression for minimum and maximum temperature gives

$$\text{minimum temperature } 50^\circ\text{C} = \frac{\dfrac{R_{13}}{R_{11} + 10{,}000 + R_{13}} \times 12\,\text{V}}{0.100\,\text{V}/^\circ\text{C}}$$

$$\text{maximum temperature } 80^\circ\text{C} = \frac{\dfrac{10{,}000 + R_{13}}{R_{11} + 10{,}000 + R_{13}} \times 12\,\text{V}}{0.100\,\text{V}/^\circ\text{C}}$$

From these two expressions, the values for the resistors R_{11} and R_{13} can be obtained:

$$R_{11} = 11\,\text{k}\Omega$$
$$R_{13} = 15\,\text{k}\Omega$$

These values allow the operating temperature to be set in the range of 50°C to 80°C.

EXAMPLE 9.14

Set up the two-position controller for a dead band of 1°C about the set point. Assume that the amplifier U1B swings equally to 10 V dc in both the positive and negative directions.

Solution

Because the controller output swing is symmetrical in the positive and negative directions, the two error bands will be identical in magnitude. Dead-band magnitude is given as

$$\begin{aligned}\text{dead band} &= 2 \times \text{error band}\\ &= 2 \times \frac{R_3}{R_2 + R_3} \times V_{sat}\end{aligned}$$

This magnitude (in volts) should be equal to the amplified transducer output corresponding to a temperature of 1°C.

But the amplified transducer output is 100 mV/°C; equating the two gives

$$\begin{aligned}100\text{ mV} &= 2 \times \frac{R_3}{R_2 + R_3} \times 10\text{V}\\ 0.1(R_2 + R_3) &= 20R_3\\ 0.1R_2 + 0.1R_3 &= 20R_3\\ 0.1R_2 &= 19.9R_3\\ R_2 &= 199R_3\end{aligned}$$

Selecting a 1-kΩ resistor for R_3 yields the R_2 resistor value of 199 kΩ. R_2 can either be a precision resistor (1% or better) or it can be a combination of a fixed resistor 190 kΩ (say) and a 10-kΩ potentiometer.

MULTIPOSITION (FLOATING) CONTROL

Multiposition control is essentially an extension of two-position control. In a two-position control system, the controller output switches between fully on and fully off states; in a multiposition control system, the controller output can have three (0%, 50%, and 100%) or more levels of output.

Figure 9.12 shows a multiposition controller with three operating states: 0%, 50%, and 100%. When the error is greater than the upper threshold limit, EH, controller output goes to 100%, and it drops to 0% whenever the error drops below the lower threshold of EL. For error conditions in between the limits of EL and EH, the controller output changes to 50%. Mutiposition control reduces the controller cycling rate compared with on-off and two-position control systems.

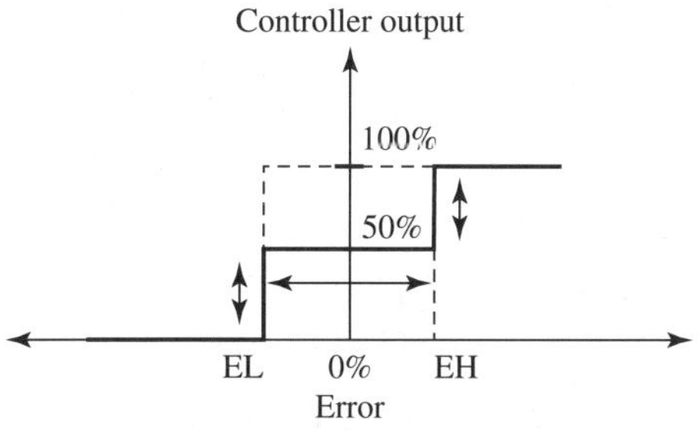

Figure 9.12 Multiposition control.

Multiposition control is not as popular as two-position control, because of difficulties in implementing it with op-amps and other conventional (mechanical, hydraulic, and pneumatic) control elements. A microprocessor-based control system is not hampered by these difficulties, and a multiposition control system can be implemented with relative ease.

PROBLEMS

1. Express the set-point value as a percentage value from the following data.

	Set-Point Range	Set-Point Value
a.	10° to 35°C	15°C
b.	1000° to 1200°C	1150°C
c.	−12 to +12 V	2.5 V

2. A commercial temperature control system has been set to operate within a temperature range of 10°C to 30°C. The display indicates the values of both the error and process temperature as a percentage value. Determine the indicator reading from the following operating data.

$$\text{SP value} = 40\%$$
$$\text{measured temperature} = 17^\circ\text{C}$$

3. Determine which of the following two velocity control systems has a greater percent error.

System A:

$$\text{operating range} = 0 \text{ to } 3000 \text{ rpm}$$
$$\text{SP speed} = 1000 \text{ rpm}$$
$$\text{measured speed} = 950 \text{ rpm}$$

System B:

$$\text{operating range} = 0 \text{ to } 600 \text{ rpm}$$
$$\text{SP speed} = 450 \text{ rpm}$$
$$\text{measured speed} = 415 \text{ rpm}$$

4. Describe three applications where on-off controllers are used.

5. What types of processes are suitable for on-off control?

6. A two-position on-off controller controls the temperature of an industrial oven. The heat switched on and off through a large electrical contactor, as dictated by the controller. It was observed that after a readjustment of the controller, the heater cycle rate increased significantly, causing a faster burnout of contactors. What is the possible cause of it?

7. A level control system is utilizing a two-position on-off controller. Water flows into the tank through a control valve, which is either fully open or fully closed. There is a constant outflow from the tank (provided there is water in the tank). It has the following specifications:

$$\text{control valve-flow rate} = 2 \text{ ft}^3/\text{min}$$
$$\text{constant outflow rate from tank} = 1 \text{ ft}^3/\text{min}$$
$$\text{tank cross-section area} = 0.5 \text{ ft}^2$$
$$\text{target (SP) level} = 5 \text{ ft}$$
$$\text{dead band} = \pm 1.0 \text{ ft}$$

a. At the start, the tank is empty and the control valve is shut off. The controller is switched on the very first time (time = 0 s). How long (minutes) will it be before the control valve is turned on?
b. Again, considering the start from an empty tank, how long (minutes) will it be before the control valve is commanded to shut off?
c. Assume that the process has been operating for a few days. What is the length of time (minutes) for which the control valve remains closed at a stretch?
d. Again, assume that the process has been operating for some time. What is the length of time (minutes) for which the control valve remains open at a stretch?
e. What is the cycle time (minutes) of the controller?
f. Draw a timing diagram showing the variation in level with respect to time. Show at least two complete cycles.
g. What will happen to the controller cycle time if the dead band is changed to ±0.5 ft.?

8. Which of the following systems are not suitable for dead-band control?
a. DC motor-speed control
b. Home-heating system
c. Oven-temperature control
d. Water-level control

9. A two-position on-off controller is used to regulate the output voltage in a low-power circuit. A large capacitor at the input of the circuit acts as a charge storage unit and cuts down the voltage variation to a minimum. Draw a fully labeled diagram showing controller action for the following operating conditions:

input voltage range = 3 to 8 V

error band = ±0.1 V

set-point range = 4.5 to 5.5 V

set-point voltage = 4.9 V

10. Draw a schematic of a two-position on-off controller to accomplish the control function of Problem 9.

10

ANALOG CONTROLLER I

INTRODUCTION

Analog controllers are the most popular type of controllers used in industry to date and have been used since the inception of control systems. Compared to an on-off controller, which is either fully on or fully off, an analog (continuous) controller can provide an output that gradually varies from 0% to 100% and back. Mechanical devices, such as gears, levers, and linkages, were originally used to implement analog controllers. Later, pneumatic devices were incorporated in complex process controllers, used for manipulation of process control valves and hydraulic and pneumatic actuators.

In this chapter all the three types of analog controllers—proportional (P), integral (I), and derivative (D)—are discussed in detail. Discussion includes the transfer function, time response, and frequency response of each of the three controller modes. Detailed explanation is provided about practical implementation of these controllers using operational amplifiers. The next chapter continues the discussion with the composite controllers: PI, PD, and PID.

PROPORTIONAL CONTROLLER (P)

In a proportional controller the controller output is directly proportional to the input (Figure 10.1). As with all controllers, the input signal is the error in controlled variable ($e = \text{SP} - \text{MV}$).

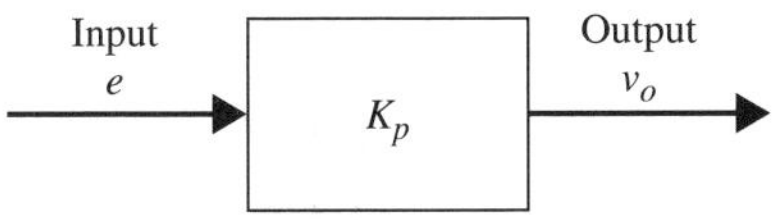

Figure 10.1 Proportional controller.

Time Equation

$$v_o = K_P e$$

where

$$e = \text{error input}$$
$$v_o = \text{controller output}$$
$$K_P = \text{controller gain}$$

Transfer Function

$$\frac{V_o(s)}{E(s)} = K_P$$

EXAMPLE 10.1

A proportional controller has a gain of 10. Determine the controller output if the input to the controller is 5 units.

Solution

$$\text{controller output } v_o = K_P \times \text{controller input}$$
$$= 10 \times 5$$
$$= 50 \text{ units}$$

Transfer Characteristic

Even though the input/output relationship of a proportional controller is expressed through a linear equation, there is a limit on how much the controller output can grow. This limitation comes from the limits of physical movement in mechanical, hydraulic, and pneumatic units and from saturation voltages in electronic controllers. Figure 10.2 shows the input/output characteristics of a proportional controller.

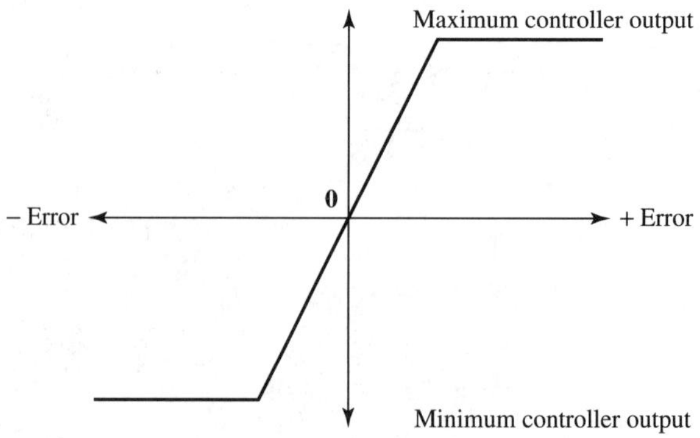

Figure 10.2 Transfer characteristic of a proportional controller.

EXAMPLE 10.2

A proportional controller has a gain of 5% of the FS output per % of error. Determine the controller output if the error is 2.5%.

Solution

The input to the controller is the error signal; thus

$$\begin{aligned}\text{controller output } v_o &= K_P \times \text{error}\\ &= \frac{5\%\ \text{FS output}}{\%\ \text{error}} \times \text{error}\\ &= \frac{5\%\ \text{FS output}}{\%\ \text{error}} \times 2.5\%\ \text{error}\\ &= 12.5\%\ \text{FS controller output}\end{aligned}$$

EXAMPLE 10.3

Determine the controller output from the following data:

$$\begin{aligned}\text{controller gain} &= 2\%\ \text{FS output}/\%\ \text{of error}\\ \text{set-point value} &= 20^\circ\text{C}\\ \text{measured value} &= 18.5^\circ\text{C}\\ \text{operating temperature range} &= 15^\circ\text{–}25^\circ\text{C}\end{aligned}$$

Solution

Controller output is directly proportional to error. Thus, % error needs to be calculated first.

$$\begin{aligned}
\%\ \text{error} &= \frac{\text{SP} - \text{MV}}{\text{range}} \times 100 \\
&= \frac{20 - 18.5}{25 - 15} \times 100 \\
&= 15\% \\
\text{controller output } v_o &= K_P \times \text{error} \\
&= \frac{2\%\ \text{FS output}}{\text{per } \%\ \text{error}} \times \%\ \text{error} \\
&= \frac{2\%\ \text{FS output}}{\text{per } \%\ \text{error}} \times 15\%\ \text{error} \\
&= 30\%\ \text{of FS controller output}
\end{aligned}$$

Proportional Band

Proportional controller gain is sometimes referred to by the term *proportional band* (PB). The relationship between PB and proportional gain is described as

$$\text{proportional band PB} = \frac{100}{\text{proportional gain } K_P}$$

Proportional Band

The **proportional band** represents the maximum percent error that will cause a change in controller output from minimum (0%) to maximum (100%).

Figure 10.3 shows the transfer characteristics of a proportional controller. Controller output changes from 0% (fully off) to 100% (fully on) for an error change of 0% to 5%. Because the minimum controller output is 0%, any error less than 0% (negative error) will result in minimum (0%) controller output. Similarly, any error greater than +5% will result in maximum (100%) controller output. So in this case,

$$\text{proportional band} = 5\%$$

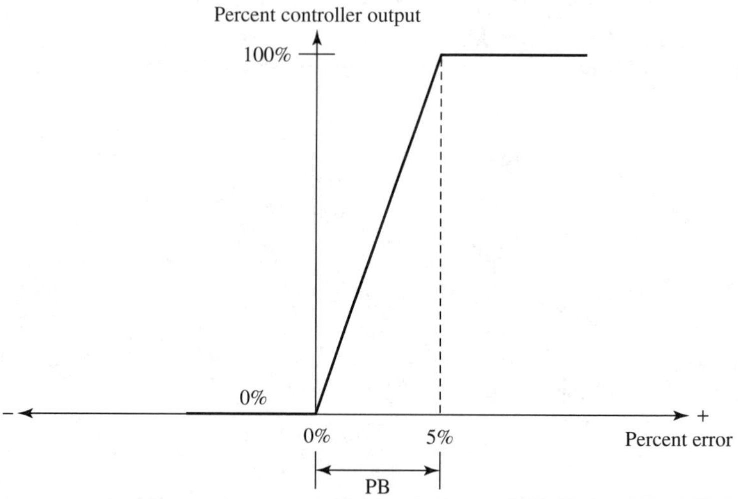

Figure 10.3 Proportional band.

The slope of the transfer characteristic is equal to the proportional gain, K_P.

$$\text{proportional gain} = \frac{100\%}{\text{PB}} = \frac{100\%}{5\%} = 20$$

Relationship Between Percent Output and Physical Controller Output

It is quite common to deal with controller output in terms of percentages. Unfortunately, real-world devices do not comprehend percentage values and require voltage, current, pressure, etc., as input. A linear relationship usually exists between percent controller output and the physical (actual) output. The physical controller output can be expressed as

$$v_a = K v_o + C$$

where

v_a = actual controller output
K = transfer gain
v_o = controller output in %
C = actual output for 0% controller output

EXAMPLE 10.4

The output of an op-amp-based controller is −10 V corresponding to 0% output and +10 V for 100% output. Determine:

a. Actual output transfer equation
b. Actual output corresponding to an 80% output signal
c. Actual output corresponding to a 40% output signal
d. Percent output corresponding to 0 V of actual output

Solution

a. The general equation for output transfer is

$$v_a = K \times v_o + C$$

Because the data are available for two points, substituting these in the equation will provide values of K and C.

Using the data for 0% output:

$$-10\text{ V} = K \times 0\% + C$$
$$C = -10\text{ V}$$

Using the data for 100% output gives

$$+10\text{ V} = K \times 100\% + (-10\text{ V})$$
$$100\%\ K = 20\text{ V}$$
$$K = 0.2\text{ V}/\%$$

Therefore, the transfer equation is

$$v_a = 0.2\text{ V}/\% \times V_o - 10\text{ V}$$

b. Substituting the data in the transfer equation, the actual output corresponding to 80% controller output is

$$v_a = 0.2\text{ V}/\% \times 80\% - 10\text{ V}$$
$$= 6\text{ V}$$

c. Similarly, the actual output corresponding to 40% controller output is

$$v_a = 0.2\text{ V}/\% \times 40\% - 10\text{ V}$$
$$= -2\text{ V}$$

d. Because the transfer characteristic is symmetrical between 0% and 100%, it can be determined intuitively that 50% output will result in an actual output of 0 V. It can also be verified by substituting the values in the transfer equation.

$$0\ \text{V} = 0.2\ \text{V}/\% \times v_o - 10\ \text{V}$$
$$v_o = 50\%$$

EXAMPLE 10.5

A controller generates a pneumatic (air-pressure) output signal. The controller's actual output ranges from 3 psi to 15 psi, corresponding to outputs of 0% and 100%, respectively. Determine:

a. Output transfer equation
b. Actual output corresponding to a 60% output signal

Solution

a. Substituting the data for 0% output in the general equation for output transfer gives

$$v_a = K \times V_o + C$$
$$3\ \text{psi} = K \times 0\% + C$$

Similarly, for 0% output,

$$15\ \text{psi} = K \times 100\% + C$$

Solving these two equations yields

$$K = 0.12\ \text{psi}/\% \quad \text{and} \quad C = 3\ \text{psi}$$

Therefore, the transfer equation is

$$v_a = 0.12\ \text{psi}/\% \times V_o + 3\ \text{psi}$$

b. Corresponding to a 60% controller output, the actual output is

$$v_a = 0.12\ \text{psi}/\% \times 60\% + 3\ \text{psi}$$
$$= 10.2\ \text{psi}$$

EXAMPLE 10.6

A proportional controller ($K_P = 5$ V of output/volt of error) is fed with the error input curve in Figure 10.4. Controller output ranges from -10 V (minimum) to $+10$ V (maximum). Determine the controller output voltage.

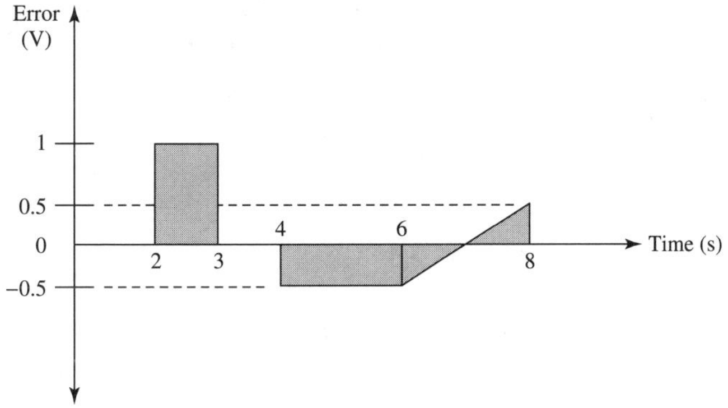

Figure 10.4 Error curve for Example 10.6.

Solution

$$\text{Controller output } v_o = K_P \times \text{error}$$
$$= 10 \text{ V/V} \times \text{error}$$

Output from 0 to 2 s:

$$v_o = 10 \text{ V/V} \times 0 \text{ V}$$
$$= 0 \text{ V}$$

Output from 2 to 3 s:

$$v_o = 10 \text{ V/V} \times 1 \text{ V}$$
$$= 10 \text{ V}$$

Output from 4 to 6 s:

$$v_o = 10 \text{ V/V} \times (-0.5 \text{ V})$$
$$= -5 \text{ V}$$

Oututput from 6 to 8 s: Error is changing linearly in the 6-s to 8-s time interval, so the controller output, which is directly proportional to error, will also change linearly during this time interval. Because the output will be a straight line, only two points need to be calculated. Suitable values are $t = 6$ s and $t = 8$ s.

Output at $t = 6$ s:

$$V_o = 10 \text{ V/V} \times (-0.5 \text{ V})$$
$$= -5 \text{ V} \qquad \text{(calculated earlier as well)}$$

Output at $t = 8$ s:

$$V_o = 10 \text{ V/V} \times 0.5 \text{ V}$$
$$= 5 \text{ V}$$

The completed output curve is shown in Figure 10.5

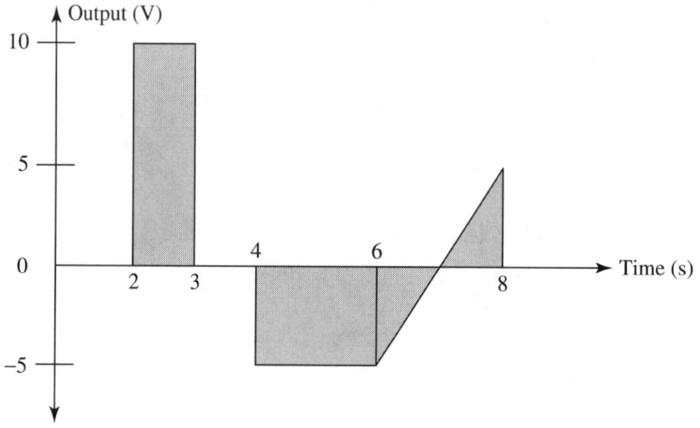

Figure 10.5 Controller output for Example 10.6.

Electronic Implementation

Schematic 1A The schematic diagram for a proportional controller is shown in Figure 10.6. It contains all the features necessary for implementing a practical proportional controller. The entire circuit can be constructed using a single quad op-amp, such as LM324.

Controller operation can be broken down into two parts: summation block and proportional controller. Amplifier U1A implements the summation block. It is

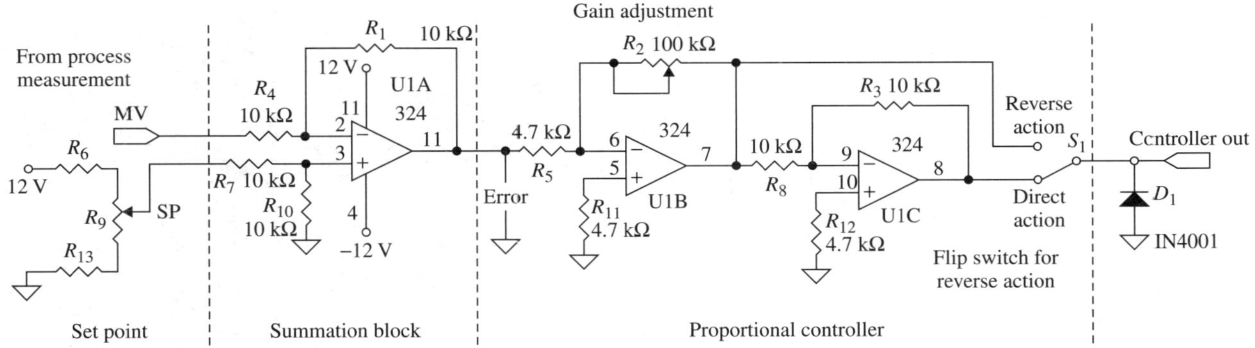

Figure 10.6 Proportional controller: schematic 1A.

set up as a difference amplifier. A set-point signal is applied to the noninverting terminal of the amplifier. The inverting amplifier terminal is fed with the measured value (MV) of the process variable. MV is the feedback signal, and thus the control loop is closed.

$$\begin{aligned}\text{output of summation block} &= \text{SP} - \text{MV}\\ &= \text{error}\end{aligned}$$

Amplifiers U1B and U1C form a proportional amplifier. Both these amplifiers are set up as inverting amplifiers. Thus the signal is inverted twice, resulting in same polarity as the error signal. Amplifier U1B provides all the necessary gain (K_P), whereas amplifier U1C has a gain of -1. Potentiometer R_2 can be set to yield the desired gain.

$$\text{proportional gain } K_P = \frac{R_2}{R_5}$$

For the values shown in the schematic diagram, a gain ranging from 1 to 20 can be achieved.

Most controllers are operated in direct action mode, where an increase in error results in a larger output from the controller, and a smaller error results in a smaller controller output. In some processes, such as a cooling system, an increase in error (cooler temperature) requires the controller output to be reduced. This calls for a reverse-acting controller. The switch S_1 allows the controller mode to be changed from direct acting to reverse acting. With the switch in the position shown, the controller output is taken from op-amp U1C and the controller acts as a direct-acting controller. For a reverse-acting controller, the switch S_1 is flipped and the final output is taken from op-amp U1B, which is inverted.

Schematic 1B Figure 10.7 shows a schematic diagram for a modified proportional controller. It again provides a complete solution for implementing a proportional controller. The design is more compact, and the entire circuit can be constructed using a single 8-pin IC containing a dual op-amp, such as LM358.

Op-amp U1A is set up as a difference amplifier and acts as the summation block. In this configuration, the set-point signal is applied to the inverting terminal, whereas the feedback signal (MV) is applied to the noninverting terminal.

$$\begin{aligned}\text{output of amplifier U1A (summation block)} &= \text{MV} - \text{SP}\\ &= -(\text{SP} - \text{MV})\\ &= -e\end{aligned}$$

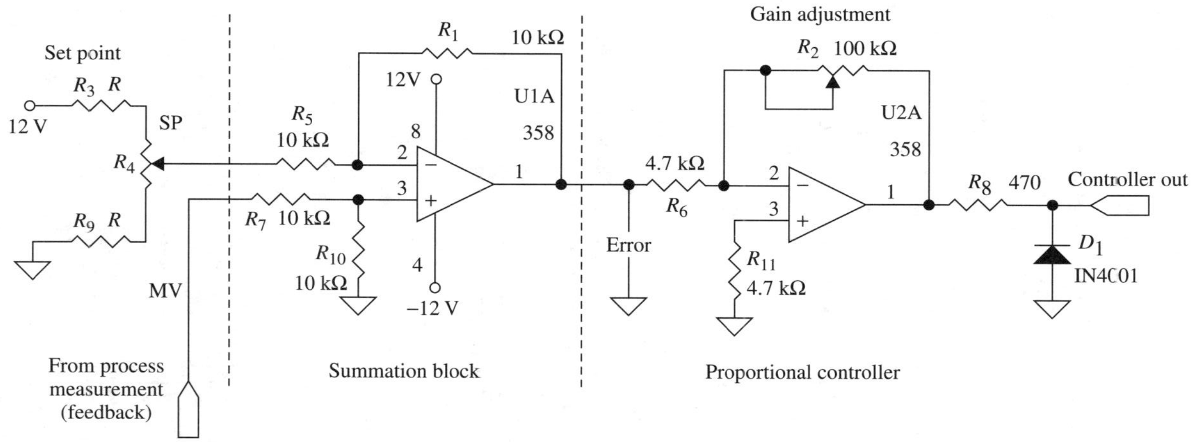

Figure 10.7 Proportional controller: schematic 1B.

Amplifier U1B is set up as an inverting amplifier. Proportional gain, K_P, is adjusted with the potentiometer located in its feedback path.

$$\begin{aligned} \text{amplifier gain} &= -\frac{R_2}{R_6} \\ &= -K_P \\ \text{output of amplifier U1B} &= -\text{input} \times K_P \\ &= -(-e) \times K_P \\ &= K_P e \qquad \text{(positive value)} \end{aligned}$$

In this design, it has been possible to eliminate one op-amp from the circuit while still maintaining full functionality. There is, however, a loss of flexibility in terms of switch selection of direct- or reverse-acting mode. Figure 10.7 is hardwired for the direct-acting mode. If the reverse-acting mode is required, it can still be achieved with a small modification: swapping the inputs SP and MV to amplifier U1A (summation block).

Schematic 1C In previous designs (Figures 10.6 and 10.7) the controller output has been limited to positive swings (0–10 V). This may be suitable for a number of applications, where only a positive (or unidirectional) controller output is required. But in some applications, such as motor velocity and position control, the electric motor may be required to rotate in either direction. In this case the controller output needs to swing in a positive as well as a negative direction.

Figure 10.8 shows a schematic diagram for a proportional controller that provides bidirectional control output. The controller has been modified from schematic 1B. The set-point potentiometer is connected to a bipolar supply, allowing either a positive or negative reference voltage to be applied. The resistor and diode pair at the controller

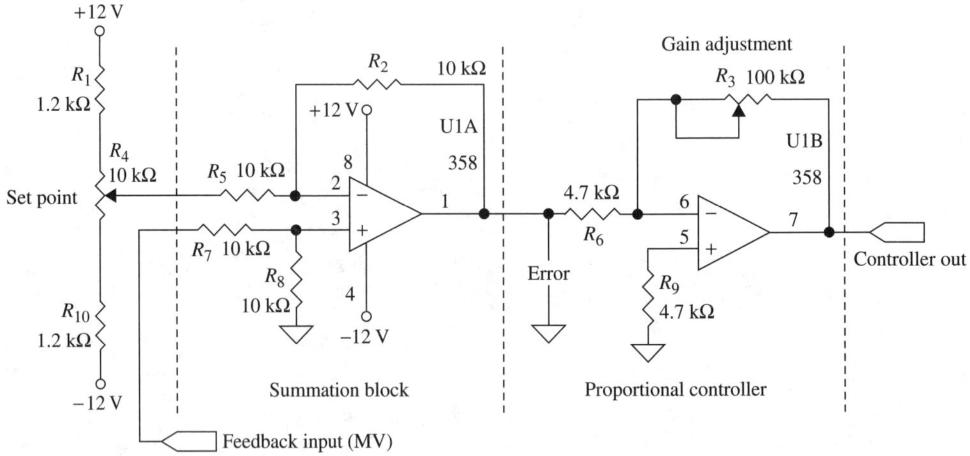

Figure 10.8 Proportional controller: bidirectional control.

output has been removed, allowing the output to swing freely from the positive to the negative region. The entire circuit can still be constructed using a single 8-pin IC containing a dual op-amp, such as LM358.

EXAMPLE 10.7

A proportional controller with a gain of 10 is subjected to the error signal in Figure 10.9. Determine the corresponding controller output.

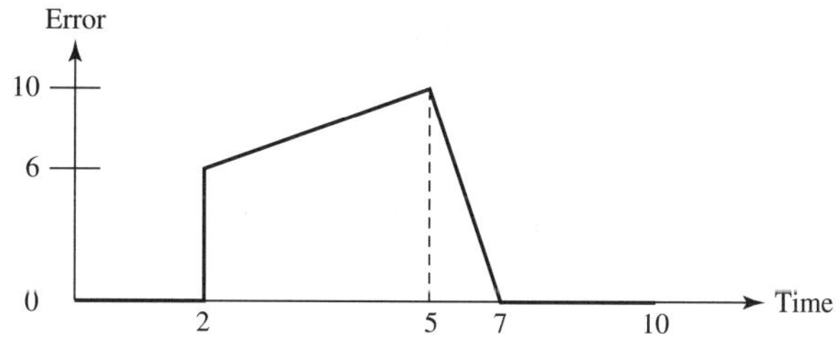

Figure 10.9 Error curve for Example 10.7.

Solution

The output $v_o(t)$ of the proportional controller is directly proportional to its error input, $e(t)$:

$$\begin{aligned} v_o &= K_P \times e \\ &= 10e \end{aligned}$$

The controller output can be calculated at $t = 0$ s, 2 s, 5 s, and 7 s. Between 2 s and 5 s, the output is rising linearly. Similarly, between 5 s and 7 s, the controller output will linearly decrease to zero.

Output at $t = 0$ s:

$$\begin{aligned} v_o &= 10 \times \text{current error (at } t = 0 \text{ s)} \\ &= 10 \times e(0) \\ &= 0 \end{aligned}$$

Output at $t = 2$ s:

$$\begin{aligned} v_o &= 10 \times e(2) \\ &= 10 \times 6 \\ &= 60 \end{aligned}$$

Output at $t = 5$ s:

$$v_o = 10 \times e(5)$$
$$= 10 \times 10$$
$$= 100$$

Output at $t = 7$ s:

$$v_o = 10 \times e(7)$$
$$= 10 \times 0$$
$$= 0$$

Figure 10.10 shows the complete controller output.

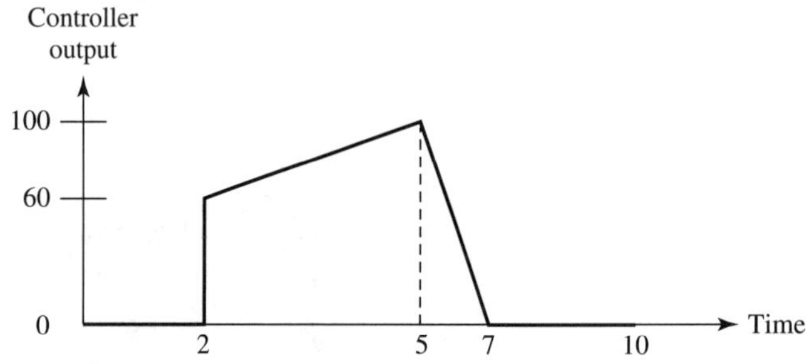

Figure 10.10 Controller output for Example 10.7.

Frequency Response of a Proportional Controller

For an ideal amplifier, the gain remains constant for all frequencies, and there is no phase difference between the input and output signals. A pure proportional controller behaves like an ideal amplifier. Its gain is assumed to remain constant for the entire (from 0 rad/s to ∞ rad/s) frequency range. It can be further assumed that there is no phase difference between input and output.

This assumption may be valid for some proportional controllers based upon mechanical elements but not for modern op-amp-based units. It is well known that the gain of all op-ams drops at higher frequencies. A phase lag also develops between output and input signals at higher frequencies. However, over the frequencies of interest, it can be safely assumed that the amplifier gain remains constant and that there is zero phase difference between the input and output signals.

$$\text{magnitude of proportional amplifier} = 20 \log(\text{gain})$$
$$= 20 \log(K_P)$$
$$\text{phase difference} = 0^\circ$$

The magnitude plot of a proportional controller is a horizontal line at a value of 20 log (K_P) dB. The phase plot is again a horizontal line passing through the 0° line.

EXAMPLE 10.8

Draw the Bode plot of a proportional controller with a gain of 1.

Solution

$$\text{magnitude } M = 20 \log(1) \text{ dB}$$
$$= 0 \text{ dB}$$
$$\text{phase angle} = 0^\circ$$

See Figure 10.11.

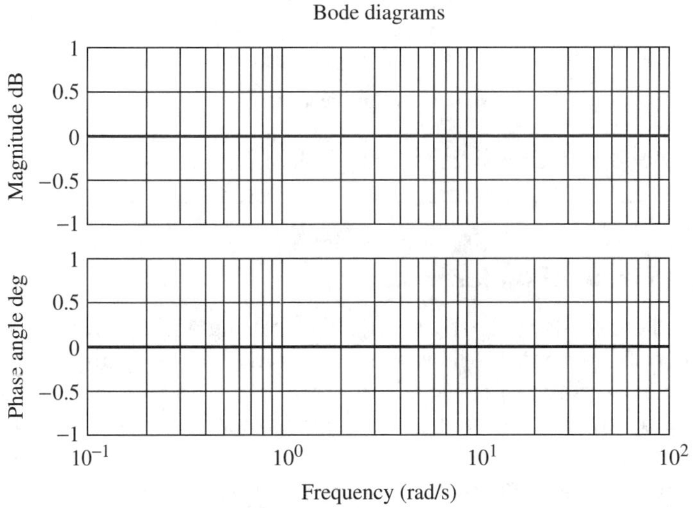

Figure 10.11 Bode plot ($K_P = 1$) for Example 10.8.

EXAMPLE 10.9

Draw the Bode plot of a proportional controller with a gain of 25.

Solution

$$\text{magnitude } M = 20\ \log(25)\ \text{dB}$$
$$= 27.96\ \text{dB}$$
$$\text{phase angle} = 0°$$

See Figure 10.12.

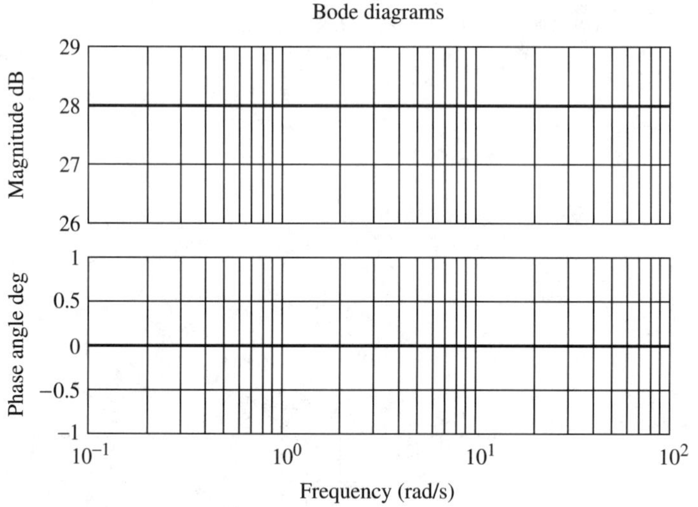

Figure 10.12 Bode plot (gain = 25) for Example 10.9.

Proportional Control of a Closed Loop

The proportional controller (Figure 10.8) is controlling a plant in a closed-loop configuration. The controller accepts the error input signal (SP – MV) and generates an output in order to drive the plant to the desired state. To keep the analysis simple, a unity feedback is employed, and it is further assumed that the plant has a transfer function of 1 ($G_1 = 1$).

If this plant is controlled in an open-loop configuration ($G_p = 1$) (Figure 10.13), then a unit signal applied at the plant input will generate a unit output signal. When

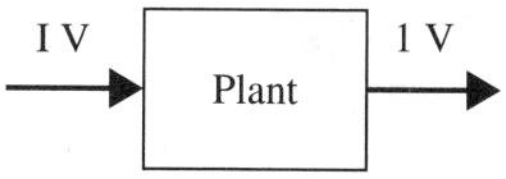

Figure 10.13 Open-loop control.

operated in a closed-loop configuration (Figure 10.14), the loop output C will depend not only upon the plant transfer function but also on the controller transfer function and the loop dynamics.

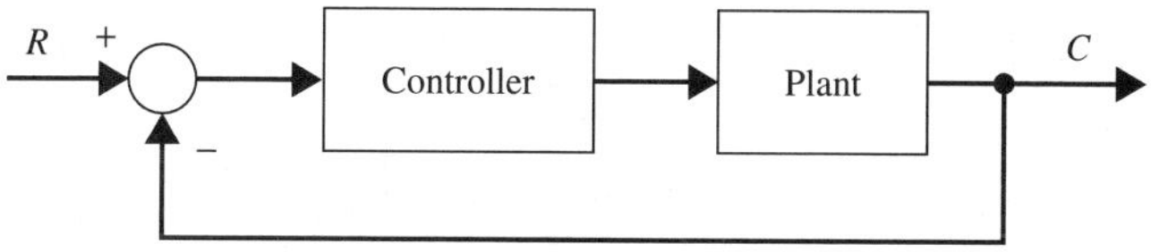

Figure 10.14 Closed-loop proportional control.

Instantaneous error in a closed-loop control system depends upon the plant dynamics and the reference (SP) value. In a servomechanism, the reference value changes frequently, whereas in a regulator setting, the reference (SP) value is more or less held constant. In these (regulator) systems, steady-state (final) error is of significance. The following discussion is based upon the steady-state error in the control of a zero-order system, where the plant gain can be expressed as a constant—e.g., $G_P = 1$.

The closed-loop transfer function (CLTF) determines the relationship between the input R and the output C.

$$\text{CLTF} = \frac{\text{forward gain}}{1 + \text{loop gain}}$$

$$= \frac{K_P G_P}{1 + K_P G_P H}$$

Let process gain $G_P = 1$ and feedback $H = 1$ (unity feedback). Then

$$\text{CLTF} = \frac{K_P}{1 + K_P}$$

$$\text{output } C = \frac{K_P}{1 + K_P} R$$

Error in a system depends upon the measured value (MV), which, in turn—because of unity feedback—is equal to C.

$$\begin{aligned}\text{absolute error} &= \text{SP} - \text{MV} \\ &= R - C \\ &= R - \frac{K_P}{1+K_P}R \\ &= \left[1 - \frac{K_P}{1+K_P}\right]R \\ &= \left[\frac{1}{1+K_P}\right]R\end{aligned}$$

$$\boxed{\text{absolute error} = \frac{R}{1+K_P}}$$

$$\begin{aligned}\text{relative error} &= \frac{\text{absolute error}}{\text{input}} \\ &= \frac{\frac{1}{1+K_P}R}{R} \\ &= \frac{1}{1+K_P}\end{aligned}$$

$$\boxed{\text{relative error} = \frac{1}{1+K_P}}$$

$$\begin{aligned}\%\ \text{error} &= \frac{\text{absolute error}}{\text{input}} \times 100 \\ &= \frac{\frac{1}{1+K_P}R}{R} \times 100 \\ &= \frac{100}{1+K_P}\end{aligned}$$

$$\boxed{\%\ \text{error} = \frac{100}{1+K_P}}$$

This is an important observation. Because K_P, which represents the proportional controller gain, is a finite value, the right-hand term (error) will also be a finite value.

Although error will become smaller with an increase in proportional gain, it will never reach zero. This is also obvious from the output equation of the proportional controller.

$$\text{controller output } v_o = K_P \times e$$

If at any time error becomes zero, then the controller output drops to zero. *In a proportional closed-loop control system, the error in the controlled variable can never be totally eliminated. Error decreases with an increase in value of the proportional gain and approaches zero for an infinitely large gain value.*

EXAMPLE 10.10

Determine the control system output, error, and percent error for the following input conditions given that the plant is of zero order and has a transfer function of 2 (Figure 10.15).

a. $R = 1$ V, $K_P = 1$
b. $R = 10$ V, $K_P = 1$
c. $R = 10$ V, $K_P = 100$

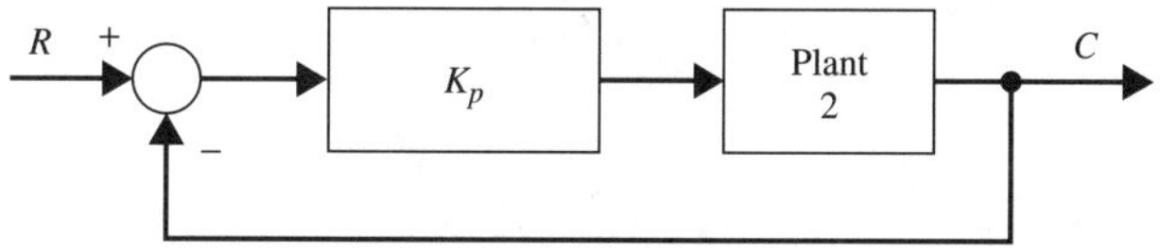

Figure 10.15 Block diagram for Example 10.10.

Solution

$$\text{control system output } C = \text{CLTF} \times \text{input } R$$

where

$$\text{CLTF} = \frac{\text{forward gain}}{1 + \text{loop gain}} = \frac{K_P \times \text{plant gain}}{1 + K_P \times \text{plant gain}}$$

$$= \frac{2K_P}{1 + 2K_P}$$

$$\text{error} = \text{input } R - \text{output } C$$

$$\%\ \text{error} = \frac{R - C}{R} \times 100$$

a.

$$\begin{aligned}
\text{output } C &= \frac{2K_P}{1+2K_P}R \\
&= \frac{2 \times 1}{1 + 2 \times 1} \times 1\text{ V} \\
&= 0.6667\text{ V} \\
\text{error} &= R - C \\
&= 1 - 0.6667\text{ V} \\
&= 0.3333\text{ V} \\
\%\text{ error} &= \frac{1 - 0.6667}{1} \times 100 \\
&= 33.33\%
\end{aligned}$$

b.

$$\begin{aligned}
\text{output } C &= \frac{2 \times 1}{1 + 2 \times 1} \times 10\text{ V} \\
&= 6.6667\text{ V} \\
\text{error} &= 10 - 6.6667\text{ V} \\
&= 3.3333\text{ V} \\
\%\text{ error} &= \frac{10 - 6.6667}{10} \times 100 \\
&= 33.33\%
\end{aligned}$$

c.

$$\begin{aligned}
\text{output } C &= \frac{2 \times 100}{1 + 2 \times 100} \times 10\text{ V} \\
&= 9.9502\text{ V} \\
\text{error} &= 10 - 9.9502\text{ V} \\
&= 0.0498\text{ V} \\
\%\text{ error} &= \frac{10 - 9.9502}{10} \times 100 \\
&= 0.498\%
\end{aligned}$$

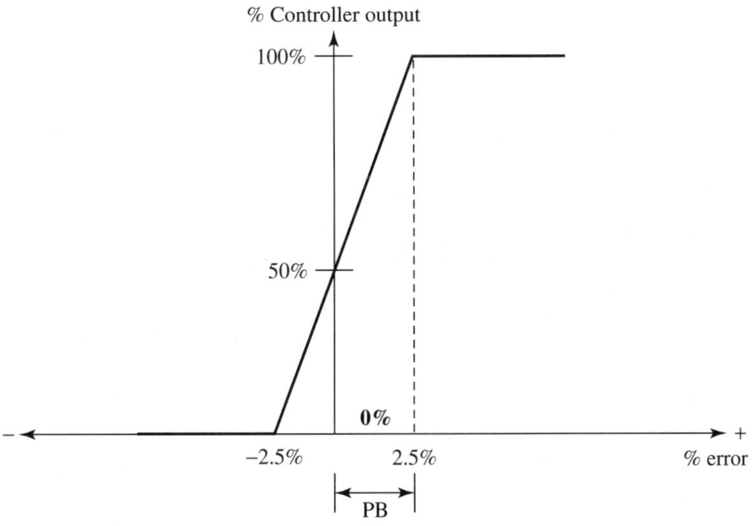

Figure 10.16 Proportional controller with an offset.

Observations

1. Error can never equal zero.
2. Percent error depends only on the proportional gain and not on the magnitude of the input signal.
3. The absolute error in (b) was larger than that in (a) because even though gain value was the same for both, the input signal was larger in (b).
4. Error becomes smaller with an increase in gain.

Steady-State Operation with Zero Error

Based upon the preceding discussions, it seems that a proportional controller will always involve an error in plant operation. This does not seem very promising for a regulator system such as home-temperature control (heating), where the set-point value (reference temperature) is essentially constant. Similarly, in the level control system of a city water tower, it is very desirable that during the steady-state operation, the water level be maintained at the desired value and without any error. To meet such a requirement, another term, *bias*, or *offset*, is added to a basic proportional controller.

$$\text{controller output } V_{\text{out}} = K_P e + \text{offset}$$

Offset allows steady-state operation, with zero error. Therefore, at zero error.

$$V_{out} = K_P \times 0 + \text{offset}$$
$$V_{out} = \text{offset}$$

Hence, offset is equal to the controller output when the process is operated at a steady-state condition and with zero error. Figure 10.16 shows a proportional controller with a proportional band of 5% and an offset value of 50%. Thus, even when there is no error in the system, the controller will generate an output of 50%.

EXAMPLE 10.11

A proportional controller is set with a gain of 20, and it is generating a 50% output at steady-state operation with no error. Determine each of the following.

a. Proportional band
b. Offset value
c. Controller equation

Solution

a.

$$\text{proportional band PB} = \frac{100}{K_P}\% = \frac{100}{20}\% = 5\%$$

b. Therefore,

$$\text{offset} = v_o \text{ at steady-state operation with zero error} = 50\%$$

c. The general equation for a proportional controller with an offset is

$$v_o = K_P e + \text{offset}$$

Substituting the value of offset (50%), the controller equation becomes

$$v_o = K_P e + 50\%$$

EXAMPLE 10.12

A proportional controller is operating with a gain of 10 (10%/% of error) and an offset of 25%. Determine controller output when an error of 1% persists in the system.

Solution

The controller output is

$$v_o = K_P e + \text{offset}$$

Substituting the values,

$$\begin{aligned} v_o &= 10 \times 1\% + 25\% \\ &= 35\% \end{aligned}$$

EXAMPLE 10.13

A level control system (Figure 10.17) is equipped with a proportional control. It is operating at a steady-state level with zero error. The output of the controller is fed to a flow control valve, thus controlling the tank-inflow rate. Water level stops changing when the inflow rate is matched to the outflow rate. The outflow rate is dictated by a downstream process and can change randomly. The controller function is to maintain the water level by manipulating the inflow rate until it matches the outflow rate.

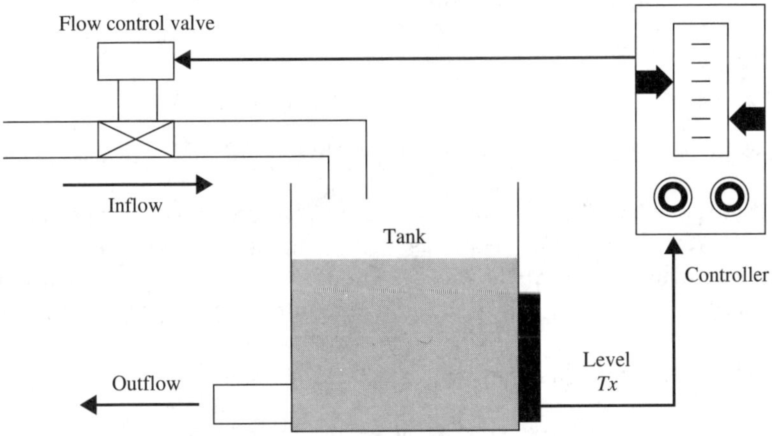

Figure 10.17 Proportional level control system for Example 10.13.

Component specifications are as follows:

$$\text{valve inflow rate} = 0.01\ \text{m}^3/\text{min/V}$$
$$\text{level transducer output} = 10\ \text{mV/cm of tank level}$$
$$\text{outflow rate range} = 0\text{–}0.1\ \text{m}^3/\text{min}$$
$$\text{tank height} = 1.5\ \text{m}$$
$$\text{set-point voltage range} = 0\text{–}10\ \text{V}$$
$$\text{controller output voltage range} = 0\text{–}10\ \text{V}$$
$$\text{proportional gain} = 5$$

Offset voltage is adjusted to provide a steady-state operation under the following conditions:

$$\text{tank level} = 60\ \text{cm}$$
$$\text{outflow rate} = 0.03\ \text{m}^3/\text{min}$$

Determine the following:

a. Inflow at steady state
b. Offset value
c. Tank level when the outflow rate changes to 0.05 m^3/min

Solution

a. Because the process is operating under steady-state conditions, there is no change in level. This can occur only if the incoming flow rate is the same as the outgoing flow rate. Thus

$$\begin{aligned}\text{inflow rate} &= \text{outflow rate}\\ &= 0.03\ \text{m}^3/\text{min}\end{aligned}$$

b. The offset value is equal to the controller output at steady-state operation with zero error. In this state, controller output is such that the flow control valve produces an inflow rate identical to the outflow rate. The flow rate through the control valve is directly proportional to the input signal or output from the controller.

$$\text{inflow rate through control valve} = 0.03\ \text{m}^3/\text{min}$$

But

$$\text{valve flow rate} = 0.01\ \text{m}^3/\text{min/V of controller output}$$

Therefore,

$$\text{controller output required to generate a flow rate of } 0.03\ \text{m}^3/\text{min} = \frac{0.03\ \text{m}^3/\text{min}}{0.01\ \text{m}^3/\text{min/V}}$$
$$= 3\ \text{V}$$

The controller outputs a 3-V signal at steady-state conditions with zero error and a tank level of 60 cm.

c. The general equation for controller output is

$$v_o = K_P e + \text{offset}$$

Substituting the values of K_P and the offset term gives

$$v_o = 5e + 3\ \text{V}$$

d. Because the outflow rate has changed, the proportional controller cannot maintain the desired (set point) level without an error. The error in the system will cause an increase in controller output, which in turn will cause the flow control valve to increase the flow rate. Eventually, a point is reached when the outflow is matched by the increased inflow rate. At this point, the error stops to change and a steady-state condition (but with an error) is reached.

$$\text{controller output required for an outflow of } 0.05\ \text{m}^3/\text{min} = \frac{\text{inflow rate}}{\text{inflow rate/V}}$$
$$= \frac{0.05\ \text{m}^3/\text{min}}{0.01\ \text{m}^3/\text{min/V}}$$
$$= 5\ \text{V}$$

But the controller output is given as

$$v_o = 5e + 3\ \text{V}$$

Therefore, upon substituting the value of controller output, the error can be determined:

$$5\ \text{V} = 5e + 3\ \text{V}$$
$$e = 0.4\ \text{V}$$
$$= 400\ \text{mV} \qquad \text{(level error)}$$

$$\text{level error (in cm) corresponding to } 0.4\ \text{V} = \frac{\text{error in volt}}{\text{level } Tx \text{ output in V/cm}}$$
$$= \frac{400\ \text{mV}}{10\ \text{mV/cm}}$$
$$= 40\ \text{cm}$$

But

$$\text{error (level)} = \text{set-point level} - \text{measured level}$$
$$40\text{ cm} = 60\text{ cm} - \text{measured level}$$
$$\text{measured level} = 60\text{ cm} - 40\text{ cm}$$
$$= 20\text{ cm} \qquad \text{(tank level)}$$

Thus, the process is operating at a 20-cm level with an outflow (and inflow) rate of 0.05 m^3/min.

EXAMPLE 10.14

In Example 10.13, a 40-cm error in operation with a flow rate of 0.05 m^3/min was obtained. This error is considered to be unacceptably large. Show what step can be taken to limit this error to 5 cm.

Solution

Here, it is going to be assumed that no hardware modification is allowed. Again, the controller output is governed by the general equation

$$v_o = K_P e + \text{offset}$$

Because the inflow rate has to be matched to an outflow rate of 0.05 m^3/min, the controller output has to be equal to 5 V. The offset term is also set by the zero error operation at the 60-cm level, with an outflow of 0.03 m^3/min.

For a reduction in error, the proportional gain term K_P needs to be increased so that a controller output of 5 V can be obtained. The level error is 5 cm.

$$\text{corresponding error value (in V)} = \text{error (in cm)} \times Tx \text{ output in V/cm}$$
$$= 5\text{ cm} \times 10\text{ mV/cm}$$
$$= 50\text{ mV}$$
$$= 0.05\text{ V}$$

Substituting the values in the preceding equation,

$$5\text{ V} = K_P \times 0.05\text{ V} + 3\text{ V}$$
$$K_P = 40$$

So, a proportional gain of 40 will limit the error to 5 cm, instead of 40 cm as was the case in the last example.

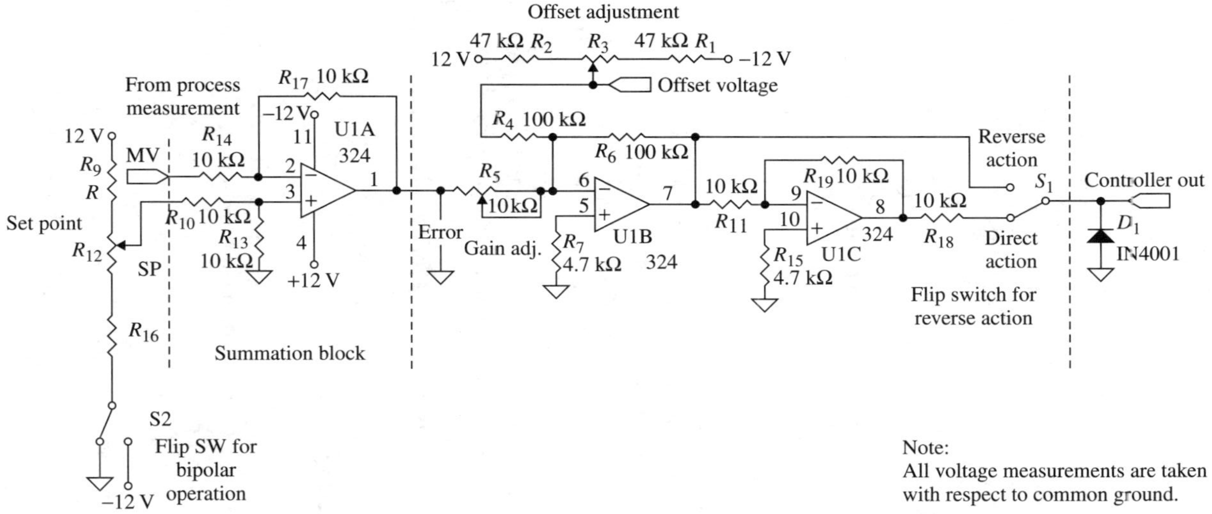

Figure 10.18 Proportional controller with offset adjustment.

Proportional Controller with Offset

Figure 10.18 shows a schematic diagram for a proportional controller incorporating offset adjustment. Resistor pair R_1, R_2 and potentiometer R_3 allow user-adjustable voltage to be added to the output of amplifier U1B.

$$\text{amplifier U1B output} = -\text{error} \times \frac{R_6}{R_5} - V_{\text{offset}} \times \frac{R_6}{R_4}$$

The output of amplifier U1B is fed to amplifier U1C, which provides the final controller output. This amplifier U1C is an inverting amplifier with a gain of −1.

$$\text{final controller output} = \text{error} \times \frac{R_6}{R_5} + V_{\text{offset}} \times \frac{R_6}{R_4}$$

Offset voltage can be adjusted for a positive or negative voltage, as required for a given configuration.

Offset Voltage Adjustment

Offset adjustment is carried out when the plant is operating at the steady-state condition. A set-point potentiometer is used to operate the plant at the desired normal operating point. Sufficient time is allowed for the plant to reach a steady-state condition. The offset potentiometer is then adjusted in a direction, which causes the error to be reduced. This adjustment is carried on until the error signal becomes equal to zero. This is the correct value of offset for this steady-state operating condition. It is important to realize that this offset setting is correct only for this operating condition, which includes the SP value, process input variable, and operating load. If any of these parameters changes, an error will develop in the system, and if these changes are permanent, readjustment of the offset value is warranted for steady-state operation under zero error.

INTEGRAL CONTROLLER (I)

In an integral controller, the controller output is not directly related to the input (error signal), but to the entire past history of error/time curve. Figure 10.19 shows a block diagram of an integral controller.

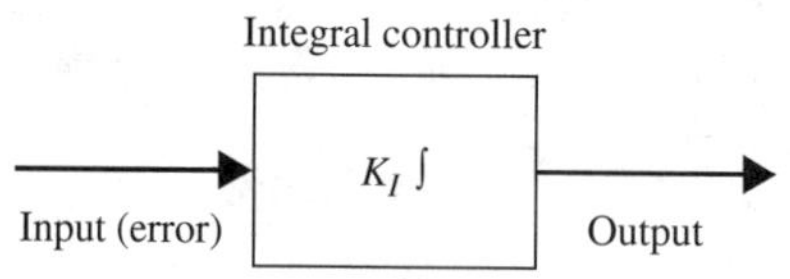

Figure 10.19 Integral controller.

Time Equation

$$v_o = K_I \int_0^T e\,dt$$

where v_o = controller output (at time T)
e = error input to controller (from time $t = 0$ to time $t = T$)
K_I = integral constant (gain or coefficient)

The integral constant is expressed in units of controller output/second/unit of error. The following shows some examples of commonly used expressions for the integral constant:

- 5
- 5/s
- 5%/s/%
- 5% output/s/% of error

The integration term is nothing but the area under the error-time curve, and the controller output is proportional to the area under the error/time curve (Figure 10.20).

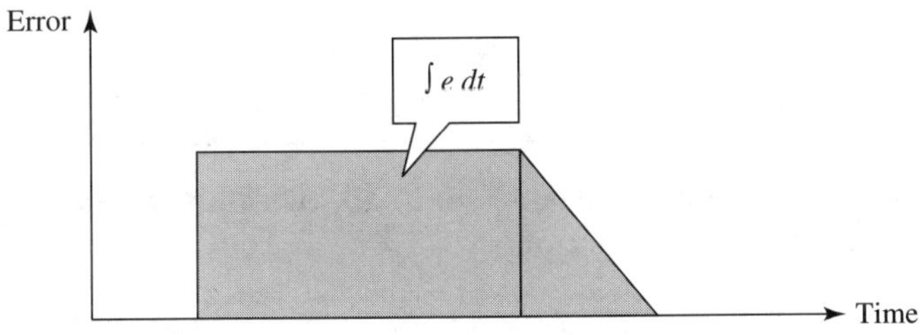

Figure 10.20 Integral as an area under the error-time curve.

Transfer Function

$$\frac{V_o(s)}{E(s)} = K_I s$$

The integral controller is not normally used by itself. It is used either with a proportional controller in a proportional + integral (P + I, or PI) configuration or with a proportional and differential controller in a full proportional + integral + derivative (P + I + D, or PID) configuration.

EXAMPLE 10.15

An integral controller has a gain of 2%/s/%. Determine the controller output after 5 s if the input to controller is constant at 5 units. Assume that the initial error is zero and the controller output initially was also zero.

Solution

Controller output is

$$v_o = K_I \int e\,dt$$

$$v_o = 2\%/\text{s}/\% \int_0^5 5\%\,dt$$

$$v_o = 2\%/\text{s}/\%\ (5\% \times 5\ \text{s})$$

$$v_o = 50\%$$

EXAMPLE 10.16

Determine the controller output from time 0 to 10 s for the following error input (Figure 10.21). The integral constant is 1.5%/s/%. The initial controller output is zero.

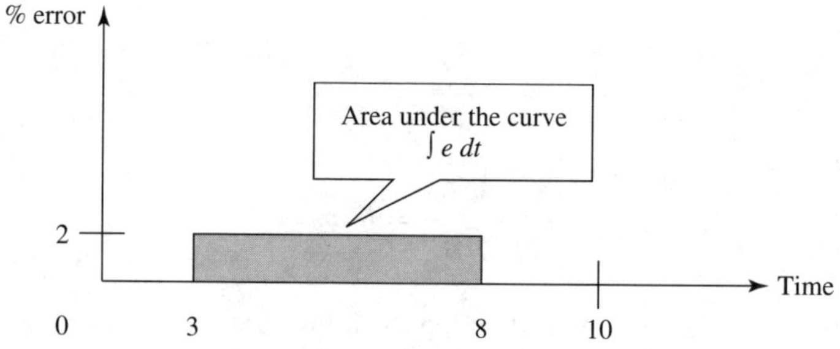

Figure 10.21 Error curve for Example 10.16.

Solution

The controller output is given by

$$v_o = K_I \int e\, dt$$

The equation can be solved either numerically or through calculation of graphical area.

a. *Numerical solution:* The numerical solution is best applied when the error curve can be defined as a function of time. It certainly applies here. The error curve from time $t = 0$ to $t = 10$ s can be broken into three parts:

$$\text{error from 0 to 3 s:} \quad e(t) = 0$$
$$\text{error from 3 to 8 s:} \quad e(t) = 2\%$$
$$\text{error from 8 to 10 s:} \quad e(t) = 0$$

Correspondingly, the output equation now becomes

$$v_o = K_I \left[\int_0^3 e\, dt + \int_3^8 e\, dt + \int_8^{10} e\, dt \right]$$

Substituting the values,

$$v_o = 1.5\%/\text{s}/\% \left[\int_0^3 0\%\, dt + \int_3^8 2\%\, dt + \int_8^{10} 0\%\, dt \right]$$

$$v_o = 1.5\%/\text{s}/\% \left[\int_3^8 2\%\, dt \right]$$

$$v_o = 1.5\%/\text{s}/\% \, [10\%\, \text{s}]$$

$$v_o = 15\%$$

Figure 10.22 shows the controller output.

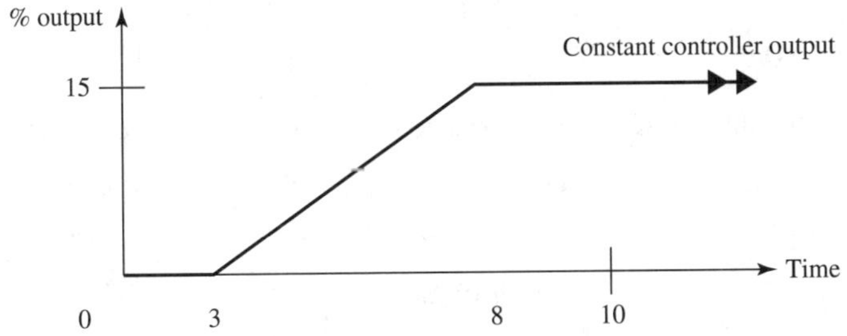

Figure 10.22 Controller output curve for Example 10.16.

b. *Graphical solution:* The graphical solution is suitable whenever the error-time graph is available. Again, observation of the error-time curve reveals that the curve can be divided into three time zones: $t = 0$ to 3 s, $t = 3$ to 8 s, and $t = 8$ to 10 s. Areas under the curve for these time zones can now be calculated. The controller output equation can now be written as

$$\text{controller output at time } t = K_I \times \text{area under the error curve between time 0 and } t$$

$$\begin{aligned} v_o \text{ (at } t = 10\text{ s)} &= K_I\ [\text{area from 0 to10 s}] \\ &= K_I\ [\text{area from 0 to 3 s} + \text{area from 3 to 8 s} \\ &\quad + \text{area from 8 to 10 s}] \\ &= K_I \times \text{area from 0 to 3 s} \\ &\quad + K_I \times \text{area from 3 to 8 s} \\ &\quad + K_I \times \text{area from 8 to 10 s} \\ &= \text{output due to error from 0 to 3 s} \\ &\quad + \text{output due to error from 3 to 8 s} \\ &\quad + \text{output due to error from 8 to 10 s} \end{aligned}$$

This result clearly indicates that the output of an integral controller is not dependent upon the instantaneous error but on the past error-time history.

$T = 0$ to 3 s:

$$\begin{aligned} \text{output due to error from 0 to 3 s} &= K_I \times \text{area from 0 to 3 s} \\ &= 1.5\%/\text{s}/\% \times 0\%\,\text{s} \\ &= 0\% \end{aligned}$$

$T = 3$ to 8 s:

$$\begin{aligned} \text{output due to error from 3 to 8 s} &= K_I \times \text{area from 3 to 8 s} \\ &= 1.5\%/\text{s}/\% \times 10\%\,\text{s} \\ &= 15\% \end{aligned}$$

$T = 8$ to 10 s:

$$\begin{aligned} \text{output due to error from 8 to10 s} &= K_I \times \text{area from 8 to 10 s} \\ &= 1.5\%/\text{s}/\% \times 0\%\,\text{s} \\ &= 0\% \end{aligned}$$

$$\begin{aligned} \text{output at 10 s} &= 0\% + 15\% + 0\% \\ &= 15\% \end{aligned}$$

It should be noted that even though error has returned to zero after 8 s, the controller continues to generate 15% of output. This will persist forever in an integral system and can be reduced to zero only if the error changes direction and becomes negative. In this case the positive area will be canceled by the negative area.

EXAMPLE 10.17

An integral controller is fed with the error-input curve in Figure 10.23. Controller output ranges from −10 V (minimum) to +10 V (maximum). Determine the controller output voltage, given that the integral constant $K_I = 0.1$ V s/V of error.

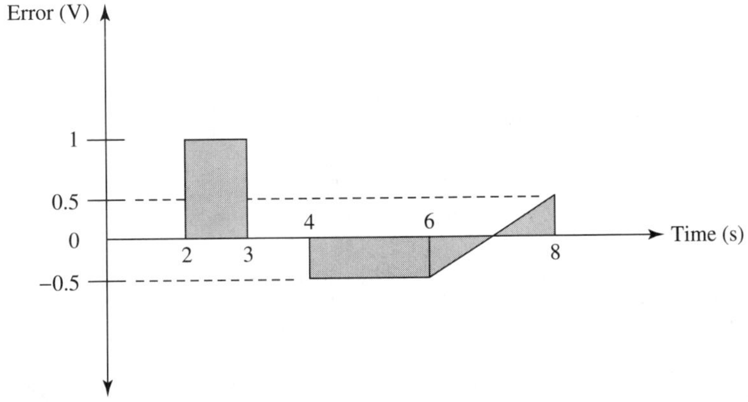

Figure 10.23 Error curve for Example 10.17.

Solution

The problem will be solved graphically. It is left to the reader to solve the problem numerically using integration.

$$\text{controller output at time } t = K_I \times \text{area under the error curve between time 0 and } t$$

Analyzing the error-time curve reveals that the problem can be divided into the six time zones:

$$
\begin{aligned}
v_o\ (\text{at } t = 8\ \text{s}) &= K_I \times \text{area under error curve from 0 to 2 s} \\
&\quad + K_I \times \text{area under error curve error from 2 to 3 s} \\
&\quad + K_I \times \text{area under error curve error from 3 to 4 s} \\
&\quad + K_I \times \text{area under error curve error from 4 to 6 s} \\
&\quad + K_I \times \text{area under error curve error from 6 to 7 s} \\
&\quad + K_I \times \text{area under error curve error from 7 to 8 s}
\end{aligned}
$$

$$
\begin{aligned}
\text{output due to error from 0 to 2 s} &= K_I \times \text{area from 0 to 2 s} \\
&= 0.1\ \text{V/s/V} \times 0\ \text{V s} \\
&= 0\ \text{V} \\
\text{output due to error from 2 to 3 s} &= K_I \times \text{area from 2 to 3 s} \\
&= 0.1\ \text{V/s/V} \times 1\ \text{V s} \\
&= 0.1\ \text{V} \\
\text{output due to error from 3 to 4 s} &= K_I \times \text{area from 3 to 4 s} \\
&= 0.1\ \text{V/s/V} \times 0\ \text{V s} \\
&= 0\ \text{V} \\
\text{output due to error from 4 to 6 s} &= K_I \times \text{area from 4 to 6 s} \\
&= 0.1\ \text{V/s/V} \times (-0.5 \times 2\ \text{V s}) \\
&= -0.1\ \text{V} \\
\text{output due to error from 6 to 7 s} &= K_I \times \text{area from 6 to 7 s} \\
&= 0.1\ \text{V/s/V} \times (-0.5 \times 1\ \text{V s}/2) \\
&= -0.025\ \text{V} \\
\text{output due to error from 7 to 8 s} &= K_I \times \text{area from 7 to 8 s} \\
&= 0.1\ \text{V/s/V} \times (0.5 \times 1\ \text{V s}/2) \\
&= 0.025\ \text{V} \\
\text{total output at 8 s} &= 0 + 0.1 + 0 + (-0.1) + (-0.025) + 0.025\ \text{V} \\
&= 0\ \text{V}
\end{aligned}
$$

The completed output curve is shown in Figure 10.24.

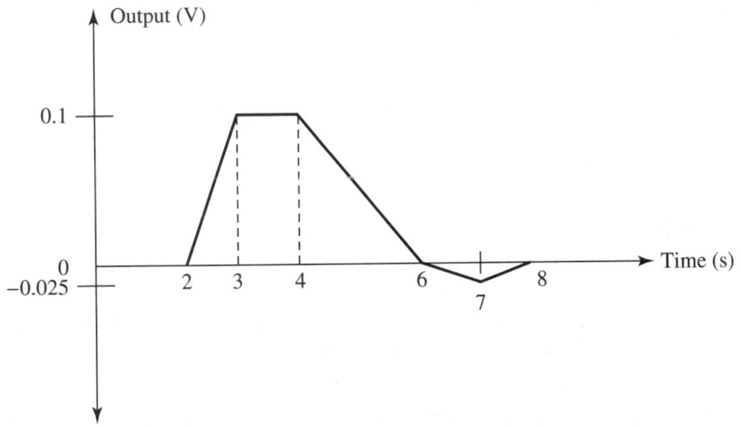

Figure 10.24 Integral controller output curve for Example 10.17.

Electronic Implementation

Figure 10.25 shows the op-amp implementation of only the integral section of a complete controller. Details of the summation block are not provided. Amplifier U1A is set up as an integrator with an output, which is of opposite polarity to that of the input (error). To reverse the inverted signal amplifier U1B is added. It is an inverting amplifier with a unity gain. Amplifiers U1A and U1B form a noninverting integral controller, resulting in a positive output for a positive error input.

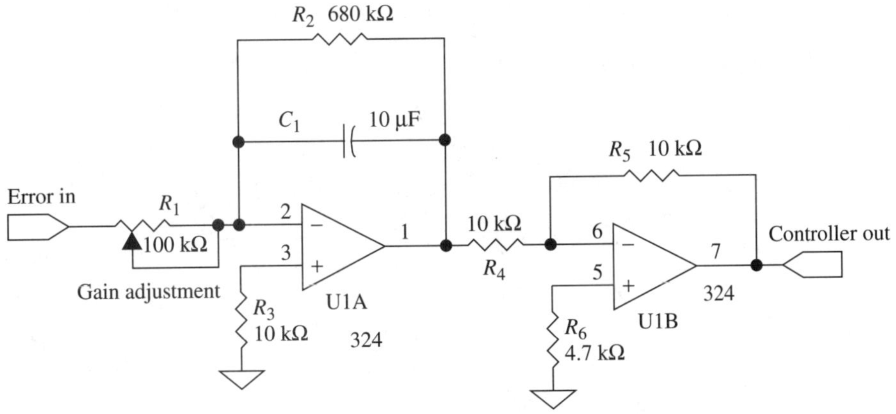

Figure 10.25 Integral controller.

The integrating amplifier U1A, along with capacitor C_1 and resistor R_1, determines the value of the integral gain.

$$\text{integral gain } K_I = \frac{1}{R_1 C_1}$$

Limiting DC Gain

A pure integrator has infinite DC gain. This would be fine for an operational amplifier with zero bias currents and offset voltages. But a real-life operational amplifier does have finite bias currents and offset voltages. These can charge the feedback capacitor C_1 and quickly saturate the op-amp. To counter this, resistor R_2, connected in parallel with the capacitor, is used. This resistor, in conjunction with resistor R_1 (potentiometer), limits the maximum gain of the integrator. A rule of thumb is to use a value for R_2 that is about 10 times greater than R_1. Another approach is to determine the lowest frequency over which integral behavior is desirable and select R_2 accordingly. For best performance, it is advisable to use an FET input op-amp for the integrating amplifier U1A.

$$\text{modified integrator gain} = -\frac{\text{feedback impedance } Z_f}{\text{input impedance } Z_i}$$

The feedback impedance Z_f is equal to the parallel combination of the feedback resistor R_2 and capacitor C_1. A unity gain inverting amplifier follows the integrating amplifier. The overall gain thus is positive, and the negative sign in the gain equation can be dropped. Therefore,

$$\frac{1}{Z_f} = \frac{1}{R_2} + Cs$$

or

$$Z_f = \frac{R_2}{R_2 Cs + 1}$$

Input impedance Z_i is equal to the input resistance R_1, or

$$Z_i = R_1$$

Therefore,

$$\begin{aligned}\text{integrator gain} &= \frac{Z_f}{Z_i} \\ &= \frac{\dfrac{R_2}{R_2 Cs + 1}}{R_1} \\ &= \frac{R_2}{R_1 R_2 Cs + R_1}\end{aligned}$$

Because this expression is derived using s terms, it is also the transfer function of a gain-limited integral controller.

$$\text{transfer function of gain-limited } I \text{ controller} = \frac{R_2}{R_1 R_2 C s + R_1}$$

At DC (zero frequency) the denominator becomes equal to R_1 and the amplifier essentially behaves as a proportional amplifier with zero phase shift. Rewriting the equation,

$$\text{transfer function} = \frac{R_2/R_1}{R_2 C s + 1}$$

The preceding transfer function can be thought of as a low-pass block with a DC gain equal to R_2/R_1 and a time constant of R_2C_1.

Frequency Response of an Ideal Integral Controller

An ideal integral amplifier has infinite DC gain and a constant phase lag of -90°. The gain rolls off at a rate of -20 dB/decade. The integral gain constant K_I defines the 0-dB crossover frequency. This frequency is numerically equal to K_I rad/s. Thus, the gain of the integral controller is dependent upon the signal frequency.

$$\begin{aligned}\text{magnitude of integrator amplifier} &= 20 \log(\text{gain}) \\ &= 20 \log(K_I/\omega) \\ \text{phase difference} &= -90^\circ\end{aligned}$$

Figure 10.26 shows the frequency response of this integral controller ($K_I = 1$).

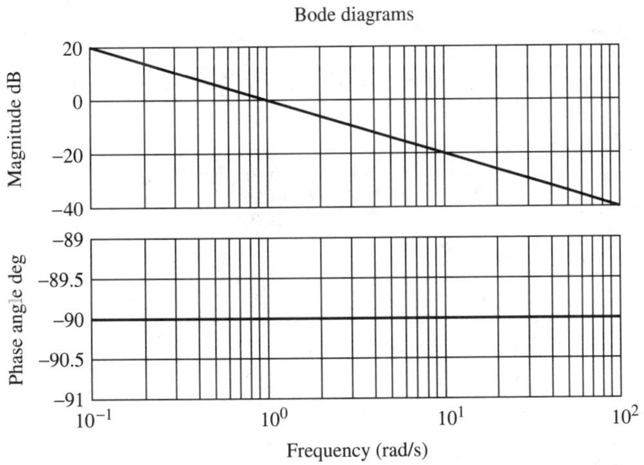

Figure 10.26 Frequency response: integral controller ($K_I = 1$).

EXAMPLE 10.18

Draw the Bode plot of an integral controller with a gain of 0.15.

Solution

Figure 10.27 shows the Bode plot of the given controller. It can be observed from the plot that it has a constant phase difference of -90°, and the crossover with the 0-dB line occurs at a frequency given by K_I (0.15 rad/s).

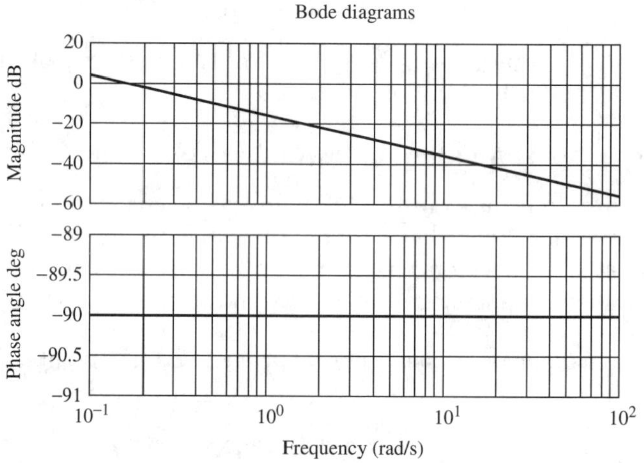

Figure 10.27 Bode plot for Example 10.18.

Frequency Response of a DC Gain-Limited Integral Controller

A DC gain-limited integral amplifier follows the characteristics of an ideal integral controller at frequencies greater than the break frequency.

- At low frequencies, it behaves as a proportional controller.
  - Maximum DC gain is given by the ratio of feedback and input resistors (R_2/R_1).
  - At low frequencies, phase angle approaches zero.
- At higher frequencies, it behaves as a pure integral controller.
  - Gain is dependent upon frequency and is given by (K_I/ω),

  where $\omega =$ frequency in rad/s

$$K_i = \frac{1}{R_1 C_1}$$

- The gain approaches zero as frequency approaches infinity.
- The phase shift remains at -90°.

- The break frequency is determined by the feedback resistor and the capacitor combination.
  - The break frequency is $1/R_2C_1$.
  - The phase angle at break frequency is -45°.
  - The gain at the break frequency is equal to R_2/R_1.

EXAMPLE 10.19

Refer to the schematic diagram in Figure 10.25. The following component values have been set up for a DC gain-limited integral controller.

$$\text{input resistor } R_1 = 10\ \text{k}\Omega$$
$$\text{feedback resistor } R_2 = 1\ \text{M}\Omega$$
$$\text{feedback capacitor } C_1 = 10\ \mu\text{F}$$

Draw the Bode plot.

Solution

$$\text{controller transfer function} = \frac{R_2}{R_1R_2Cs + R_1} = \frac{100}{1{,}000{,}000\,s + 1}$$

The Bode plot of the controller transfer function is shown in Figure 10.28, and the following characteristics can be observed from it:

$$\text{DC gain} = \frac{R_2}{R_1} = 40\ \text{dB}\ (100)$$
$$\text{gain roll-off} = -20\ \text{dB/decade}$$
$$\text{DC phase shift} = 0^\circ$$
$$\text{high-frequency phase shift} = -90^\circ$$
$$\text{break frequency} = \frac{1}{R_2C_1} = 0.1\ \text{rad/s}$$

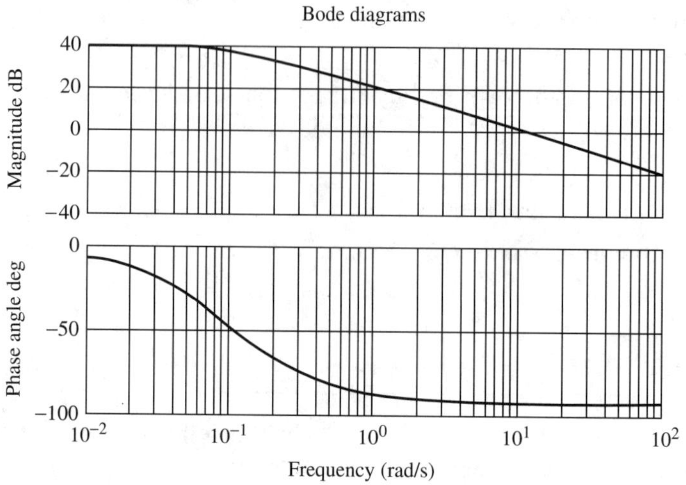

Figure 10.28 Bode plot for Example 10.19.

EXAMPLE 10.20

Refer to the schematic diagram in Figure 10.25. An integral controller is required to have a maximum DC gain of 50, and it is imperative that the phase angle does not drop below −45° at a frequency of 1 rad/s. Determine suitable values for the components, and draw the frequency response.

Solution

The DC gain is controlled by the ratio of feedback and input resistor. Therefore,

$$\frac{R_2}{R_1} = 50 \qquad \text{(a)}$$

The phase angle at break frequency is −45°, and it drops below −45° at lower frequencies. Thus, the break frequency is 1 rad/s. But,

$$\text{break frequency} = \frac{1}{R_2 C_1}$$

Therefore,

$$\frac{1}{R_2 C_1} \geq 1 \quad \text{(rad/s)} \qquad \text{(b)}$$

or,

$$R_2C_1 \leq 1$$

There are three unknowns and only two equations. Start with a common value for capacitor C_1; let it be 1μF. Thus

$$1 \times 10^{-6} R_2 \leq 1$$
$$R_2 \leq 1\ m\Omega$$

Considering DC gain, from expression (a),

$$R_1 = \frac{R_2}{50}$$
$$R_1 = \frac{10^6}{50} = 2\ k\Omega$$

The frequency response plot is shown in Figure 10.29.

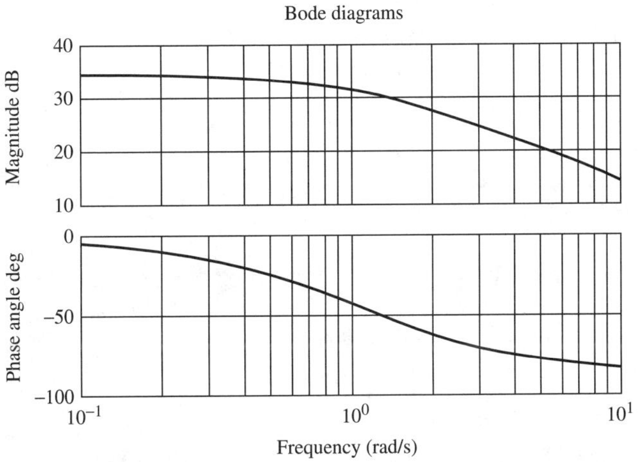

Figure 10.29 Bode plot for Example 10.20.

DERIVATIVE CONTROLLER (D)

A differential controller is also known as a rate, or anticipatory, controller. Its behavior is a marked departure from proportional and integral controllers. The output of a derivative controller is not dependent upon the present or past error, but on the

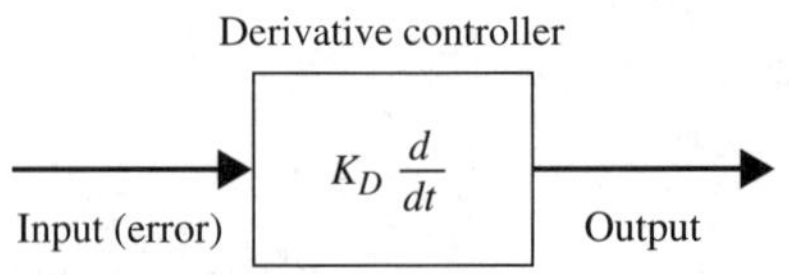

Figure 10.30 Derivative controller.

rate at which the error is changing. Figure 10.30 shows the block diagram of a derivative controller.

Time Equation

$$v_o = K_D \frac{de}{dt}$$

where v_o = controller output (at time t)
de/dt = rate of change of error at time t
K_D = derivative constant (gain or coefficient)

The derivative constant is expressed in units of controller output × seconds × unit of error rate. The following are the valid forms of expression for the derivative constant:

- 5
- 5 s
- 5% s/%/s
- 5% output s/%/s of error

Transfer Function The transfer function of derivative controller is given by

$$\frac{V_o(s)}{E(s)} = K_D s$$

A derivative controller normally cannot be used by itself, as is the case with an integral controller. It is either used with a proportional controller in a proportional + derivative (P + D, or PD) configuration or with a proportional and an integral controller in a full proportional + integral + derivative (P + I + D, or PID) configuration.

EXAMPLE 10.21

A derivative controller has a gain of 10 s. Determine the controller output if the controller is subjected to a constant error of 5. Assume that initial controller output is zero (Figure 10.31).

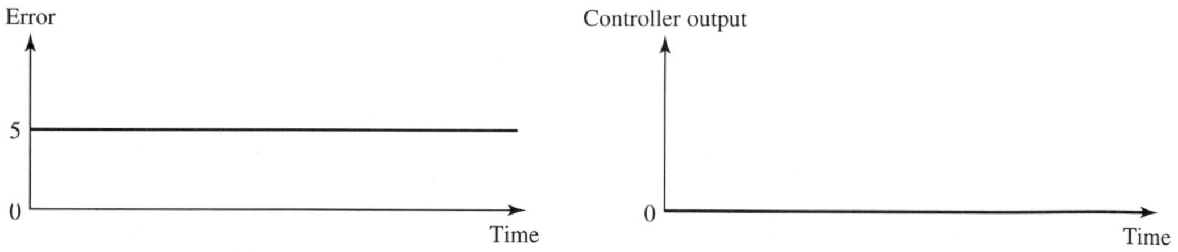

Figure 10.31 Error and output curves for Example 10.20.

Solution

The controller output is

$$v_o = K_D \frac{de}{dt}$$

$$v_o = 10\ \text{s} \frac{de}{dt}$$

Because the error is constant at 5, the rate of change of error is zero.

$$v_o = 10\ \text{s} \times 0/\text{s}$$

$$v_o = 0$$

There is no output from the derivative controller, even though there is a large error present all the time. The derivative controller does not respond to instantaneous error but generates an output only when error is changing.

EXAMPLE 10.22

Determine the controller output when it is subjected to the given error input (Figure 10.32). Assume that initial controller output is zero and controller gain $K_D = 2\%$ s.

Solution

The controller output is

$$v_o = K_D \frac{de}{dt}$$

$$v_o = 2\%\ \text{s} \frac{de}{dt}$$

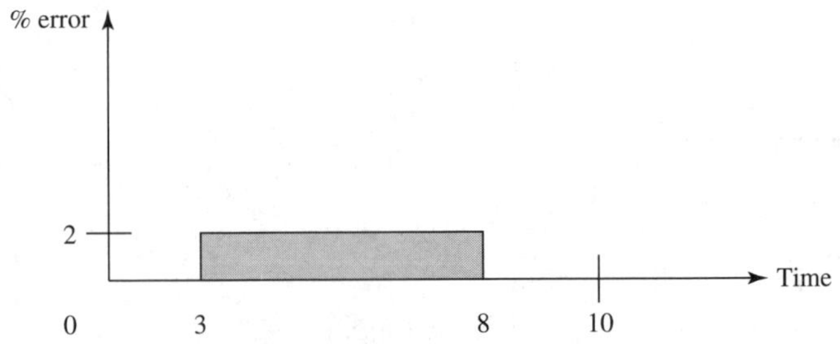

Figure 10.32 Error curve for Example 10.22.

The controller operation can be broken down into three time zones: 0 to 3 s, 3 to 8 s, and 8 s onwards.

***T* = 0 to 3 s:** The controller output is

$$v_o = 2\%\ \mathrm{s}\frac{de}{dt}$$

The error during this duration is zero; thus,

$$\frac{de}{dt} = 0$$

$$v_o = 0\%$$

***T* = 3 to 8 s:** At $t = 3$ s, the error abruptly (instantaneously) changes from 0% to 2%, and the slope of the error with respect to time approaches infinity:

$$\frac{de}{dt} = \frac{2}{0} \to \infty$$

Thus, controller output is

$$v_o = 2\%\ \mathrm{s}\frac{de}{dt}$$

$$v_o \to \infty$$

As the instant time crosses the $t = 3$-s boundary ($t = 3 + \varepsilon$ s),[1] the error becomes constant at 2%. Thus the rate of change of error drops from near infinity to zero. This will force the controller output back to zero.

[1] ε is an infinitesimaly small number.

The net effect of this sudden change in error results in the controller output appearing as a vertical, positive-going spike (fast rise-up and instant fallback).

During the 3- to 8-s interval, error remains constant, i.e., *de/dt* is zero. The controller output from $3+\varepsilon$ s to $8-\ \varepsilon$ s is 0%. At $t=8$ s again the error changes from positive 2% to 0%. Thus results in slope of error curve to approach negative infinity. Immediately, afterwards error becomes a constant 0% and slope of error curve drops to zero. It will result in a negative-going spike.

***T* = 8 s onward:** The error remains constant at 0% after $t=8$ s, and *de/dt* is zero. Controller output is

$$v_o = 0\%$$

The controller output–time curve can now be drawn (Figure 10.33).

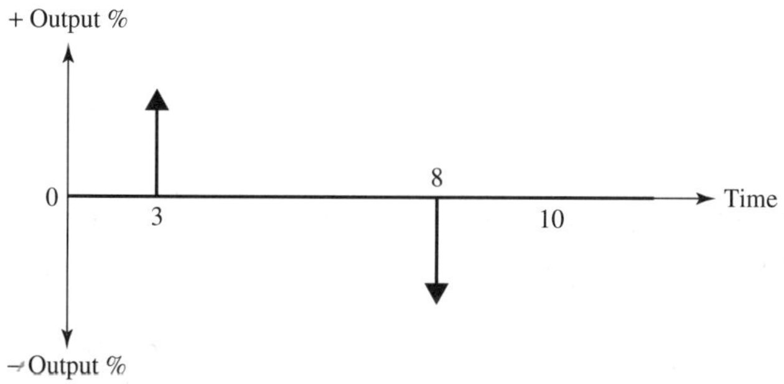

Figure 10.33 Controller output curve for Example 10.22.

EXAMPLE 10.23

Determine the controller output when it is subjected to the given error input Figure 10.34. The controller output at 0 s is zero and the derivative gain K_D is 5% s.

Solution

The controller output is

$$v_o = K_D \frac{de}{dt}$$

$$v_o = 5\%\ \text{s} \frac{de}{dt}$$

We can break the operation down into four time zones: 0 to 2 s, 2 to 4 s, 4 to 6 s, and 6 s onward.

$T = 0$ to 2 s: During this time frame, the error is constant at zero, so the slope of error curve is also equal to zero. Therefore, controller output is

$$v_o = 0\%$$

The error during this duration is zero; thus, $de/dt = 0$.

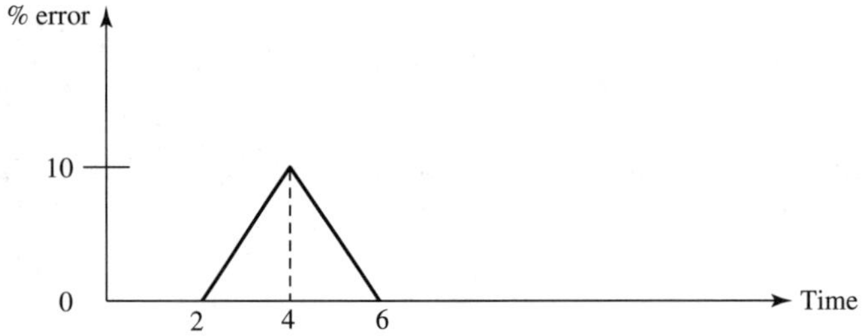

Figure 10.34 Error curve for Example 10.23.

$T = 2$ to 4 s: At $t = 2$ s, the error starts to increase at a constant rate. The change from 0% to 10% takes 2 s to complete.

$$\text{slope of error curve } \frac{de}{dt} = \frac{10\%}{(4-2)\text{ s}} = 5\%/\text{s}$$

$$\text{controller output } v_o = 5\text{ s} \times 5\%/\text{s}$$

$$v_o = 25\%$$

$T = 4$ to 6 s: At $t = 4$ s, the error starts to decrease at a constant rate, dropping to 0% from 10% in 2 s.

$$\text{slope of error curve } \frac{de}{dt} = \frac{-10\%}{(6-4)\text{ s}} = -5\%/\text{s}$$

$$\text{controller output } v_o = 5\text{ s} \times (-5\%/\text{s})$$

$$v_o = -25\%$$

$T = 6$ s onward: The error remains constant at 0% after $t = 6$ s, and de/dt is zero. The controller output is

$$v_o = 0\%$$

The controller output–time curve (Figure 10.35) can now be drawn.

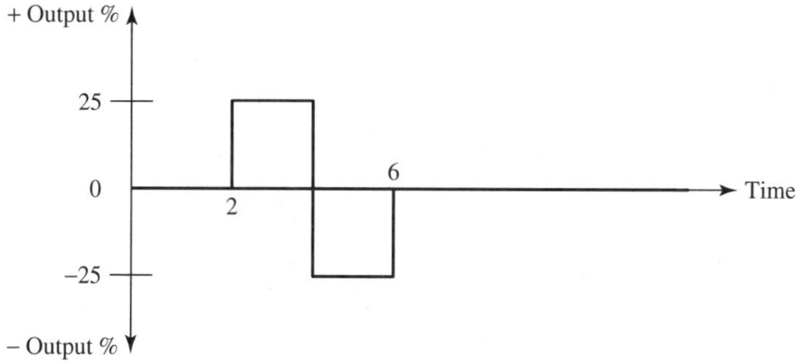

Figure 10.35 Controller output for Example 10.23.

Electronic Implementation

Figure 10.36 shows a schematic diagram for implementing a derivative controller. Only the derivative amplifier section is shown. A summation block (not shown) precedes the derivative amplifier and generates an error signal from set-point and feedback inputs. Amplifier U1A acts as an inverting derivative amplifier. Unity gain inverting amplifier U1B inverts the output of U1A. Amplifier pairs U1A and U1B together form a noninverting derivative amplifier. Capacitor C_1 and resistor R_1 control the value of derivative gain, K_D.

$$\text{derivative gain } K_D = R_1 C_1$$

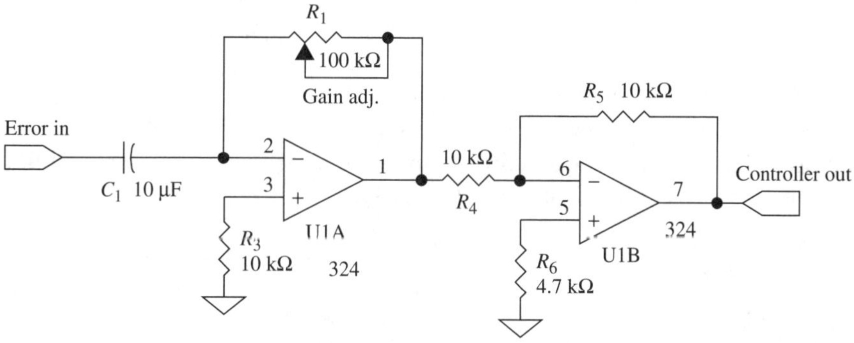

Figure 10.36 Schematic diagram: derivative controller.

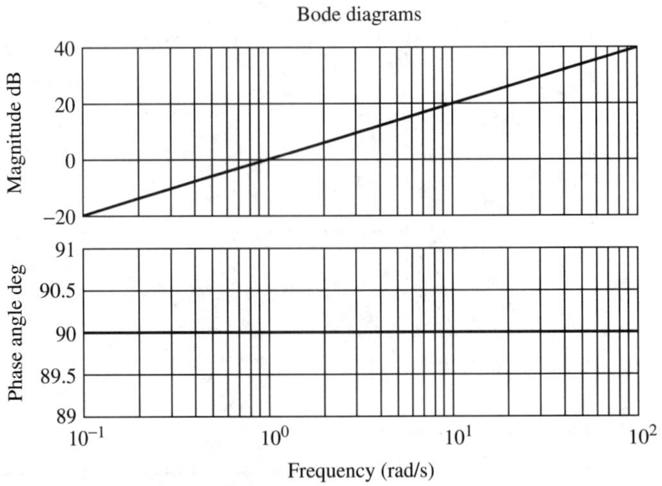

Figure 10.37 Derivative controller: normalized Bode plot.

The gain of the derivative amplifier is frequency-dependent. It is zero for DC signals and becomes infinite as the frequency approaches infinity.

Frequency Response of a Derivative Controller

An ideal derivative controller has zero DC gain and a constant phase shift of +90°. Controller gain increases at a rate of +20 dB/decade. The derivative gain constant K_D defines the 0-dB crossover frequency. This frequency is numerically equal to $1/K_D$ rad/s.

$$\text{controller gain} = K_D\omega$$

$$\text{magnitude} = 20 \log (K_D\omega)$$

$$\text{phase difference} = +90^\circ$$

Figure 10.37 shows the frequency response of an ideal derivative controller with a derivative constant K_D of 1 (normalized Bode plot).

EXAMPLE 10.24

Draw the Bode plot of an ideal derivative controller with a gain of 0.2.

Solution

The crossover frequency ω_c is related to the derivative constant. Therefore,

$$\text{crossover frequency } \omega_c = \frac{1}{K_D}$$

$$\omega_c = \frac{1}{0.2}$$

$$= 5 \text{ rad/s}$$

The magnitude plot is a straight line with a constant slope of +20 dB/decade, and it will intersect the 0-dB line at a frequency of 5 rad/s.

The phase angle of the derivative amplifier (block) is a constant +90°. With this information, the frequency response plot (Figure 10.38) can now be drawn.

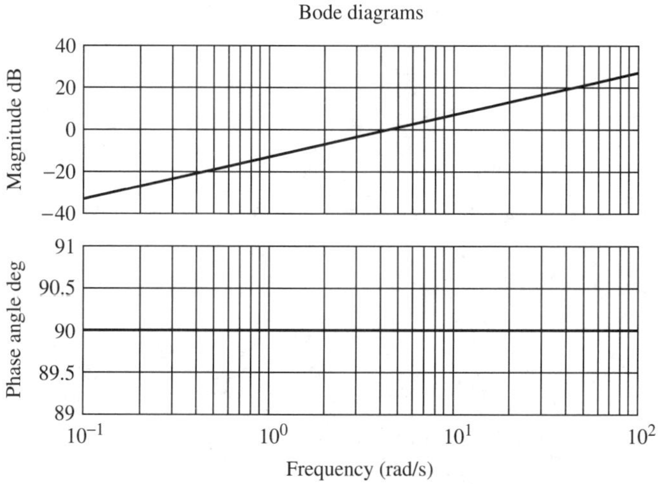

Figure 10.38 Bode plot for Example 10.24.

Limiting High-Frequency Gain

Gain of a pure derivative controller increases with an increase in frequency and becomes infinite as frequency approaches infinity. Obviously, a real operational amplifier has a limited bandwidth, and its gain starts to roll off at very high frequencies. But with the continued development in semiconductor technology, the gain-bandwidth product of operational amplifiers has increased, resulting in an improved, higher-frequency response. But this improved high-frequency response carries a hidden cost.

An electromagnetic noise signal is always present in an industrial environment. The noise can be induced by any one of several conditions: switching of large electromagnetic relays and motors, electrostatic discharge, or simply because of electrical storm activity. The noise signal occupies the high end of the frequency spectrum. Since the gain of a derivative controller increases linearly with frequency, the noise signal tends to be greatly amplified. In other words, a pure derivative controller works as an excellent noise amplifier. To reduce the controller sensitivity to noise, amplifier gain needs to be limited. Without using a complex filtering scheme, it can be achieved by adding a resistor in series with the input capacitor. This *RC* network changes the amplifier characteristics, resulting in a limited high-frequency gain.

A schematic diagram of the modified derivative controller is shown in Figure 10.39. Resistor R_2 in conjunction with resistor R_1 (potentiometer) limit the maximum amplifier gain. As the frequency increases, the impedance of the input capacitor becomes smaller. Because the amplifier gain is controlled by the ratio of feedback and input impedance, with input impedance becoming smaller with an increase in frequency, the amplifier gain continues to increase. The resistor R_2 in series with the capacitor sets a limit on the minimum value of input impedance.

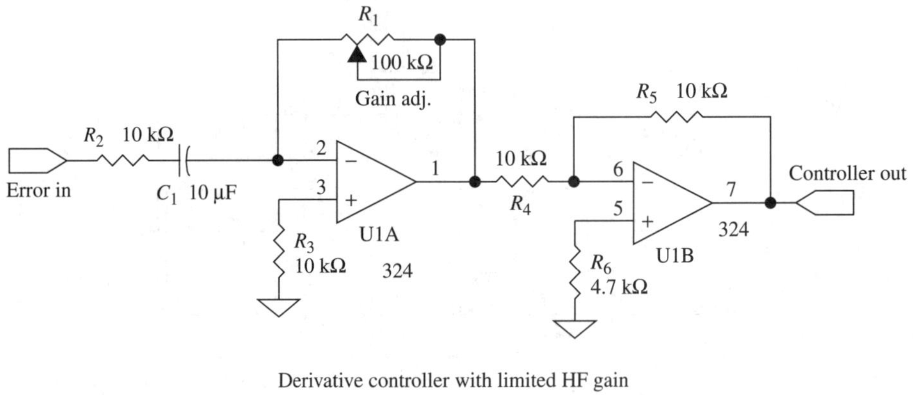

Figure 10.39 Modified derivative controller.

The value of resistor R_2 depends upon the maximum desirable high-frequency gain. One rule of thumb is to use a value for R_2 that is about 1/10 of resistor R_1. In another approach, the highest frequency over which derivative behavior is desirable is determined and R_2 is selected accordingly.

$$\text{modified derivative gain} = -\frac{\text{feedback impedance } Z_f}{\text{input impedance } Z_i}$$

Input impedance Z_f is equal to the series combination of resistor R_2 and capacitor C_1. The derivative section is followed by a unity gain inverting amplifier, resulting in the output signal being of the same polarity as the input signal.

$$Z_i = R_2 + \frac{1}{C_1 s}$$

$$Z_i = \frac{R_2 C_1 s + 1}{C_1 s}$$

Feedback impedance Z_f is equal to the feedback resistance R_1.

$$Z_f = R_1$$

Therefore,

$$\text{derivative gain} = \frac{Z_f}{Z_i}$$

$$= \frac{R_1}{\dfrac{R_2 C_1 s + 1}{C_1 s}}$$

$$= \frac{R_1 C_1 s}{R_2 C_1 s + 1}$$

which is also the transfer function of a high-frequency, gain-limited derivative controller.

$$\text{transfer function of gain-limited derivative controller} = \frac{R_1 C_1 s}{R_2 C_1 s + 1}$$

At high frequencies, the amplifier essentially behaves as a proportional amplifier with zero phase shift. The preceding transfer function can be thought of as a derivative block followed by a first-order log block with a time constant equal to R_2C_1.

A high-frequency gain-limited derivative controller follows the characteristics of an ideal derivative controller, except when operating above the break frequency.

- At low frequencies, it behaves as a pure derivative controller.
  - Gain is dependent upon frequency and is given by $(K_D\,\omega)$

    where
    $$\omega = \text{frequency in rad/s}$$
    $$K_D = R_1 C_1$$

  - Gain approaches zero as frequency approaches zero (DC).
  - As the frequency approaches zero, the phase shift becomes $+90^\circ$.

- At high frequencies, it behaves as a proportional controller.
  - Maximum DC gain is given by the ratio of feedback and input resistors (R_1/R_2).
  - At high frequencies, the phase angle approaches zero.
- The break frequency is determined by a combination of the input resistor and capacitor.
  - Break frequency $= 1/R_2C_1$.
  - The phase angle at break frequency is $+45°$.
  - The gain at break frequency is equal to R_1/R_2.

EXAMPLE 10.25

Refer to the schematic diagram in Figure 10.39. The following component values have been set up for a high-frequency, gain-limited derivative controller.

$$\text{input resistor } R_2 = 10\ \text{k}\Omega$$
$$\text{feedback resistor } R_1 = 100\ \text{k}\Omega$$
$$\text{feedback capacitor } C_1 = 10\ \mu\text{F}$$

Draw the Bode plot.

Solution

$$\text{controller transfer function} = \frac{R_1C_1s}{R_2C_1s+1}$$

Substituting the values, the transfer function expression becomes

$$\text{TF} = \frac{0.1s}{s+1}$$

The system is a cascade combination of two blocks: a first-order lag ($\tau = 1$) and a lead (derivative) block ($\tau = 0.1$). The corresponding crossover frequencies are

First-order lag block:	1 rad/s
Derivative block:	10 rad/s

At low frequencies the magnitude plot will be a rising line with a slope of +20 dB/decade. The high-frequency gain is given by the ratio R_1/R_2, which for the given values is 0.1 (−20 dB).

The composite Bode plot (Figure 10.40) can now be drawn.

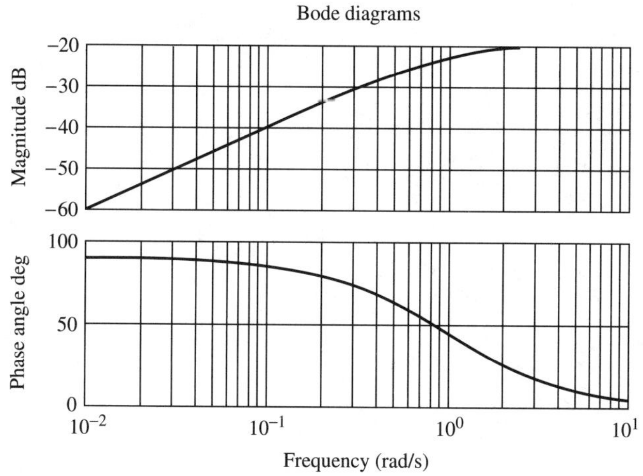

Figure 10.40 Modified derivative controller for Example 10.25.

The following observations can be made from the plot:

- HF gain = 0.1 (R_1/R_2)
- HF magnitude = −20 dB
- gain increase rate = +20 dB/decade
- low-frequency phase shift = +90°
- high-frequency phase shift = 0°
- break frequency = 10 rad/s $(1/R_2C_1)$

EXAMPLE 10.26

The minimum break frequency for a derivative controller used in materials testing is 500 Hz. The maximum allowed high-frequency gain is 10. Determine the component values and draw the frequency response.

Solution

The break frequency is controlled by the combination of the input capacitor and resistor.

$$\text{break frequency} = \frac{1}{R_2C_1}$$

Therefore, break frequency is

$$\frac{1}{R_2C_1} \geq 500 \quad \text{(rad/s)} \qquad \text{(a)}$$

The maximum value of the gain, which occurs at high frequency, is controlled by the ratio of feedback to the input resistor. Therefore,

$$\frac{R_1}{R_2} = 10 \qquad \text{(b)}$$

Because there are three unknowns and only two equations, select the capacitor C_1 to be 2.2 μF. Then

$$\frac{1}{R_2 2.2 \times 10^{-6}} \geq 500 \quad \text{(rad/s)}$$

or

$$R_2 \leq 909\ \Omega$$

A common value of 910 Ω can be used for R_2.
From the maximum HF gain (Equation(a)),

$$\begin{aligned} R_1 &= 10\ R_2 \\ &= 9100\ \Omega \end{aligned}$$

Thus, the component values are

$$\begin{aligned} \text{input resistor } R_2 &= 910\ \Omega \\ \text{feedback resistor } R_1 &= 9100\ \Omega \\ \text{input capacitor } C_1 &= 2.2\ \mu\text{F} \\ \text{transfer function } &= \frac{R_1C_1s}{R_2C_1s+1} \\ &= \frac{0.02002s}{0.002002s+1} \end{aligned}$$

The frequency response plot can now be drawn (Figure 10.41).

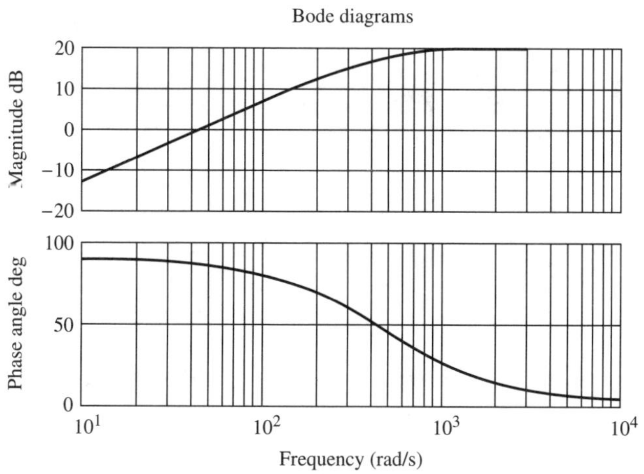

Figure 10.41 Bode plot for Example 10.26.

EXAMPLE 10.27

A derivative controller is fed by a sinusoidal signal with a peak amplitude of 5 mV and a frequency of 10 Hz. Determine the magnitude (dB) of the output signal, if controller gain $K_D = 0.25$.

Solution

Fundamental approach: The output of a derivative controller is the derivative of the input signal.

$$\text{Input signal } v_{\text{in}} = A \sin 2\pi f t$$

Here

$$A = \text{peak amplitude} = 5 \text{ mV} = 0.005 \text{ V}$$
$$f = \text{frequency} = 10 \text{ Hz}$$

Substituting the value $v_{\text{in}} = 0.005 \sin 62.832t$ gives

$$\begin{aligned}
\text{controller output } v_o &= K_D \frac{dv_{\text{in}}}{dt} \\
&= 0.25 \frac{d(0.005 \sin 62.832t)}{dt} \\
&= 0.07854 \cos 62.832t
\end{aligned}$$

The output signal leads the input signal by 90° and has a peak amplitude of 0.07854 V.

$$\begin{aligned} \text{signal amplification (gain)} &= \frac{\text{output}}{\text{input}} \\ &= \frac{0.07854 \text{ V}}{0.005 \text{ V}} \\ &= 15.71 \\ &= 23.92 \text{ dB} \end{aligned}$$

Using a Bode plot: The same information can be derived from a Bode plot of this controller.

$$\begin{aligned} \text{controller transfer function} &= K_D s \\ &= 0.25s \end{aligned}$$

The Bode plot for this transfer function is shown in Figure 10.42.

$$\begin{aligned} \text{frequency in rad/s } \omega &= 2\pi f \\ &= 2\pi \times 10 \\ &= 62.832 \text{ rad/s} \end{aligned}$$

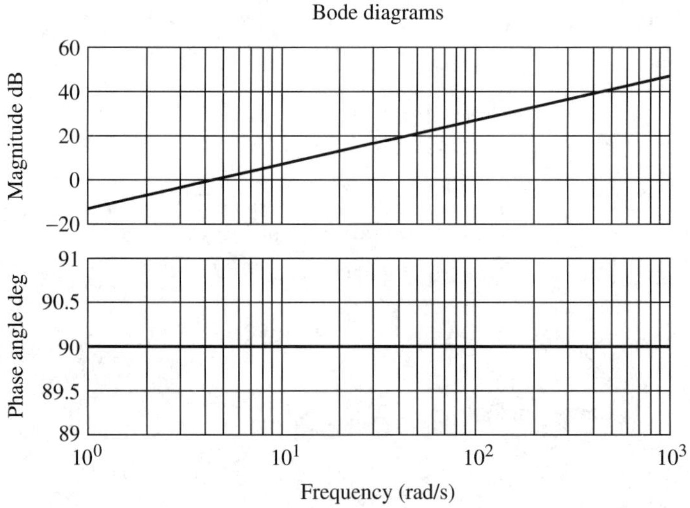

Figure 10.42 Bode plot for Example 10.27.

It can be observed from the magnitude plot that frequency at 62.832 rad/s is about 25 dB and the phase shift is $+90°$. It also illustrates a problem in extracting the numerical data from the plot. It is generally difficult to pinpoint a frequency on a logarithmic scale, and precise determination of magnitude may be difficult.

Using mathematical calculations: An exact value can be calculated from the transfer function. The s term can be replaced with $j\omega$, where ω represents the frequency under consideration.

$$\begin{aligned} \text{TF} &= 0.25s \\ &= 0.25\,j\omega \end{aligned}$$

But,

$$\begin{aligned} \omega &= 2\pi f \\ &= 63.832 \text{ rad/s} \end{aligned}$$

Substituting the value of ω in the transfer function expression gives

$$\begin{aligned} \text{TF} &= 0.25\,j \times 63.832 \\ &= 15.708\,j \end{aligned}$$

This represents a phasor with a magnitude (gain) of 15.708 and an angle of $+90°$. Thus

$$\begin{aligned} \text{output signal peak amplitude} &= 15.70796 \times 0.005 \text{ V} \\ &= 0.07854 \text{ V} \end{aligned}$$

PROBLEMS

1. A P-mode controller has $K_P = 1$. Determine the following:

a. Magnitude (dB) at a frequency of 1 rad/s
b. Gain at a frequency of 1 rad/s
c. Phase difference between input and output at a frequency of 1 rad/s.
d. Magnitude (dB) at a frequency of 10 rad/s
e. Gain at a frequency of 10 rad/s
f. Phase difference between input and output at a frequency of 10 rad/s.

Hint: Sketch the frequency response (Bode plot) of a proportional controller ($K_P = 1$) (shown below).

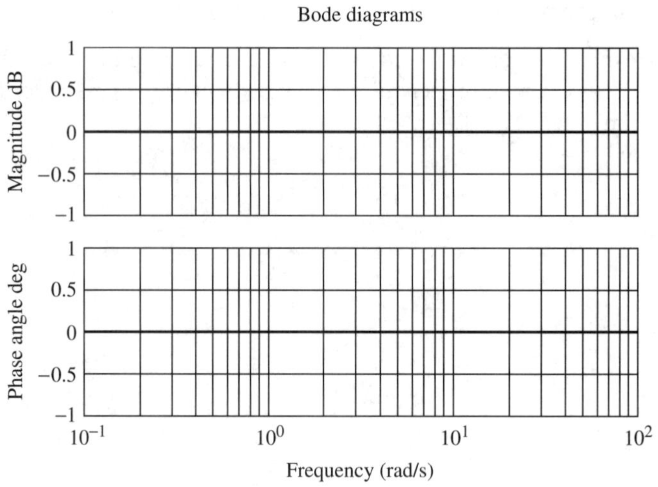

2. An I-mode controller has the coefficient $K_I = 1$. Determine the following:
   a. Magnitude (dB) at a frequency of 1 rad/s
   b. Gain at a frequency of 1 rad/s
   c. Phase difference between input and output at a frequency of 1 rad/s.
   d. Magnitude (dB) at a frequency of 10 rad/s
   e. Gain at a frequency of 10 rad/s
   f. Phase difference between input and output at a frequency of 10 rad/s.
3. A D-mode controller has the coefficient $K_D = 1$. Determine the following:
   a. Magnitude (dB) at a frequency of 1 rad/s
   b. Gain at a frequency of 1 rad/s
   c. Phase difference between input and output at a frequency of 1 rad/s.
   d. Magnitude (dB) at a frequency of 10 rad/s
   e. Gain at a frequency of 10 rad/s
   f. Phase difference between input and output at a frequency of 10 rad/s.
4. Repeat Problems 1–3 with $K_P = 0.5$, $K_I = 0.1$, and $K_D = 10$.
5. An integral controller has $K_I = 0.25$. Determine the frequency (Hz) at which the gain equals 0.1.
6. A derivative controller has $K_D = 25$. Determine the frequency (Hz) at which the gain equals 10.
7. Write the time equation and transfer function for each of the following controller configurations.
   a. Proportional controller with $K_P = 16$
   b. Integral controller with $K_I = 0.08$
   c. Derivative controller with $K_D = 2.5$

8. An analog controller is subjected to the following error input. Calculate the controller output and draw a diagram showing the controller output with respect to time for each of the following cases:

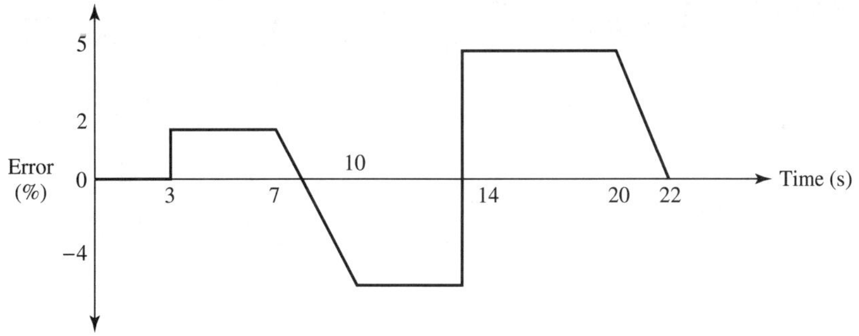

a. Proportional controller, $K_P = 2$
b. Proportional controller, $K_P = 50$
c. Integral controller, $K_I = 1$
d. Derivative controller, $K_D = 4$

9. Draw a schematic diagram for a proportional controller complete with offset adjustment. Select the component values to allow for a proportional gain ranging from 0 to 15.

10. A proportional controller is controlling a first-order lag process. Explain what effect controller gain has on the process step response.

11. Explain why a derivative controller cannot be used alone (without any proportional or integral mode) to control a process.

12. Explain the term *integral windup*. What are the consequences of integral windup? Is it possible to avoid it or at least minimize its effect?

13. What is the procedure used to reduce the sensitivity of a derivative controller to signal noise?

14. What effect does the proportional gain have on the stability of a control loop?

11

ANALOG CONTROLLER II

INTRODUCTION

This chapter continues with the coverage of analog controllers. In Chapter 10, only the single modes of analog controllers—proportional, integral, and derivative—were covered. Here the combinations of the various controller modes, including the time and frequency responses, are discussed. The chapter includes complete schematic diagrams for implementing these controllers.

COMPOSITE CONTROLLER MODES

Of the three controllers (P, I, and D), the proportional (P) controller is the only one that can be used by itself. The other two controllers, integral and derivative, are not used individually. Each of these two controllers is always used in combination with at least the proportional mode. The following composite controller modes are commonly used:

- PI controller
- PD controller
- PID controller

The composite controllers can be implemented in parallel or serial configuration, depending on how the error signal is transferred to each controller mode.

Serial Implementation

In serial implementation, the proportional controller always precedes the integral and/or derivative controller. Figure 11.1 shows a serial implementation of a full three-mode controller. The error signal is first fed to a proportional controller, which then feeds it to integral and derivative controllers. In this configuration, the proportional gain directly affects the overall gain. This configuration was very popular earlier, because it could be constructed with a minimum amount of hardware. In recent years parallel implementation has gained more popularity.

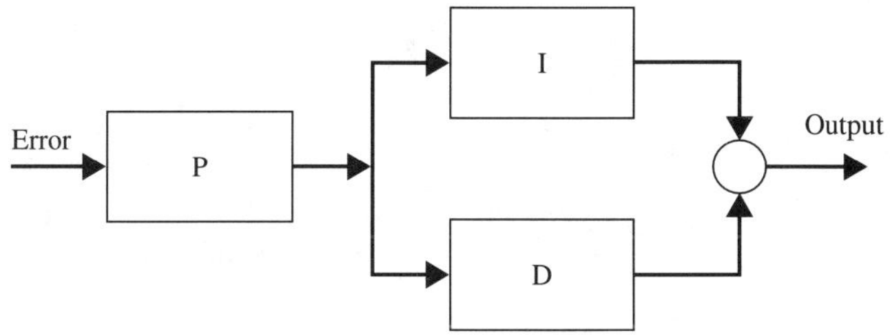

Figure 11.1 Composite controller: serial implementation.

Parallel Implementation

As the name implies, parallel implementation requires the member controllers to be placed in a parallel configuration. Figure 11.2 shows a composite controller working either as a proportional and integral (P + I) or proportional and derivative (P + D) controller. In each case the proportional controller is placed in parallel to the other (integral or derivative) controller.

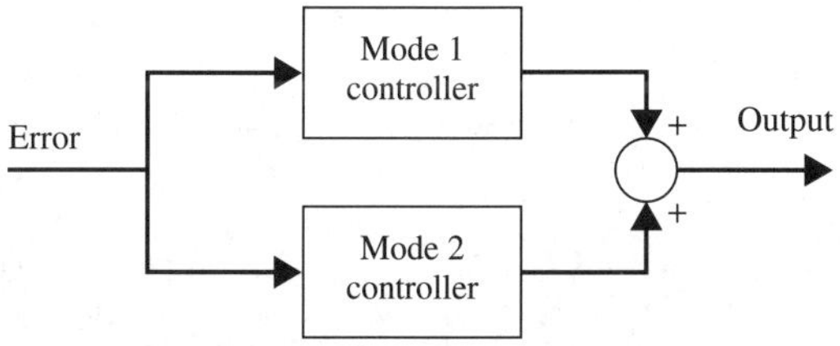

Figure 11.2 Composite controller: parallel implementation.

Each controller is fed with the same error signal, and the output from each controller is algebraically added to form a composite output. It is fairly easy to implement and analyze. Each controller mode can be treated individually, and controller gain can be adjusted (tuned) without affecting other controller(s). *Note:* In this text, only the parallel implementation is discussed. It is the most popular configuration. Nearly all microcontrollers used as digital controllers rely on parallel implementation. In any case, the operating principles remain the same, and the individual effects of gain constants K_P, K_I, and K_D on control-loop performance also remain the same.

PI CONTROLLER

The PI mode controller (Figure 11.3) is commonly used in slow-to-moderate-speed processes. It consists of a proportional controller in parallel with an integral controller.

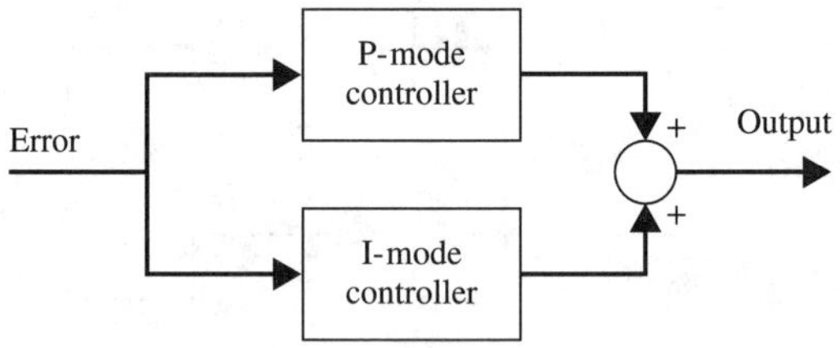

Figure 11.3 Proportional + integral (PI) controller.

Time Equation

$$\text{controller output} = \text{proportional controller output} + \text{integral controller output}$$

$$v_o = K_P e + K_I \int e\,dt$$

Transfer Function Taking the Laplace transform of both sides of the time equation gives

$$\text{output } V(s) = K_P E(s) + \frac{K_I E(s)}{s}$$

$$V(s) = \frac{K_P s + K_I}{s} E(s)$$

$$\text{TF} = \frac{\text{controller output}}{\text{error input}}$$

$$\text{TF} = \frac{V_o(s)}{E(s)} = \frac{K_P s + K_I}{s}$$

$$\boxed{\text{TF} = \frac{K_P s + K_I}{s}}$$

This form of the transfer function is convenient to remember as well as suitable for entry into a program such as MATLAB. It can also be represented as a combination of two blocks, each represented in pure form. This form provides more insight into the frequency response of the PI controller.

Rearranging the terms,

$$\frac{V(s)}{E(s)} = \frac{K_I\left(\frac{K_P}{K_I}s + 1\right)}{s}$$

$$\frac{V(s)}{E(s)} = \frac{\frac{K_P}{K_I}s + 1}{\frac{1}{K_I}s} = \frac{\tau_1 s + 1}{\tau_2 s}$$

$$\boxed{\mathrm{TF} = \frac{\tau_1 s + 1}{\tau_2 s}}$$

From this form of the transfer function, it can clearly be seen that a PI controller consists of an integral block in cascade with a first-order lead block.

EXAMPLE 11.1

Determine the output of a PI controller when subjected to the error input in Figure 11.4. The initial controller output is zero, and it is given that $K_P = 10$ and $K_I = 1.5$.

Solution

$$\text{controller output } v_o = \text{output due to proportional controller} + \text{output due to integral controller}$$

$$v_o = K_P e + K_I \int e\,dt$$

Proportional controller: The proportional controller will generate an output only when there is nonzero error.

$$\begin{aligned}\text{controller output } v_P &= K_P e \\ &= 10 \times 2 \\ &= 20\end{aligned}$$

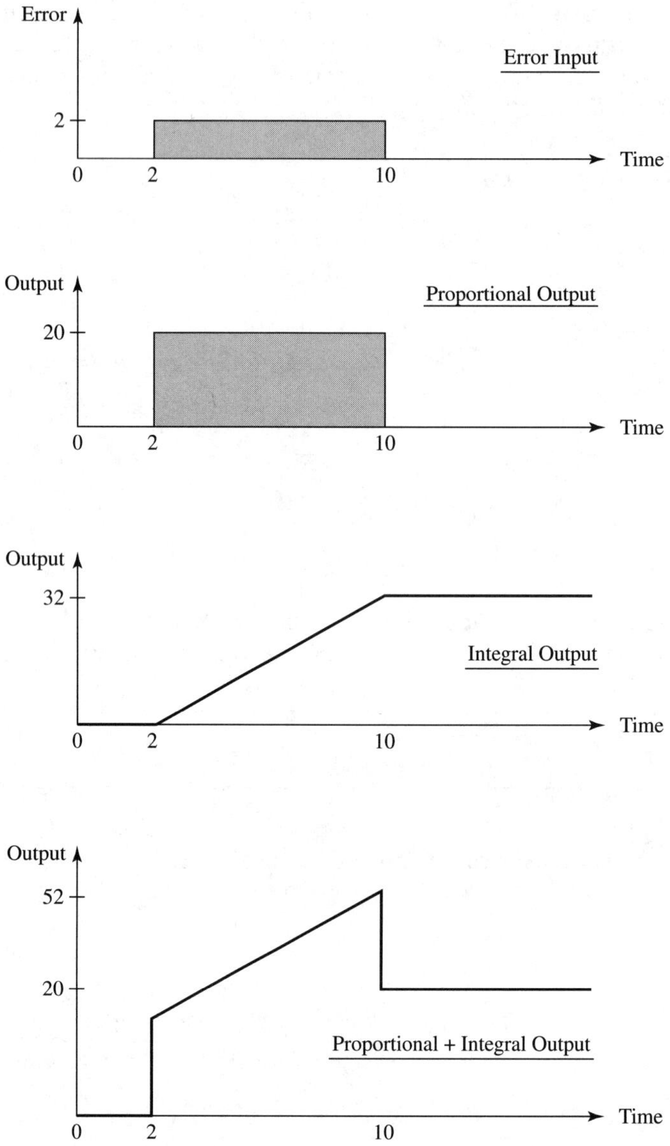

Figure 11.4 Diagrams for Example 11.1.

Integral controller: The output of an integral controller is based upon the area under the error-time curve. Controller output is

$$v_I = K_I \int e\,dt$$

At time $t = 2$ s:

$$\text{controller output} = 0$$

At time $t = 10$ s: controller output is

$$
\begin{aligned}
v_I &= 2\int_0^{10} e\,dt \\
v_I &= 2\int_0^{2} e\,dt + 2\int_2^{10} e\,dt \\
v_I &= 2\int_0^{2} 0\,dt + 2\int_2^{10} 2\,dt \\
&= 0 + 32 \\
&= 32
\end{aligned}
$$

Therefore, total controller output at $t = 10$ s is

$$
\begin{aligned}
v_P + v_I &= 20 + 32 \\
&= 52
\end{aligned}
$$

Electronic Implementation of a PI Controller

Figure 11.5 shows a schematic diagram for implementing a PI controller. Details of the summation block are not provided. The error signal is coupled to the input of U1A and U1C. Amplifier U1A serves as a proportional controller, and U1C acts as an integrator. The summing amplifier U1B takes the output of both the units and produces an output that is an algebraic sum of the two.

Proportional controller gain is set through the potentiometer R_2, and integral gain is set through potentiometer R_3.

$$\text{proportional gain } K_P = \frac{R_2}{R_1}$$

$$\text{integral gain } K_I = \frac{1}{R_3 C_1}$$

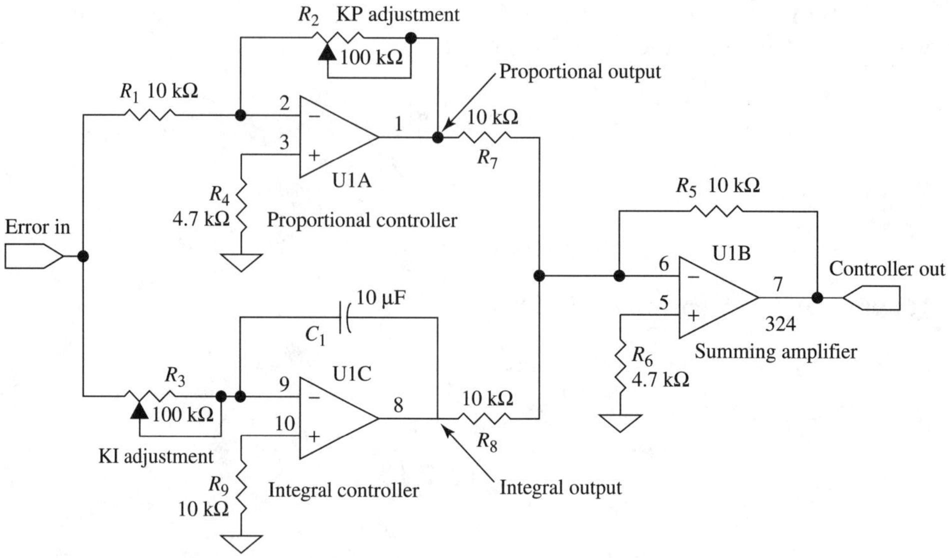

Figure 11.5 PI controller: schematic diagram.

Frequency Response of a PI Controller

A PI controller carries the characteristics of both integral and proportional elements. At low frequencies, the controller essentially behaves as an integral controller, whereas at higher frequencies, its behavior mimics a proportional controller. Its characteristics are as follows:

1. Break frequency is given by $\omega_b = 1/\tau_1\,(K_I/K_P)$ rad/s.
2. Low-frequency ($<K_I/K_P$) gain has a slope of -20 dB/decade.
3. Steady-state gain after the break frequency is K_P ($20 \log K_P$ dB).
4. The phase angle at break frequency is -45°.
5. The lower-frequency phase angle approaches -90°.
6. The higher-frequency phase angle approaches 0°.

EXAMPLE 11.2

Draw the Bode plot for a PI controller, and verify its response characteristics. The controller coefficients are $K_P = 0.1$ and $K_I = 0.01$.

Solution

The PI controller transfer function is

$$\mathrm{TF} = \frac{\frac{K_P}{K_I}s + 1}{\frac{1}{K_I}s}$$

Substituting the values of gain coefficients,

$$\frac{V(s)}{E(s)} = \frac{\frac{0.1}{0.01}s + 1}{\frac{1}{0.01}s}$$

$$\frac{V(s)}{E(s)} = \frac{(10s + 1)}{100\,s}$$

The lead block has a time constant of 10 s, which results in a crossover frequency of 0.1 rad/s. The integral block has a time constant of 100 s, resulting in a crossover frequency of 0.01 rad/s. The composite frequency response of this PI controller can now be drawn. Figure 11.6 shows an exact plot.

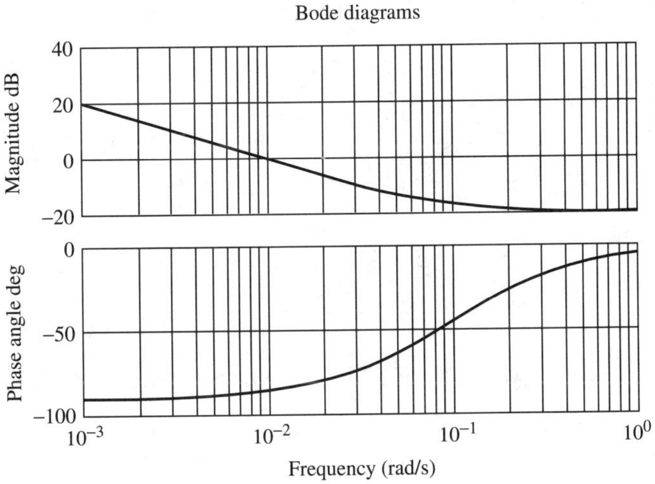

Figure 11.6 Composite Bode plot for Example 11.2.

The following response characteristics can be readily verified from the plot:

$$\text{overall break frequency } \omega_b = \frac{1}{\tau_1} = \frac{K_I}{K_P}$$

$$= \frac{0.01}{0.1} \text{rad/s}$$

$$\text{low-frequency gain droop rate} = -20\,\text{dB/decade}$$

$$\text{steady-state gain after break frequency} = K_P$$

$$= 0.1\ (-20\,\text{dB})$$

$$\text{phase angle at break frequency} = -45^\circ$$

The lower-frequency phase angle approaches -90°.
The higher-frequency phase angle approaches 0°.

EXAMPLE 11.3

Draw the Bode plot of a PI controller with the following coefficients: $K_P = 1$ and $K_I = 0.01$.

Solution

Substituting the values of the gain coefficients,

$$\frac{V(s)}{E(s)} = \frac{(100s + 1)}{100s}$$

The following response characteristics can be predicted:

$$\text{overall break frequency } \omega_b = \frac{K_I}{K_P}$$

$$= 0.01 \text{ rad/s}$$

$$\text{low-frequency gain droop rate} = -20\,\text{dB/decade}$$

$$\text{steady-state gain after break frequency} = K_P$$

$$= 1\ (0\,\text{dB})$$

$$\text{phase angle at break frequency} = -45^\circ$$

The lower-frequency phase angle approaches $-90°$.
The higher-frequency phase angle approaches $0°$.
These values can be verified from the (exact) Bode plot (Figure 11.7).

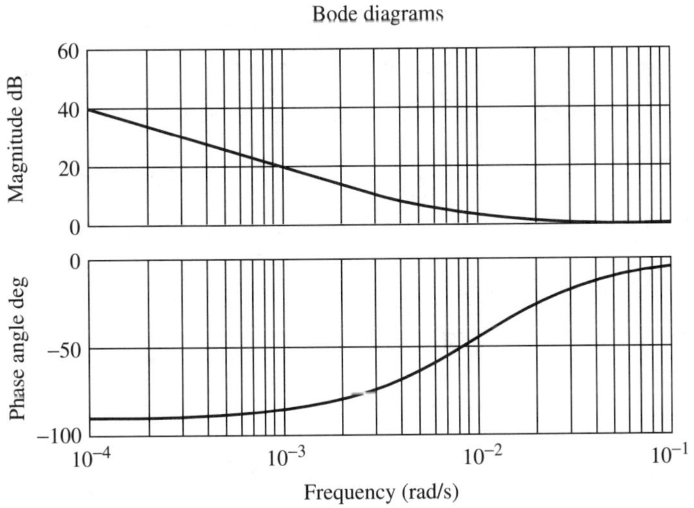

Figure 11.7 Bode plot for Example 11.3.

EXAMPLE 11.4

A sinusoidal signal ($0.1 \sin 100t$) is fed to a PI controller. Determine the amplitude and the phase angle of the output signal, given that $K_P = 10$ and $K_I = 0.01$.

Solution

The signal gain and phase difference between input and output can be determined from the Bode plot of the controller. Substituting the values of the gain coefficients,

$$\frac{V(s)}{E(s)} = \frac{\frac{10}{0.01}s + 1}{\frac{1}{0.01}s}$$

$$\frac{V(s)}{E(s)} = \frac{(1000\,s + 1)}{100\,s}$$

The following response characteristics can be calculated based upon the values of the proportional and integral gain coefficients:

$$\text{overall break frequency } \omega_b = \frac{K_I}{K_P}$$
$$= 0.001 \text{ rad/s}$$
$$\text{low-frequency gain droop rate} = -20 \text{ dB/decade}$$
$$\text{steady-state gain after break frequency} = K_P$$
$$= 10 \text{ (20 dB)}$$

The exact Bode plot of this PI controller is shown in Figure 11.8. From the given data:

$$\text{input signal amplitude} = 0.1 \text{ V}$$
$$\text{input signal frequency} = 100 \text{ rad/s}$$

From the plot:

$$\text{gain at 100 rad/s} = 20 \text{ dB}$$
$$= 10$$
$$\text{phase shift} = 0^\circ \quad \text{(approximately)}$$
$$\text{output signal amplitude} = \text{input amplitude} \times \text{gain}$$
$$= 0.1 \text{ V} \times 10$$
$$= 1 \text{ V}$$

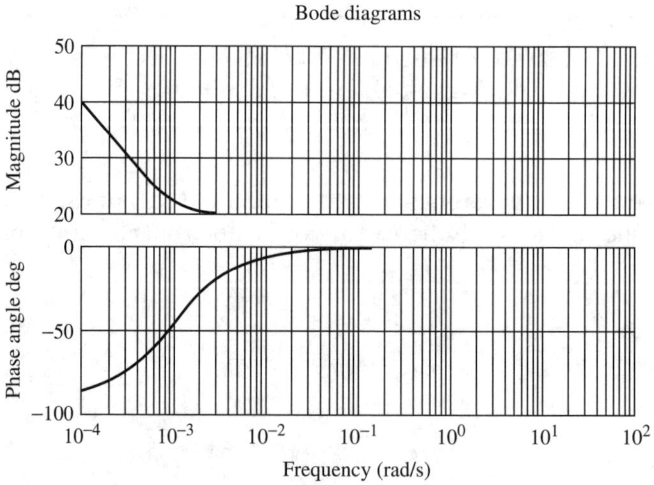

Figure 11.8 Bode plot ($K_P = 10$, $K_I = 0.01$) for Example 11.4.

The output signal has an amplitude of 1 V and is in phase with the input signal. The frequency of 100 rad/s is far greater than the break frequency, and thus at this frequency, the PI controller is simply behaving as a proportional controller.

EXAMPLE 11.5

The Bode plot of a PI controller used in a control system is given in Figure 11.9. It is desired to reduce the high-frequency gain in half while not affecting the break frequency. What steps should be taken?

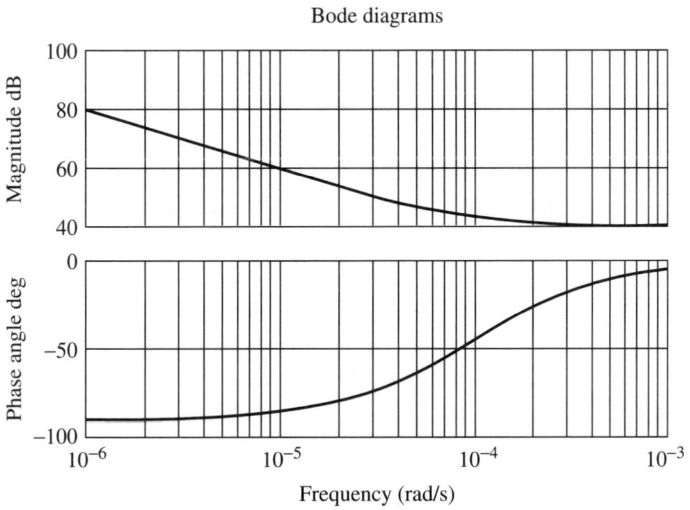

Figure 11.9 Bode plot of the original system of Example 11.5.

Solution

Based upon the standard characteristics of PI controllers, the following data can be extracted from the plot.

$$\begin{aligned}\text{proportional gain constant } K_P &= \text{high-frequency gain}\\ &= 40\text{ dB}\\ &= 100\\ \text{break frequency } \frac{K_I}{K_P} &= 0.0001\text{ rad/s}\end{aligned}$$

From these values, the integral gain constant can be calculated.

$$\begin{aligned} K_I &= 0.0001 \times K_P \\ &= 0.0001 \times 100 \\ &= 0.01 \end{aligned}$$

In order to reduce the high-frequency gain, the value of the proportional gain constant needs to be reduced. Therefore,

$$\begin{aligned} \text{new value of } K_P &= \frac{\text{old value}}{10} \\ &= \frac{100}{10} \\ &= 10 \end{aligned}$$

A change in the proportional constant to 10 will result in a change in break frequency from the current value of 0.0001 rad/s. To counteract this, the integral gain constant needs to be reduced proportionally.

Since the break frequency is equal to K_I/K_P,

$$\frac{K_I}{K_P} = 0.0001 \text{ rad/s}$$

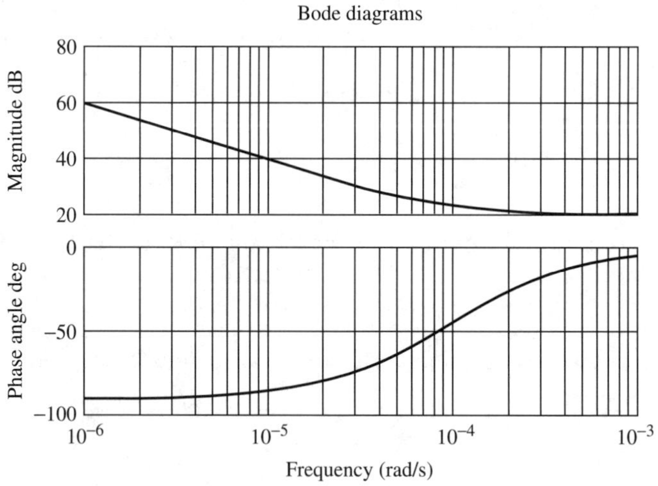

Figure 11.10 Bode plot: modified system for Example 11.5.

or

$$K_I = 0.0001 \times 10$$
$$= 0.001$$

By setting the proportional and integral coefficients to these values, the desired system response can be obtained, as shown in the Bode plot of the modified system (Figure 11.10).

PD CONTROLLER

Similar to the PI mode, a combination of proportional and derivative mode is also used in industrial process control. This controller consists of a proportional controller in parallel with a derivative controller (Figure 11.11).

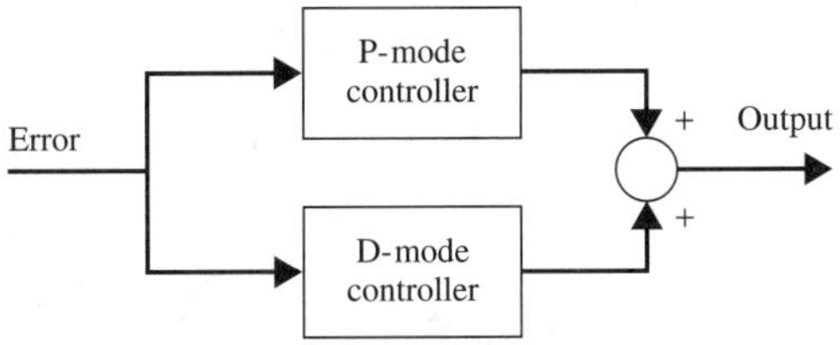

Figure 11.11 PD controller.

Time Equation

$$\text{controller output} = \text{proportional controller output} + \text{derivative controller output}$$

$$v_o = K_P e + K_D \frac{de}{dt}$$

Transfer Function Taking the Laplace transform of both sides of the time equation gives

$$\text{output } V(s) = K_P E(s) + sK_D E(s)$$
$$V(s) = K_P E(s) + sK_D E(s)$$
$$\text{TF} = \frac{V(s)}{E(s)} = sK_D + K_P$$

$$\boxed{\text{TF} = (K_D s + K_P)}$$

TF can again be expressed as a product of two pure blocks. Rearranging the terms,

$$\text{TF} = K_P\left(\frac{K_D}{K_P}s + 1\right) = K_P(\tau s + 1)$$

$$\boxed{\text{TF} = K_P(\tau s + 1)}$$

This equation indicates that a PD controller consists of a proportional block followed by a first-order lead block.

EXAMPLE 11.6

Determine the output of a PD controller when subjected to the error input in Figure 11.12. The initial controller output is zero, and it is given that $K_P = 10$ and $K_D = 1$.

Solution

The controller output is

$$\begin{aligned} v_o &= \text{output due to proportional controller} \\ &\quad + \text{output due to derivative controller} \\ v_o &= K_P e + K_D \frac{de}{dt} \end{aligned}$$

Proportional controller: The proportional controller will generate an output that is proportional to

$$\begin{aligned} v_P &= K_P e \\ &= 10 \times 2 \\ &= 20 \end{aligned}$$

Derivative controller: The output of the derivative controller is based upon the slope of error-time curve:

$$v_D = K_D \frac{de}{dt}$$

At time $t = 2$ s: The error changes instantaneously with the slope approaching infinity. This will cause the controller output to approach infinity. But immediately afterward the error becomes constant with zero slope, and, controller output will drop

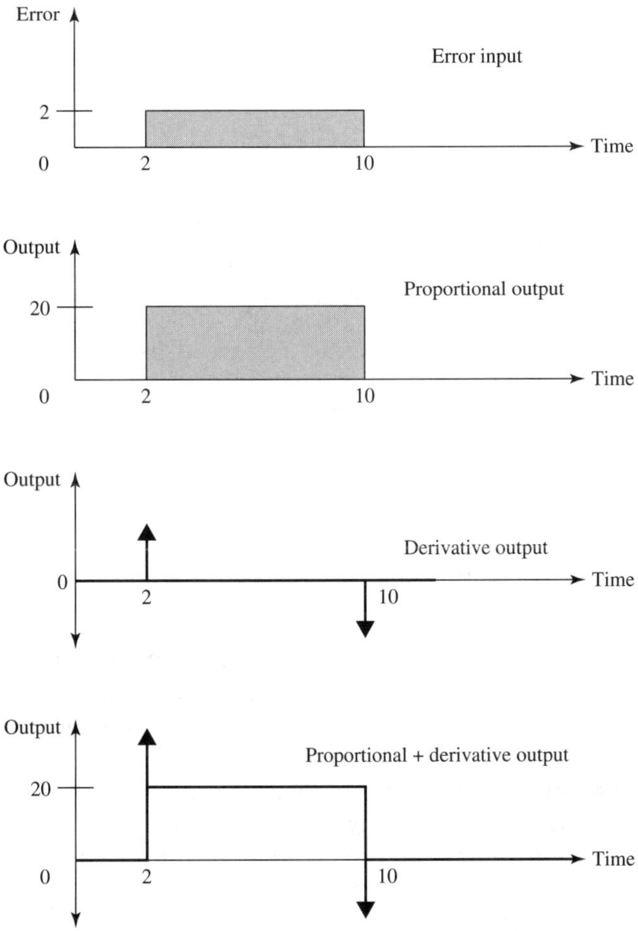

Figure 11.12 Input and output curves for Example 11.6.

to zero. As a consequence, the derivative section will simply generate a spike at this instant.

At time $t = 10$ s: Similarly, controller output will be a negative-going spike. Thus, the overall output will be the same as the output of the proportional controller, with spikes at 2-s and 10-s intervals.

Electronic Implementation of a PD Controller

Figure 11.13 shows a schematic diagram for implementing a proportional and derivative (PD) controller. Details of the summation block are excluded from the schematic

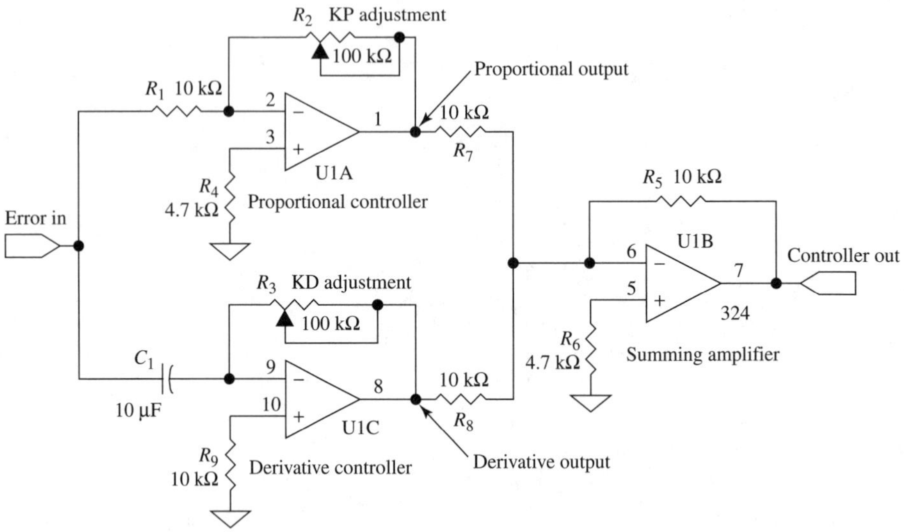

Figure 11.13 Schematic diagram of a PD controller.

(refer to the schematics of proportional controllers in Chapter 10). It is very similar to the schematic of a PI controller. The main difference lies in replacement of the integral amplifier with a derivative amplifier (U1C). The error signal is coupled to the input of amplifiers U1A and U1C. U1A acts as the proportional controller, and U1C acts as the derivative unit. Summing amplifier U1B takes the output of both the units and produces an output that is an algebraic sum of the two.

The proportional controller gain is set through potentiometer R_2 and is

$$K_P = \frac{R_2}{R_1}$$

The derivative gain, K_D, is adjusted through the potentiometer R_3:

$$K_D = R_3 C_1$$

EXAMPLE 11.7

Select the component values to provide a derivative gain of 0.05 and a proportional gain of 7.5.

Solution

Refer to the schematic diagram in Figure 11.13.

Proportional gain: The proportional gain is

$$K_P = \frac{R_2}{R_1}$$

Therefore,

$$\frac{R_2}{R_1} = 7.5$$
$$R_2 = 7.5\,R_1$$

Select a 10-kΩ resistor for R_1. Thus, $R_2 = 75\,\text{k}\Omega$. A 100-kΩ potentiometer (as shown in the schematic diagram) can be used for resistor R_2, and it can be set to 75 kΩ. A potentiometer will also prove useful during controller tuning as opposed to using a fixed 75-kΩ resistor.

Derivative gain: The derivative gain is

$$K_D = R_3 C_1$$
$$R_3 C_1 = 0.05$$

Select a 1-μF capacitor for C_1. Thus,

$$R_3 C_1 = \frac{0.05}{1 \times 10^{-6}}$$
$$R_3 = 50\,\text{k}\Omega$$

A fixed 50-kΩ resistor can be used. A second choice is to use a 100-kΩ potentiometer and set it to 50 kΩ.

Frequency Response of a PD Controller

Because the transfer function of a PD controller is made of a proportional block in cascade with a first-order lead element, it exhibits the characteristics of these elements. At low frequencies, the controller essentially behaves as a proportional controller, whereas at higher frequencies its behavior follows that of a first-order lead element. Its characteristics, in general, are as follows:

1. Break frequency is given by $\omega_b = K_P/K_D$ rad/s.
2. High-frequency ($>\omega_b$) gain has a slope of +20 dB/decade.
3. Lower-frequency ($\ll\omega_b$) gain approaches K_P ($20 \log K_P$ dB).

4. The phase angle at the break frequency is $+45°$.
5. The lower-frequency phase angle approaches $0°$.
6. The higher-frequency phase angle approaches $+90°$.

EXAMPLE 11.8

Draw the Bode plot for a PD controller, and verify its response characteristics. The controller coefficients are $K_P = 0.1$ and, $K_D = 0.01$.

Solution

The PD controller transfer function is

$$\mathrm{TF} = K_P\left(\frac{K_D}{K_P}s + 1\right)$$

Substituting the values of the gain coefficients,

$$\begin{aligned}\mathrm{TF} &= 0.1\left(\frac{0.01}{0.1}s + 1\right)\\ &= 0.1(0.1s + 1)\end{aligned}$$

An exact frequency response of this PD controller is shown in Figure 11.14. The following response characteristics can be verified from the plot:

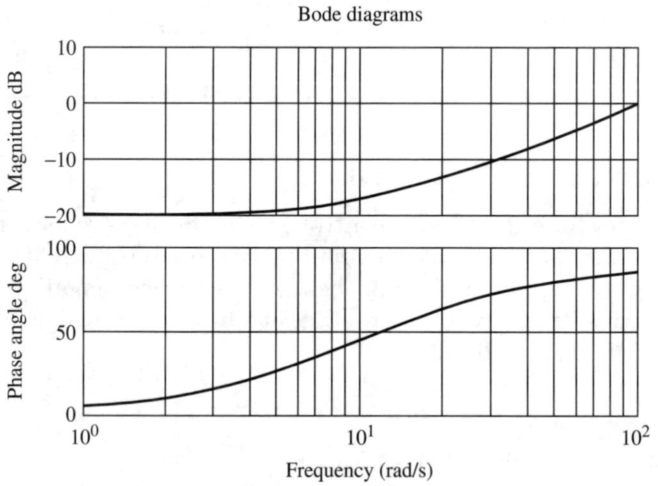

Figure 11.14 Frequency response for the PD controller in Example 11.8.

$$\text{break frequency } \omega_b = \frac{K_P}{K_D}$$
$$= 10 \text{ rad/s}$$
$$\text{high-frequency } (>\omega_b) \text{ gain change rate} = +20 \text{ dB/decade}$$
$$\text{lower-frequency } (\ll\omega_b) \text{ gain approaches } K_P = 0.1\ (-20 \text{ dB})$$
$$\text{phase angle at break frequency} = +45^\circ$$

The lower-frequency phase angle approaches 0°.
The higher-frequency phase angle approaches +90°.

EXAMPLE 11.9

Draw the Bode plot of a PD controller with the following coefficients: $K_P = 1$ and $K_D = 0.01$.

Solution

Substituting the values of the gain coefficients into TF,

$$\frac{V(s)}{E(s)} = \left(\frac{0.01}{1}s + 1\right)$$
$$= (0.01s + 1)$$

The frequency response of this PD controller is shown in Figure 11.15.

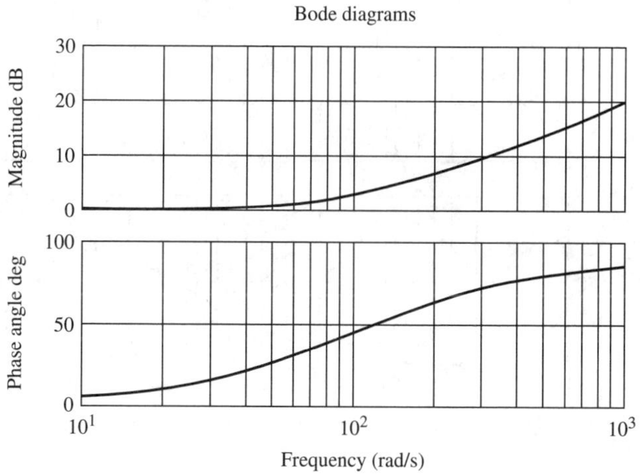

Figure 11.15 Bode plot for Example 11.9.

EXAMPLE 11.10

Determine gain and phase angle for a PD controller at the following frequencies, given that $K_P = 10$ and $K_D = 0.1$.

a. 1 rad/s
b. 10 rad/s
c. 1 KHz

Solution

This information can be obtained from the frequency response plot. Substituting the gain coefficients in the transfer function expression gives

$$\mathrm{TF} = 10(0.01s + 1)$$

The frequency response plot is shown in Figure 11.16. Signal gain and phase angle can be read from the plot.

a.

$$\text{magnitude} = 20\ \text{dB}$$
$$\text{phase angle} = 0^\circ \qquad \text{(approximately)}$$
$$\text{gain (calculated)} = 10$$

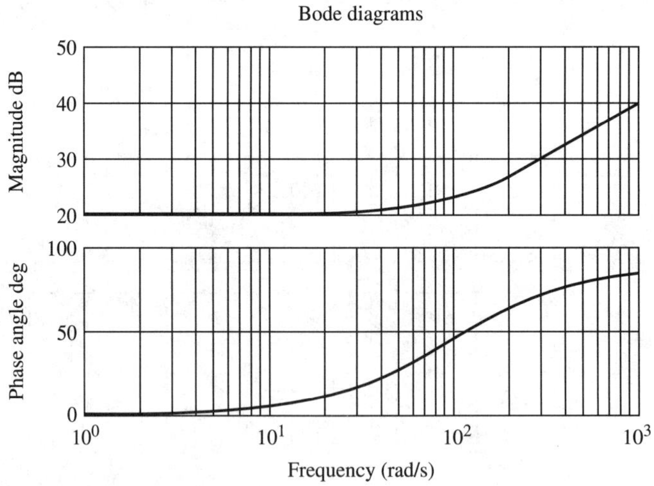

Figure 11.16 Bode plot for Example 11.10.

b.

$$\text{magnitude} = 20\,\text{dB}$$
$$\text{phase angle} = 5^\circ \qquad \text{(approximately)}$$
$$\text{gain (calculated)} = 10$$

c.

$$\text{frequency in radian/s} = 2\pi \times 1000\,\text{rad/s}$$
$$= 6284\,\text{rad/s}$$
$$\text{magnitude} = 55\,\text{dB} \qquad \text{(estimated)}$$
$$\text{phase angle} = 90^\circ \qquad \text{(approximately)}$$
$$\text{gain (calculated)} = 562.3$$

EXAMPLE 11.11

A PD controller is subjected to a sinusoidal voltage of 2.5 mV peak amplitude and a frequency of 100 Hz. Determine the precise value of the output signal amplitude and phase angle, given that $K_P = 22.5$ and $K_D = 0.25$.

Solution

The Bode plot of the PD controller with the given values of K_P and K_D is shown in Figure 11.17. It is difficult to determine the precise values of gain and phase angle from the graph, especially for a fractional frequency value. Numerical calculations can, however, be used to obtain accurate results.

In this case, the first form of the transfer function is more convenient to use.

$$\text{controller TF} = (K_D\, s + K_P)$$
$$= (0.25s + 22.5)$$

Replacing s with $j\omega$, where ω is the frequency in radians/second, gives a phasor quantity of $0.25j\omega + 22.5$. We know

$$\omega = 100\ \text{Hz}$$
$$= 628.319\ \text{rad/s}$$

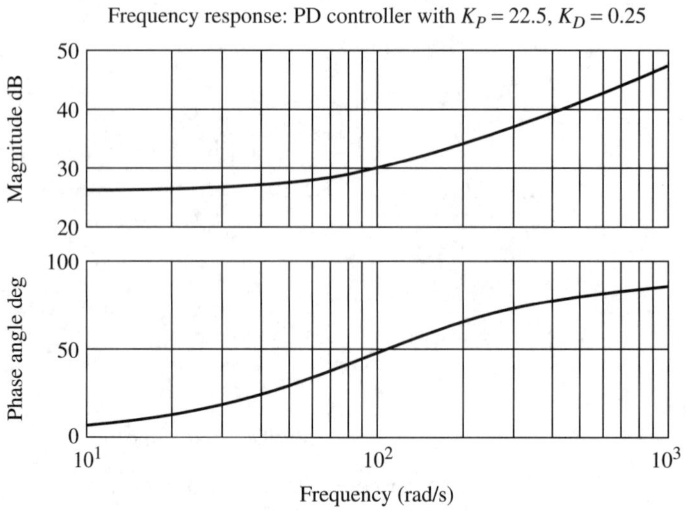

Figure 11.17 Bode plot for Example 11.11.

Substituting the value of ω, the phasor becomes

$$(157.0796j + 22.5)$$

Converting to polar form gives

$$158.683\angle 81.85$$

The signal gain is 159.683 and the phase angle is 81.85° at a frequency of 100 Hz.

$$\text{output amplitude} = 158.683 \times 2.5\,\text{mV}$$
$$= 396.7\,\text{mV}$$

PID CONTROLLER

The PID controller is a full three-term (proportional + integral + derivative) controller. It is the most versatile of all controller modes and is commonly used. A PID controller can be made to act as a single mode (P, I, or D) controller or as any possible combination

(PI or PD). An operational mode (P, I, or D) can be removed by setting its coefficient to zero. Figure 11.18 shows the parallel implementation of a PID controller.

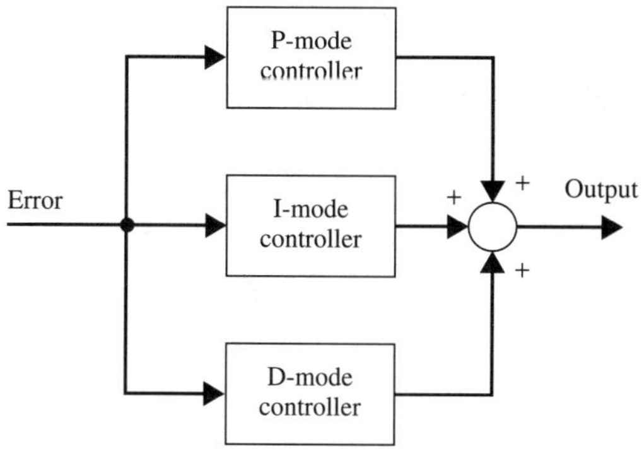

Figure 11.18 PID controller.

Time Equation

controller output = proportional controller output + integral controller output + derivative controller output

$$v_o = K_P e + K_I \int e\,dt + K_D \frac{de}{dt}$$

Transfer Function Taking the Laplace transformation of both sides of the equation,

$$V(s) = K_P E(s) + \frac{K_I E(s)}{s} + sK_D E(s)$$

$$V(s) = \left[K_P + \frac{K_I}{s} + sK_D\right] E(s)$$

Rearranging the terms, the transfer function is

$$\frac{V(s)}{E(s)} = \frac{K_D s^2 + K_P s + K_I}{s}$$

$$\boxed{\text{TF} = \frac{K_D s^2 + K_P s + K_I}{s}}$$

This form is convenient to remember and for entering into MATLAB, although greater insight can be obtained from the following form:

$$\frac{V(s)}{E(s)} = \frac{K_I\left(\frac{K_D}{K_I}s^2 + \frac{K_P}{K_I}s + 1\right)}{s}$$

$$\mathrm{TF} = \frac{\left(\frac{K_D}{K_I}s^2 + \frac{K_P}{K_I}s + 1\right)}{\frac{1}{K_I}s}$$

$$\boxed{\mathrm{TF} = \frac{A_2 s^2 + A_1 s + 1}{\tau s}}$$

where

$$A_2 = \frac{K_D}{K_P}$$

$$= \frac{1}{\omega_n^2}$$

$$A_1 = \frac{K_P}{K_I}$$

$$= \frac{2\zeta}{\omega_n}$$

$$\omega_n = \text{natural frequency of oscillation}$$

$$\zeta = \text{damping ratio}$$

$$\tau = \text{integral time constant}$$

$$= \frac{1}{\text{(integral) break frequency}}$$

$$= \frac{1}{K_I}$$

From this form of the transfer function, it is evident that a PID controller can be considered as a combination of an integral element in cascade with a second-order lead element.

EXAMPLE 11.12

A PID controller is subjected to the given error input curve (Figure 11.19). Determine the controller output given that $K_P = 10$, $K_I = 1$, and $K_D = 2$.

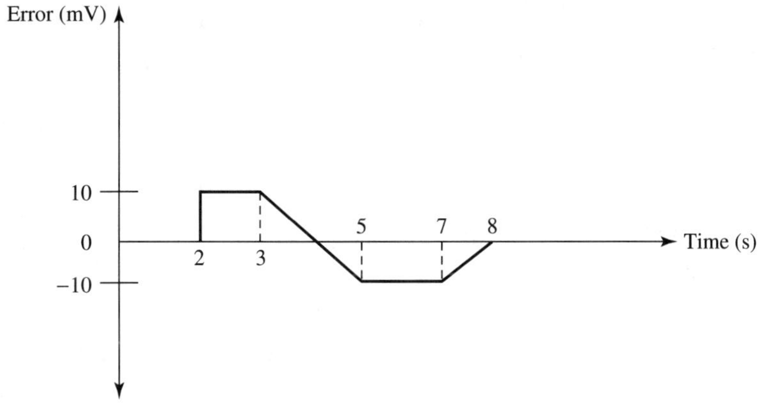

Figure 11.19 Error curve for Example 11.12.

Solution

$$\text{controller output} = v_P \quad \text{(proportional controller output)}$$
$$+ v_I \quad \text{(integral controller output)}$$
$$+ v_D \quad \text{(derivative controller output)}$$

Proportional controller output: Controller output is directly proportional to the error input:

$$v_P = K_P \times \text{error}$$
$$= 10e$$

Output from 0 to 2 s:

$$v_P = 10 \times 0\,\text{mV}$$
$$= 0\,\text{mV}$$

Output from 2 to 3 s: It will be constant, because the error is constant at 10 mV.

$$v_P = 10 \times 10\,\text{mV}$$
$$= 100\,\text{mV}$$

Output from 3 to 5 s: The error is changing linearly during this interval, so controller output, which is directly proportional to error, will also change linearly during this time interval. Since the output will be a straight line, only two points need to be calculated. Suitable values will be $t = 3$ s and $t = 5$ s.

$$\text{output } (t = 3\text{ s}) = v_P = 100\text{ mV} \qquad \text{(calculated earlier)}$$
$$\text{output } (t = 5\text{ s}) = v_P = 10 \times (-10\,\text{mV})$$
$$= -100\,\text{mV}$$

Output from 5 to 7 s: It will be constant as the error is constant at −10 mV.

$$v_P = 10 \times (-10\,\text{mV})$$
$$= -100\,\text{mV}$$

Output from 7 to 8 s: The controller output will change linearly from −100 mV and reach zero at $t = 8$ s.

Figure 11.20 shows the output of the proportional controller section.

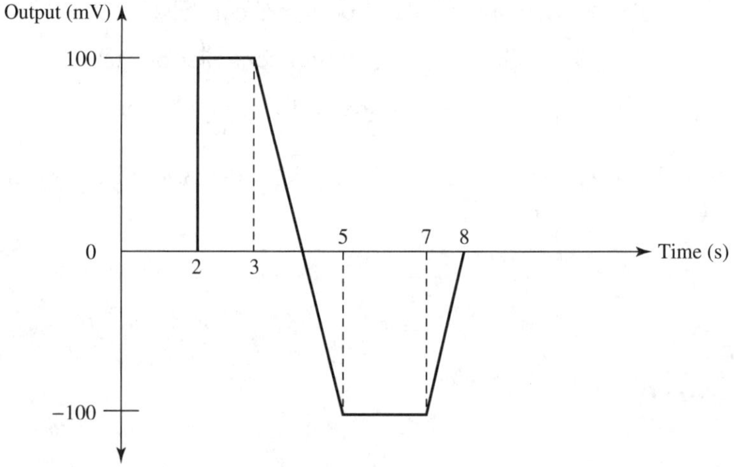

Figure 11.20 Proportional controller output for Example 11.12.

Integral controller output: The controller output is proportional to the area under the error curve from the start:

$$\begin{aligned} v_I &= K_I \int e\,dt \\ &= K_I \times \text{area under the error curve (until this time)} \end{aligned}$$

Output from 0 to 2 s: The area under the curve is zero, thus,

$$v_I = 0\,\text{mV}$$

Output from 2 to 3 s: The area under the curve is increasing linearly and so output will rise linearly from a zero initial value.

$$\begin{aligned} \text{output } (t = 3\text{ s}) = v_I &= 1 \times \text{area under the curve from 0 to 3 s} \\ &= 1 \times (10 \times 1)\,\text{mV} \\ &= 10\,\text{mV} \end{aligned}$$

Output from 3 to 4 s: Error drops to zero during this interval. But this still results in a net gain in the area under the curve. Controller output will increase in a nonlinear fashion.

$$\begin{aligned} \text{output } (t = 4\text{ s}) = v_I &= 1 \times \text{area under the curve from 0 to 4 s} \\ &= 1 \times \left(10 \times 1 + 10 \times \frac{1}{2}\right)\text{mV} \\ &= 15\,\text{mV} \end{aligned}$$

Output from 4 to 5 s: The error increases in the negative direction during this interval. This reduces the net positive area. Controller output will decrease in a nonlinear fashion.

$$\begin{aligned} \text{output } (t = 5\text{ s}) = v_I &= 1 \times \text{net area under the curve from 0 to 4 s} \\ &= 1 \times \left(10 \times 1 + 10 \times \frac{1}{2} - 10 \times \frac{1}{2}\right)\text{mV} \\ &= 10\,\text{mV} \end{aligned}$$

Output from 5 to 7 s: Constant negative error generates more negative area, resulting in a linear reduction in controller output.

$$\text{output } (t = 7\text{ s}) = v_I = 1 \times \text{net area under the curve from 0 to 7 s}$$
$$= 1 \times \left(10 \times 1 + 10 \times \frac{1}{2} - 10 \times \frac{1}{2} - 10 \times 2\right) \text{mV}$$
$$= -10\,\text{mV}$$

Output from 7 to 8 s: The controller output will drop further in a nonlinear fashion.

$$\text{output } (t = 8\text{ s}) = v_I = 1 \times \text{net area under the curve from 0 to 8 s}$$
$$= 1 \times \left(10 \times 1 + 10 \times \frac{1}{2} - 10 \times \frac{1}{2} - 10 \times 2 - 10 \times \frac{1}{2}\right) \text{mV}$$
$$= -15\,\text{mV}$$

Output beyond $t = 8$ s: As the error has returned to zero, there is no change in area under the error curve. The integral controller will continue to generate the output of −15 mV.

Figure 11.21 shows the output of the integral controller section.

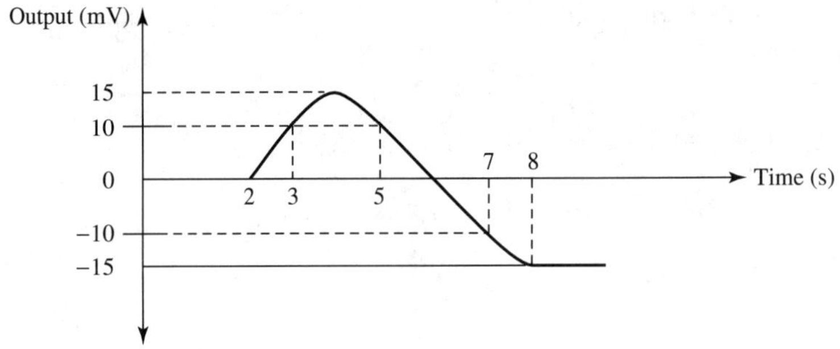

Figure 11.21 Integral controller output for Example 11.12.

Derivative controller output: The controller output is proportional to the rate of change of error.

$$v_D = K_D \frac{de}{dt}$$
$$= K_D \times \text{slope of error curve}$$
$$= 2 \times \text{slope of error curve}$$

Output from 0 to 2 s: The error is constant at 0, so the slope is zero.

$$\begin{aligned} v_D &= 2 \times \text{slope of error curve} \\ &= 2 \times 0 \\ &= 0 \end{aligned}$$

Output at 2 s: The error suddenly changes to 10 units; thus, the slope is infinite.

$$\begin{aligned} v_D &= 2 \times \text{slope of error curve} \\ &= 2 \times \infty \\ &\to \infty \end{aligned}$$

It will result in a positive-going spike.

Output from 2 to 3 s: The error is constant, and so the slope is zero.

$$\begin{aligned} v_D &= 2 \times \text{slope of error curve} \\ &= 2 \times 0 \\ &= 0 \end{aligned}$$

Output from 3 to 5 s: The error drops at a constant rate during this interval.

$$\begin{aligned} v_D &= 2 \times \text{slope of error curve} \\ &= 2 \times \frac{(\text{error at 5 s} - \text{error at 3 s})}{(5-3)\,\text{mV}} \\ &= 2 \times \frac{(-10-10)}{2\,\text{mV}} \\ &= -20\,\text{mV} \end{aligned}$$

Output from 5 to 7 s: The error is constant during this interval and so the slope is zero.

$$\begin{aligned} v_D &= 2 \times \text{slope of error curve} \\ &= 2 \times 0 \\ &= 0 \end{aligned}$$

Output from 7 to 8 s: The error starts to change in the positive direction at a constant slope.

$$
\begin{aligned}
v_D &= 2 \times \text{slope of error curve} \\
&= 2 \times \frac{(\text{error at } 8\text{ s} - \text{error at } 7\text{ s})}{1\text{ mV}} \\
&= 2 \times (0 - (10))\text{ mV} \\
&= 20\text{ mV}
\end{aligned}
$$

Output beyond $t = 8$ s: The error has returned to zero, and the slope of the error curve is also equal to zero. Thus, there will be no (zero) output from the derivative controller.

Figure 11.22 shows the output of the derivative controller section.

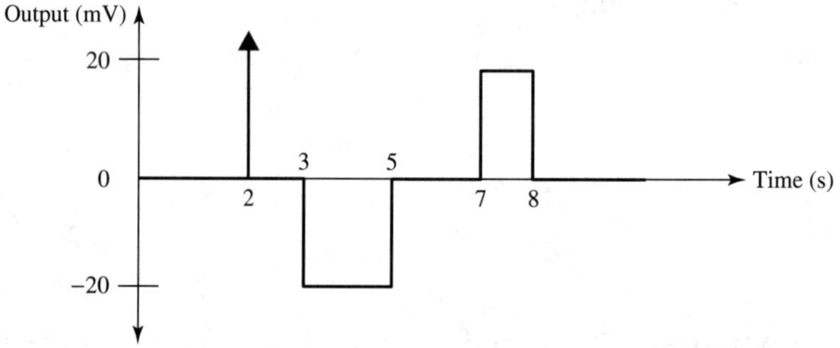

Figure 11.22 Derivative controller output for Example 11.12.

PID controller output: Final output will be the algebraic sum of all three outputs.

$$v_o = v_P + v_I + v_D$$

From 0 to 2 s:

$$
\begin{aligned}
v_o &= 0 + 0 + 0 \\
&= 0
\end{aligned}
$$

At 2 s:

$$
\begin{aligned}
v_o &= 0 + \infty + 0 \\
&= \infty \qquad \text{(positive spike)}
\end{aligned}
$$

At 3^- s (just before $t = 3$ s):

$$\begin{aligned} v_o &= 100 + 10 + 0 \\ &= 110\,\text{mV} \end{aligned}$$

At 3^+ s (just after $t = 3$ s):

$$\begin{aligned} v_o &= 100 + 10 - 20 \\ &= 90\,\text{mV} \end{aligned}$$

At 4 s:

$$\begin{aligned} v_o &= 0 + 15 - 20 \\ &= -5\,\text{mV} \end{aligned}$$

At 5^- s:

$$\begin{aligned} v_o &= -100 + 10 - 20 \\ &= -110\,\text{mV} \end{aligned}$$

At 5^+ s:

$$\begin{aligned} v_o &= -100 + 10 + 0 \\ &= -90\,\text{mV} \end{aligned}$$

At 7^- s:

$$\begin{aligned} v_o &= -100 - 10 + 0 \\ &= -110\,\text{mV} \end{aligned}$$

At 7^+ s:

$$\begin{aligned} v_o &= -100 - 10 + 20 \\ &= -90\,\text{mV} \end{aligned}$$

At 8^- s:

$$\begin{aligned} v_o &= 0 - 15 + 20 \\ &= +5\,\text{mV} \end{aligned}$$

At 8^+ s:

$$v_o = 0 - 15 + 0$$
$$= -15\,\text{mV}$$

The controller output will remain at −15 V beyond the 8-s interval. Figure 11.23 shows the composite output of this PID controller when subjected to the given error curve.

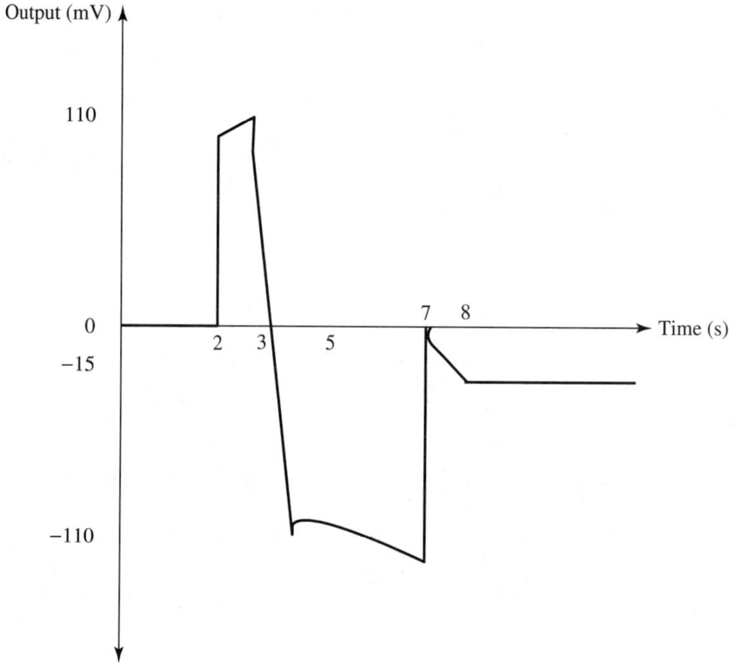

Figure 11.23 PID controller: composite output for Example 11.12.

Electronic Implementation

Figure 11.24 shows a schematic diagram of a full three-term PID controller. The details of the summation block are not provided. The error signal is commonly applied to the inputs of all three amplifiers, U1A, U1B, and U1D. U1A is a proportional amplifier. U1B is an integrator, and U1D acts as a derivative amplifier. U1C is an inverting summing amplifier. It takes output from all three amplifiers and generates an output that is an algebraic sum of the three. All three amplifiers, U1A, B, and D, are inverting in nature. The summing amplifier U1C adds the three signals together

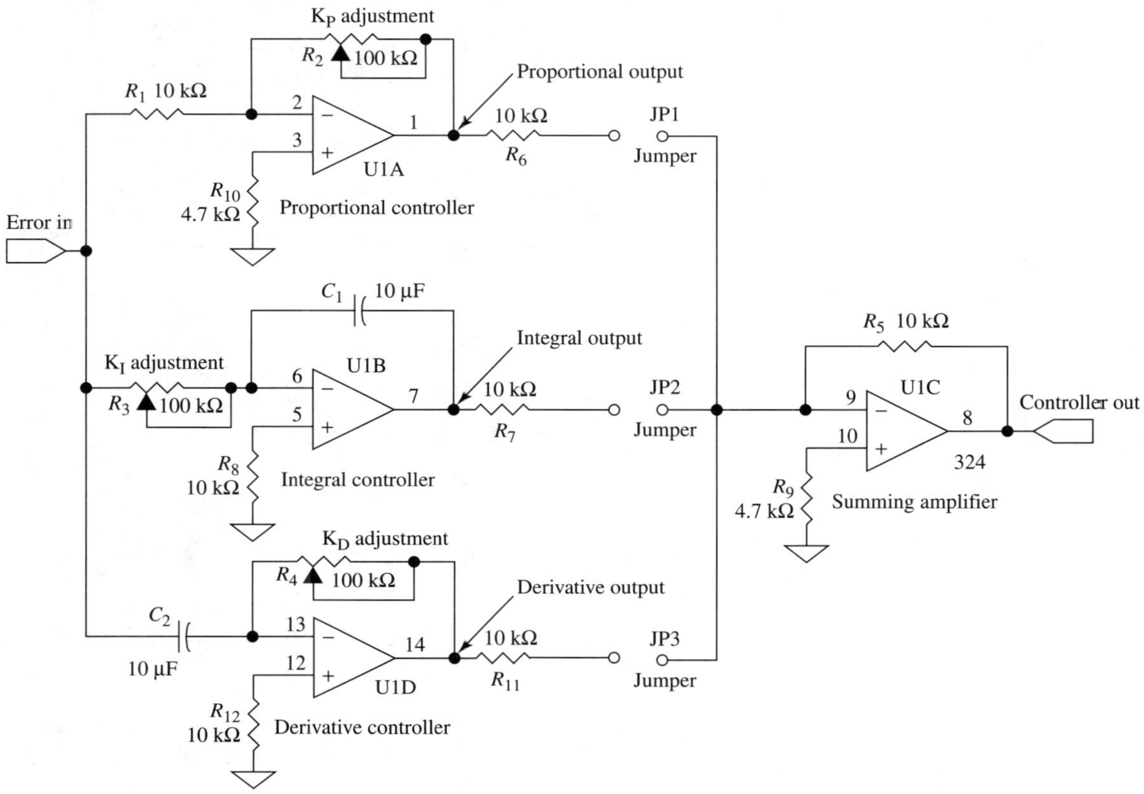

Figure 11.24 Schematic diagram: PID controller.

and further inverts the sum signal. Hence, the overall controller output has the same polarity as that of the error input signal.

Proportional gain K_P is controlled by the resistor R_1 and the potentiometer R_2.

$$K_P = \frac{R_2}{R_1}$$

Integral gain K_I is set through a combination of the capacitor C_1 and the potentiometer R_3.

$$K_I = \frac{1}{R_3 C_1}$$

Derivative gain K_D is controlled by the combination of the potentiometer R_4 and capacitor C_2.

$$K_D = R_4 C_1$$

Changing the Operating Mode

The full three-term controller can be changed to work either as a single-mode controller (P, I, or D) or as a two-mode controller (PI or PD). This can be accomplished in two ways.

- *Tuning out the gain coefficients:* The gain coefficient of the mode, which is not included, is set to an insignificant (practically zero) value. For example, to change the operating mode to a PI controller, the derivative coefficient K_D is tuned out. This is accomplished by setting the potentiometer R_4 to a near-zero value.
- Unfortunately, this procedure cannot be used to eliminate the integral gain. The integral gain is controlled by the inverse of the product of capacitor C_1 and potentiometer R_3. In order to achieve a zero integral gain the product of C_1 and R_1 has to be infinite. In the circuit (Figure 11.24), it can be done by either setting R_3 to infinite value, or removing C_1 or R_3 from the circuit—not a practical solution for soldered-in components.
- *Jumper setting:* The three jumpers, JP1, 2, and 3, provide an easier method for changing the mode. For a full three-mode operation, all three jumpers are left in place. If a particular mode is to be eliminated, the corresponding jumper is removed. For example, to make the unit work in a single proportional-only mode, jumper JP1 is left on and the other two jumpers, JP2 and JP3, are removed. Similarly, for the PD mode of operation, jumper JP2 is removed, leaving JP1 and JP3 on.

Frequency Response of a PID Controller

As evident from the transfer function, a PID controller reflects the characteristics of an integral element in cascade with a second-order lead element. A typical frequency response of a PID controller is shown in Figure 11.25.

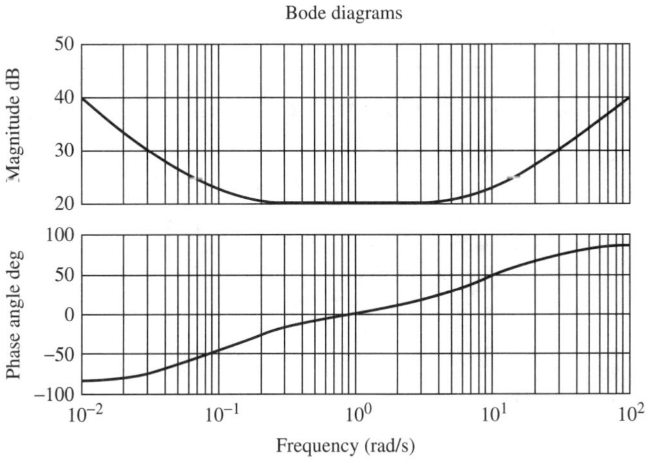

Figure 11.25 Typical frequency response: PID controller.

In general, the gain coefficients are selected so that the break frequency due to the integral mode, ω_a, is less than the break frequency, ω_b, due to the derivative mode ($\omega_a < \omega_b$). With such a setting, the frequency response can essentially be divided into three parts:

The low-frequency region is controlled by integral gain.
The midfrequency region is controlled by proportional gain.
The high-frequency region is controlled by derivative gain.

Such a PID controller has the following characteristics:

Break Frequency

1. There are two break frequencies: ω_a and ω_b.
2. The lower break frequency, ω_a, is given by $\omega_a = K_I/K_P$ (rad/s).
3. The upper break frequency, ω_b, is given by $\omega_b = K_P/K_D$ (rad/s).

Gain

1. The lower-frequency ($\ll\omega_a$) gain has a slope of -20 dB/decade.
2. The upper-frequency ($\gg\omega_b$) gain has a slope of $+20$ dB/decade.
3. The midfrequency ($\omega_a \ll \omega \gg \omega_b$) gain is given by K_P (dB).

Phase Angle

1. The midfrequency ($\omega_a < \omega > \omega_b$) phase angle ranges from $-45°$ to $+45°$.
2. The lower-frequency ($\ll\omega_a$) phase angle approaches $-90°$.
3. The upper-frequency ($\gg\omega_b$) phase angle approaches $+90°$.

EXAMPLE 11.13

Draw a frequency response plot of a PID controller with the following coefficient setting:

$$\text{proportional gain } K_P = 10$$
$$\text{integral gain } K_I = 0.1$$
$$\text{derivative gain } K_D = 1$$

Solution

The PID controller transfer function is

$$\mathrm{TF} = \frac{K_D s^2 + K_P s + K_I}{s}$$

Substituting the values in the transfer function equation gives

$$\mathrm{TF} = \frac{s^2 + 10s + 0.1}{s}$$

It can be seen from the plot (Figure 11.26) that there are two distinct break frequencies.

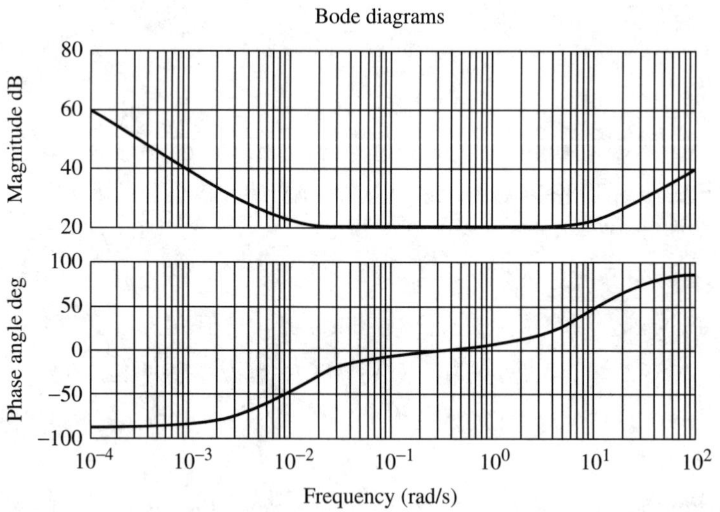

Figure 11.26 Bode plot: PID controller for Example 11.13.

$$\text{lower frequency} = \frac{K_I}{K_P}$$
$$= \frac{0.1}{10} = 0.01 \text{ rad/s}$$
$$\text{upper frequency} = \frac{K_P}{K_D}$$
$$= \frac{10}{1} = 10 \text{ rad/s}$$
$$\text{gain at midfrequency} = 20 \log K_P \text{ (dB)}$$
$$= 20 \log 10 = 20 \text{ dB}$$

The phase angle starts at -90°, becomes zero at midfrequency, and increases to $+90^\circ$ at high frequency.

EXAMPLE 11.14

Determine the exact magnitude and phase angle of a PID controller at the following frequencies, given that $K_P = 12.5$, $K_I = 0.35$, and $K_D = 1.25$.

a. 0.1 Hz
b. 0.1 Hz
c. 100 Hz

Solution

The controller transfer function is

$$\text{TF} = \frac{K_D s^2 + K_P s + K_I}{s}$$

Substituting the values of gain coefficients gives

$$\text{TF} = \frac{1.25s^2 + 12.5s + 0.35}{s}$$

The gain and phase angle can be determined from the phasor obtained by replacing s with $j\omega$.

$$\frac{1.25(j\omega)^2 + 12.5\, j\omega + 0.35}{j\omega} = 12.5 + J\,\frac{1.25\omega^2 - 0.25}{\omega}$$

a. Converting to radians,

$$\omega = 2\pi 0.001 \text{ rad/s} = 0.006283 \text{ rad/s}$$

Substituting the value in the phasor,

$$12.5 + j\frac{1.25(0.006283)^2 - 0.25}{0.006283} = 12.5 - 39.7809j$$
$$= 41.698\angle -72.56$$

The controller gain is 41.698 and the output signal lags the input by an angle of 72.56°.

b. Converting to radians,

$$\omega = 2\pi 0.1 \text{ rad/s} = 0.6283 \text{ rad/s}$$

Substituting the value in the phasor,

$$12.5 + j\frac{1.25(0.6283)^2 - 0.25}{0.6283} = 12.5 + 0.3875j$$
$$= 12.506\angle 1.78$$

The controller gain is 12.506 and the output signal leads the input by an angle of 1.78°.

c. Converting to radians,

$$\omega = 2\pi 100 \text{ rad/s}$$
$$= 628.318 \text{ rad/s}$$

Substituting the value in the expression for phasor and simplifying gives

$$785\angle 89.09$$

The controller gain is 785 and the output signal leads the input by an angle of 89.09°.

The controller gain is large at lower frequencies, drops at midfrequency, and rises again at the higher frequencies. The phase angle starts with a lag (typical of an integral element) and changes to a lead (typical of a derivative element) at high frequencies.

PROBLEMS

1. Write the time equation and transfer function for each of the following controller configurations.
   a. PI controller with coefficients $K_P = 10$ and $K_I = 0.1$
   b. PD controller with coefficients $K_P = 22$ and $K_I = 1.25$
   c. PI controller with coefficients $K_P = 0.1$ and $K_I = 0.01$
   d. PID controller with coefficients $K_P = 14$, $K_I = 0.25$, and $K_D = 1.5$
2. Transfer functions of some analog controllers are given here. Identify the controller configuration and the controller gain in each case.
   a. Controller TF $= \dfrac{10s + 1}{s}$
   b. Controller TF $= \dfrac{10}{s}$
   c. Controller TF $= \dfrac{s + 5}{4s}$
   d. Controller TF $= \dfrac{s}{1.25}$
   e. Controller TF $= 5s + 10$
   f. Controller TF $= \dfrac{3s^2 + 8s + 0.5}{s}$
3. A PI controller has the coefficients $K_P = 0.1$ and $K_I = 0.01$.
   a. Determine the frequency (rad/s) at which the magnitude becomes 0 dB.
   b. Determine the frequency (rad/s) at which the magnitude is 20 dB.
   c. Determine the output signal for an input signal of 0.1 V (pk-pk) and a frequency of 100 rad/s.
4. The coefficients for a PD controller are $K_P = 1$ and $K_D = 0.1$.
   a. Determine the frequency (rad/s) at which the magnitude becomes 0 dB.
   b. Determine the frequency (rad/s) at which the magnitude is 20 dB.
   c. Determine the signal for an input signal of 0.25 V (pk-pk) and a frequency of 100 rad/s.
5. A PI ($K_P = 10$ and $K_I = 0.01$) controller is subjected to a constant error signal of 0.1 V DC. Determine the output voltage at the following time intervals. Assume that the initial controller output is zero and the controller saturation voltage is ± 10 V.
   a. 2 s
   b. 20 s
   c. 1 h
6. An analog controller is subjected to the error input shown in the accompanying figure. Calculate the controller output and draw a diagram showing the controller output with respect to time for each of the following cases:

a. PI controller with $K_P = 10$ and $K_I = 0.9$
b. PD controller with $K_P = 4.8$ and $K_D = 1.6$
c. PID controller with $K_P = 12$, $K_D = 2.25$, and $K_I = 0.4$

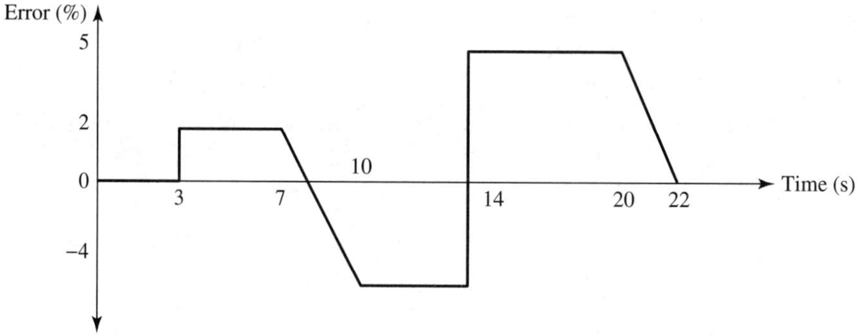

7. Write down the time equation and transfer function for a PID controller with the controller coefficients of $K_P = 10$, $K_I = 0.5$, and $K_D = 2$.

8. Draw a schematic diagram for a PI controller. Select the component values to allow for a proportional gain in the range of 0 to 10 and integral gain in the range of 0.1 to 2.

9. A process controller is operating as a full three-term (PID) controller. Due to poor shielding and some large electromechanical equipment in the near vicinity, a significant amount of noise is superimposed on the transducer feedback signal. Explain what effect this will have on each of the three controller modes.

10. Explain during what portion of its entire frequency range each of the three modes of a PID controller is most effective.

11. Draw a Bode plot showing the magnitude and phase response of a typical PID controller.

12. A PID controller is represented by the following time equation:

$$\text{output } v = 10\,e + 0.1 \int e\,dt + 2.5\frac{de}{dt}$$

Determine the controller output for each of the following frequencies:
a. 0.01 rad/s
b. 1 rad/s
c. 60 Hz

13. Draw a complete schematic of a PID controller. Select the components to allow for the following gain settings:

$$\begin{aligned} \text{proportional coefficient} &= 1 \text{ to } 20 \\ \text{integral coefficient} &= 0.05 \text{ to } 2 \\ \text{derivative coefficient} &= 0.25 \text{ to } 4 \end{aligned}$$

14. Draw the frequency response of a PID controller given by the following time equation:

$$\text{output } v = 5e + 1.2 \int e\,dt + 4\frac{de}{dt}$$

12

DIGITAL CONTROLLER

INTRODUCTION

The advent of the microprocessor has revolutionized the entire measurement and control industry. Originally, hydraulics, pneumatics, and mechanical linkages were used as control and sensing elements, and it took a lot of ingenuity on the part of control engineers to implement continuous control with these elements. Later these were replaced with electronic analog units using transistors and operational amplifiers. Electronic analog units opened the door for comprehensive PID control and some nonlinear control. But the proliferation of digital controllers opened the doors wide, permitting nearly endless control schemes to suit practically any type of system.

This chapter covers the application of the digital computer as a controller. It discusses the pros and cons of applying a digital controller. The subject matter focuses on using a microprocessor/microcontroller as a replacement for an analog controller while still implementing conventional control algorithms, such as two-position and PID control.The *Z*-transform and some of the algorithms suitable only for digital controllers are not discussed. The basic structure of a digital controller and its implementation are explained. Development of the digital PID control algorithm is discussed in detail. Software and hardware details for the implementation of the digital controller algorithm on a PC as well as on microcontrollers are also provided.

DIGITAL CONTROLLER

The main function of a digital controller is the same as that of an analog controller: to control the (closed-loop) response of a system. A digital controller differs from its analog counterpart mainly in the fact that a digital controller is essentially a computer (or a microcontroller). Because a digital computer cannot accept an analog signal, it cannot be directly interfaced to feedback transducers or set-point potentiometers. Similarly, a digital computer cannot generate a real-world analog signal, such as 3–15 psi of air pressure, 0–10 V DC signal, or 4–20 mA current signal, so it cannot be directly interfaced to final control elements. An analog-to-digital (A/D) converter is needed for accepting analog input signal and a digital-to-analog (D/A) converter is needed for interfacing to final control elements (Figure 12.1).

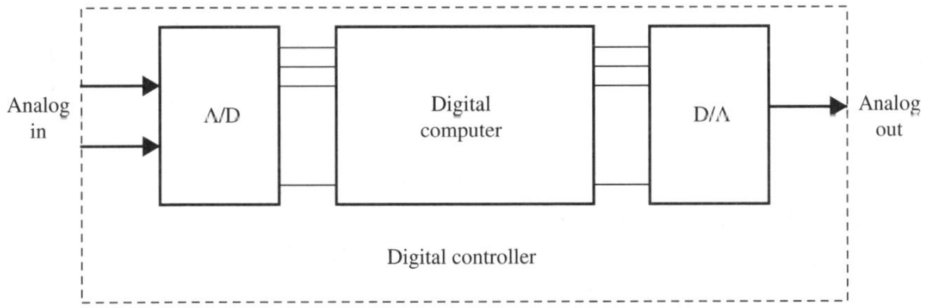

Figure 12.1 Structure of a digital controller.

Figure 12.2 shows a closed-loop control system with a digital controller. As shown in the diagram, the conventional analog controller has been replaced with computer software executing the control algorithm. The only purpose of the summation block

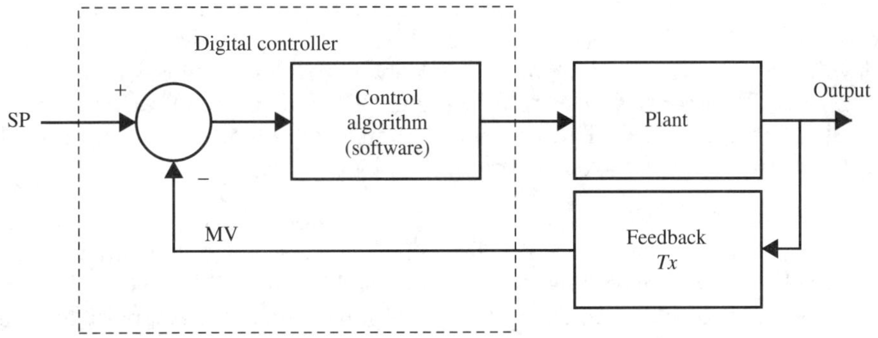

Figure 12.2 Digital controller in a closed loop.

is to generate the error signal as a difference of SP and MV values. It is a very trivial task for a digital computer, which can accept the values of SP and MV through analog and digital converters and internally generate the error signal. Thus, in practice, the summation block is replaced by a line of computer code, which takes the numerical (digitized) set-point data and subtracts measured values from that data. Figure 12.3 shows this updated configuration for a digital controller in closed-loop setup.

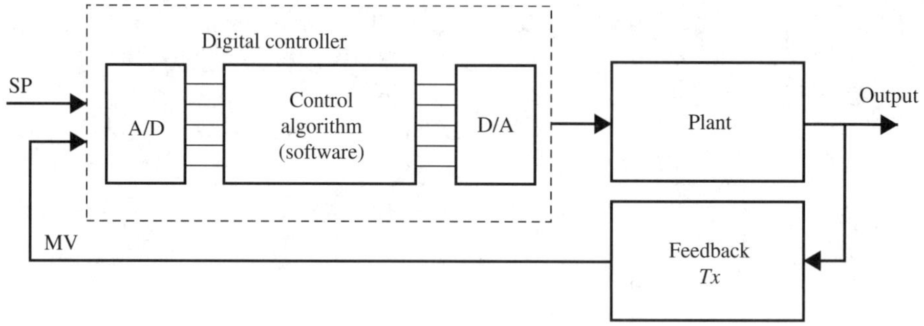

Figure 12.3 Digital control loop configuration.

The digital controller accepts the values of SP and MV as digitized by the A/D converter. It processes this information according to its internal programmed software and generates an appropriate output through the D/A converter. The software is coded to execute a control algorithm suitable for the closed-loop system under control. The flowchart in Figure 12.4 describes the controller operation.

Controller Sequence of Operation

As shown in the flowchart in Figure 12.4, the digital controller first reads the output from the A/D converter channel connected to the SP input and then reads the A/D channel connected to the process variable MV. From these two measurements the present value of the error is determined. A new controller output is generated based upon this error. This controller output value is then fed to the D/A converter, and an analog output is generated. The controller then completes any outstanding housekeeping tasks and returns to repeat the control loop.

Housekeeping Tasks: In a number of systems, a general-purpose computer such as a desktop PC may act as a digital controller, whereas in others, a dedicated computer or a microcontroller may be used. Besides acting as digital controller, a PC will be required to respond to keyboard input, update graphic displays, and also respond to any network communication. Even a dedicated microcontroller may be required to update a graphic display and respond to serial communication from a supervisory computer. In both

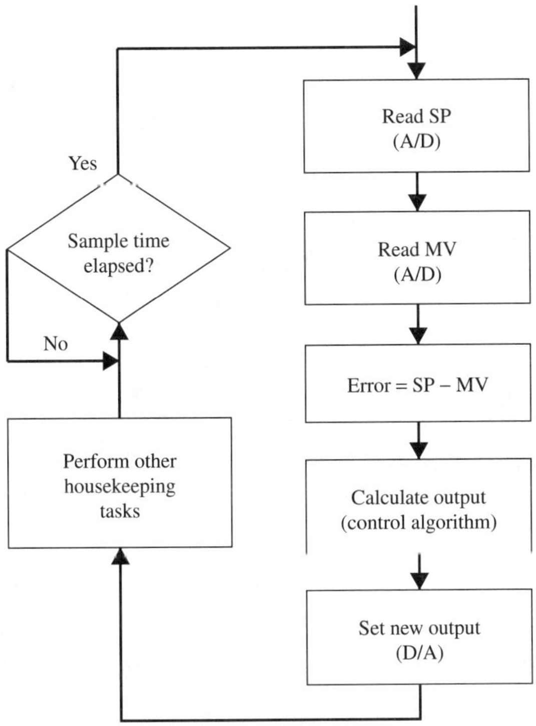

Figure 12.4 Flowchart of digital controller operation.

cases the additional tasks are labeled as *housekeeping tasks*, and a finite amount of computer time is spent on responding to these tasks.

Sample Time

Sample time is the time taken for a computer to complete all the tasks in a loop, or the time interval with which the computer will repeat a task.

A closed-loop control system incorporating a digital controller is analyzed using *Z*-transforms. Laplace transform analysis is not applicable to a digital controller. *Z*-transforms are necessitated for a process system that has a relatively fast response time (short time constant), and comparable to loop execution time (sample time). Sample time is process-dependent and can range from a few microseconds to several seconds. If the loop execution time is very small compared to the process response time, sample time considerations (and *Z*-transforms) in analysis are not warranted. Also, for simple control strategy, such as on-off control, no consideration needs to be made for the sampling interval, and the control loop is repeated as fast as possible. The decision block representing the sample time in the flowchart ensures that the control loop is repeated only at every sampling time interval.

Considerations for Digital Control

Equipped with A/D and D/A, a digital controller performs like any analog controller (on-off, P, PI, PID, etc.) and also allows some very unique control strategies to be implemented. The following factors should be considered when incorporating digital controllers for closed-loop control.

Truncation Error A truncation error is caused by the inability of digital data to fully represent the (analog) information in a digital domain. There are three sources of truncation error.

- A digital computer performs calculations based upon its internal data width (8, 16, or 32 bits). Because of limited bandwidth, the calculations are always rounded to the number of bits used. This creates some residual error. This is normally not a concern, especially with 16- or 32-bit calculations, but it can lead to a large accumulated error in integral sums. To counteract the truncation error, calculations are sometimes carried out using long integers or floating-point format. Unfortunately, these extended calculations increase the calculation time and tend to slow the rate of algorithm execution.
- An A/D converter is another source of error. An 8-bit A/D has only 8 bits with which to represent the entire analog input range. Converter truncation error is equivalent to the converter resolution.

$$\text{truncation error (resolution)} = \frac{\text{input range}}{2^N}$$

where N = number of bits.

EXAMPLE 12.1

Determine the truncation error of the following A/D.

$$\text{input range} = 0 \text{ to } 10 \text{ V}$$
$$\text{number of bits} = 8$$

Solution

The truncation error equals

$$\text{converter resolution} = \frac{10}{2^8}\text{ V}$$
$$= 0.0391 \text{ V}$$

The converter will not be able to detect any change in input voltage, which is smaller than this value (0.391 V).

Truncation error due to the A/D converter can be reduced by using higher-resolution A/D converters, but the reduction comes at the expense of increased cost and complexity.

- A D/A converter is also a source of truncation error, similar to an A/D converter. Because of limited resolution, a D/A converter is not capable of generating a continuous analog output ranging from zero to full-scale output. The output appears to have a staircase pattern. Each step of this pattern is equivalent to the converter resolution.
- The majority of microcontrollers do not carry a D/A converter on board, so a D/A function is either provided through a clever use of on-board PWM along with an external low-pass filter, or an external D/A is interfaced to it. The D/A truncation error is the same as the D/A converter resolution.

$$\text{truncation error (resolution)} = \frac{\text{input range}}{2^N}$$

Execution Time The digital controller accepts the digitized data related to SP and MV, performs a calculation, generates an output, and returns to repeat the process. Because of the finite execution speed of the CPU, a small but finite time delay exists between successive calculations of outputs. The total time (sample time) taken before the loop can be repeated, depends on the following factors:

- A/D conversion time
- CPU calculation time
- D/A update time

Once the measurements of SP and MV are done and calculations have begun, the computer is not able to monitor the process. For all practical purposes, during this period (of computer calculations) the control loop acts as an open loop. The inputs may change during this interval (sample time), but they cannot be detected by the digital controller, because it is busy doing calculations. Only when the controller has completed all the calculations and sent the digitized output to the D/A converter, it can return and take another set of measurements.

It is important to select a processor with reasonable operating speed, but the language used for execution of control algorithms and the type of variables used also have a significant effect on loop time. Finally, the control algorithm used has to be carefully considered.

DIGITAL TWO-POSITION CONTROL

A digital controller lends itself very easily to two-position control. Figure 12.5 shows the transfer characteristic of a two-position controller.[1] The controller output changes between 0% and 100%, depending upon the error and direction of movement. It should be noted that error is represented on an absolute scale. If desired, error can be represented as a percentage, but additional calculations would be required, resulting in longer execution time. A two-position controller is commonly used as a digital thermostat for residential heating and/or cooling. The temperature error band in these controllers is set in terms of degrees Celsius (or degrees Fahrenheit).

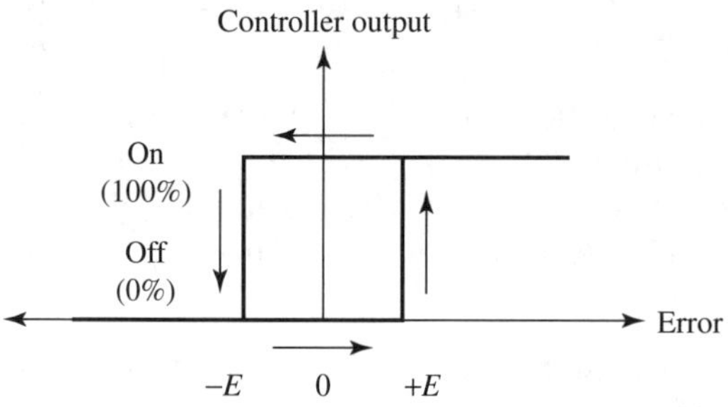

Figure 12.5 Digital two-position control.

Sequence of Operation

The sequence of operation for a controller can be described by the following pseudocode.

```
WHILE (TRUE)
      Read Set Point
      Read Measured Value
      Calculate Error

      IF Error > +E THEN
              Set Output = ON (100%)
      ELSE IF Error < -E THEN
              Set Output = OFF (0%)
      'ELSE  IF (Error > -E) AND (Error < +E) THEN
      '       Do Not Change Output
      ENDIF
WEND
```

[1] Review Chapter 9.

Microcontroller Implementation of a Two-Position Controller

Figure 12.6 shows a schematic diagram for a microcontroller-based on-off controller. The process is very similar to an analog two-position controller (Chapter 9). Motorola's MC68HC11 microcontroller is used as the digital controller, although there are several microcontrollers that can be used here. The application is essentially similar to the one covered in Chapter 9. It controls the temperature of a small oven.

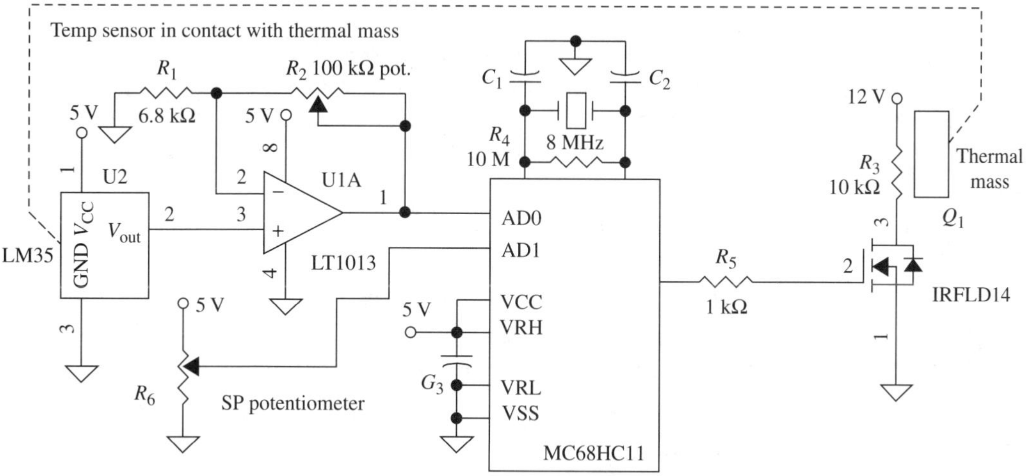

Figure 12.6 Microprocessor-based two-position controller.

Circuit Description

Digital Controller: A commercially available HC11 microcontroller module, 'MOTU 1' [11], has been chosen. It houses Motorola's MC68HC811E2 microcontroller but can also accommodate any 52-pin PLCC version. The microcontroller module replaces the analog controller described in Chapter 9. Several versions of the MC68HC11 microcontroller are available. Some of these can be used in the single-chip mode, whereas other devices are designed for use in an expanded system with external memory. Two of the very popular devices are MC68HC811E2FN and MC68HC711E9CFN. Both of these are designed for use in a single-chip mode. The current application can be run in any one of these two modes. If desired, the application can also be run on Motorola's HC11 EVB evaluation board. The program has been coded for the single-chip version MC68HC811E2 [8, 18, 19]. Its main features are as follows:

- 2 kilobytes of EEPROM code space
- 256 bytes of RAM

- 8 channels of 8-bit A/D converters
- Counters/timers
- Serial communication
- Parallel I/O ports

The diagram does not show the connections for every single pin but covers the interface needed for the current application. For more details, refer to several excellent books covering the details of HC11 programming and interfacing.

Only two of the eight available A/D channels are used in this application. A nice feature about HC11 A/D conversion is that once microcontroller is configured, it can continuously convert 4 (or 8) analog input signals and place the results in an internal register file (internal special-purpose RAM). The converter resolution is controlled by the differential voltage present between two reference terminals, V_{ref_H} and V_{ref_L}:

$$\text{resolution} = \frac{V_{\text{ref}_H} - V_{\text{ref}_L}}{2^8}$$

It is possible to set V_{ref_H} to below 5 V and V_{ref_L} to above 0 V, thus narrowing the operating range and improving the measurement resolution. Here, however, a range of 5 V is used, resulting in a measurement resolution of about 19.5 mV.

Temperature Sensor: LM35 from National Semiconductor is used as the temperature sensor. It has a temperature range of −50°C to more than 100°C. A bipolar supply is needed if a negative temperature is to be measured. In this application a temperature control range of 5°C to 45°C is assumed, eliminating the need for a bipolar supply. LM35 generates a linear output of 10 mV/°C.

Interfacing an LM35 directly to HC11 will result in a measurement resolution of about 2°C. It means that the control cannot be exercised to better than 2°C. An amplifier can be used to amplify the LM35 output and increase the measurement resolution. U1A (LT1013, from Linear Technology) is a noninverting amplifier and is adjusted for a gain of 10. Thus, temperature-sensor output is amplified to 100 mV/°C, resulting in a measurement (and control) resolution of about 0.2°C. The LT1013 is a superior-quality operational amplifier and provides low offset, low drift, and an output that approaches rail-to-rail supply voltages. Amplification of the temperature transducer output results in a temperature range of 5°C to 45°C.

Output Generation: An IRLD14 from International Rectifier controls the power to the electric heater. It is a power MOSFET, which can be controlled through logic 0–5 V. It has a low ON-resistance and contains a integrated diode for bypassing switching transients. The MOSFET is turned on or off by the logic level present on one of the output lines from 68HC11.

Set-Point Input: Set-point input is provided through a potentiometer. The potentiometer is connected across 0- and 5-V supplies and provides a full temperature control range (0°C to 50°C). Keypads are commonly used in microprocessor-based control systems for entering the control parameters. In some applications, serial communication from a remote host computer can be used to enter the set-point data to the controller. The same link can also be used to set the error band and other operating parameters.

Software Description Software is the heart of controller operation. The hardware provides a means of accepting the input information (SP and MV) and generating an output, but it is the software that controls the mode of operation. With a different final control element (MOSFET in this example) and a suitable control algorithm, the same controller (68HC11 here) can be made to function in a different mode of control, such as PID, lag, lead, FIR, or IIR.

The software for on-off control is given in Program Listing 12.1. It can be divided into the following tasks:

1. System initialization
2. Loop execution
3. Housekeeping

System Initialization: Before the 68HC11 can start executing the control loop, the various peripheral registers need to be properly configured and memory allocated for variables. The initialization routine does just that. After setting up the stack pointer, it sets a continuous conversion mode for the first four A/D channels, and the pin Port B.0, is set to low. After initialization, the microcontroller enters an indefinite loop, which contains the loop execution and housekeeping tasks.

Loop Execution: Having initialized the peripherals, the microcontroller enters the control loop. For on-off control operation, the software is fairly simple. Two channels of A/D results are read from the A/D result registers, ADR1 and ADR2. A comparison is made to determine if the process value is within the dead band. If it is within the band, the output is not changed; that is, if it is LOW, it is left at LOW, and if it is HIGH, it is left at that. If this is not the case, the output is set to LOW or HIGH, depending upon the error (SP – MV) being too high or too low. The user can modify the set-point value at any time, because it is read at the beginning of every control loop. The error band is set (coded) in software, but if desired, an additional potentiometer can be installed to manually adjust this value.

Housekeeping: Having executed the control loop, the microcontroller is free to perform any housekeeping tasks. No task is set up here, but this is the place to add such a task. The program returns to the indefinite loop and continues to perform the on-off control function.

Program Listing 12.1

```
*************************************************************
*                   Dec 19,1999                             *
*                   S.Gupta                                 *
*                   Two_Pos.ASM                             *
*                   MC68HC811E2                             *
*************************************************************

; This program is designed for 68HC11 to act as an on-off controller
; with hysteresis
; Operation:
; ---------
; ADR1  ; MV, Temperature data
        ; 100 mv/deg C
        ; 0.200 v (2 deg C) to 4.8 v (48 deg C)
; ADR2  ; SP, Temperature value
        ; 100 mv/deg C
        ; 0.0 v (0 deg C) to 5.0 v (50 deg C)
        ; Allow only 5 to 45 deg C control range

; Output
        ; Port B.0 is used for output
;       *************************************************************

;       ===============================================================
;                           Constants
;       ===============================================================
EBAND   EQU     $64
; Allow for an error of 1 deg C about SP.
; 1°C is equivalent to +100 mv of amplified tx output.
; Change this value to suit.
; Use an additional potentiometer to set this error band, dynamically,
; or use serial port communication for modification of parameters.
;       ---------------------------------------------------------------

;       ===============================================================
;                           Registers
;       ===============================================================
PORTB   EQU     $04
PORTE   EQU     $0A
ADCTL   EQU     $30
ADR1    EQU     $31
ADR2    EQU     $32
OPTION  EQU     $39
;       ---------------------------------------------------------------

;       ===============================================================
;                           Variables
;       ===============================================================
```

```
          ORG $0
PORTB_DATA       RMB     1
MV               RMB     1
;        -----------------------------------------------------------------

;        =================================================================
;                                VECTOR  EQUATES
;        =================================================================
         ORG     $FFFE
         FDB     START
;        -----------------------------------------------------------------

;        |||||||||||||||||||||||||||||||||||||||||||||||||||||||||||||||||
;                                Program
;        |||||||||||||||||||||||||||||||||||||||||||||||||||||||||||||||||
         ORG $F800
START:
         LDS     #$FF
         LDX     #$1000          ; Offset for control registers.
         JSR     INITIALIZE      ; Initialize the peripherals

AGAIN:
         JSR     CONTROL         ; Execute the control algorithm
         JSR     HOUSEKEEPING    ; Do any additional work
         BRA     AGAIN           ; Repeat the loop
;        -----------------------------------------------------------------
;        -----------------------------------------------------------------


;        =================================================================
;                                Sub INITIALIZE
;        =================================================================
; This routine will initialize all the necessary peripheral and
reset the output pin.
;

INITIALIZE:

         ; A/D Converter
         ; ------------
         CLRA
         STAA    PORTB_DATA      ; Reset port B data in memory
         STAA    PORTB,X         ; Set all outputs to Low

         BSET    OPTION,X,$80    ; ADPU = 1 to power up A/D
         BCLR    OPTION,X,$40    ; CSEL = 1, Use E Clock
         LDAB    #$30            ; Select Continuous conversion
         STAB    ADCTL,X         ; On Channel AD0 - AD3
         RTS
;        -----------------------------------------------------------------
```

```
;       ==================================================================
;                                Sub CONTROL
;       ==================================================================;
; It is the heart of the controller program
;
CONTROL:
        LDAB    ADR1,X          ; Read process temperature (MV)
        LDAA    ADR2,X          ; Read desired temperature (SP)
        PSHA                    ; Save a copy of current SP

        SBA                     ; ACCA = SP - MV
        BEQ     EXIT_CONTROL    ; If equal then exit
        BHI     CHECK_POSITIVE  ; If it is positive then find if outside
                                ;  error band.
        ADDA    #EBAND          ; ACCA = SP - MV □
        BPL     EXIT_CONTROL    ; It is within -ve error band

        LDAA    PORTB_DATA      ; Read the current port B value from
memory
        ANDA    #$FE            ; Reset PortB.0
        STAA    PORTB_DATA      ; Update the data
        STAA    PORTB_DATA      ; Update the data
        BRA     EXIT_CONTROL

CHECK_POSITIVE:
        SUBA    #EBAND          ; ACCA = (SP-MV) - EBAND
        BHI     EXIT_CONTROL    ; No it is within  error band

        LDAA    PORTB_DATA      ; Read the current port B value from
memory
        ORAA    #$01            ; Set PortB.0
        STAA    PORTB_DATA      ; Update the data
        STAB    PORTB,X         ; Output the data

EXIT_CONTROL:
        RTS
;       ------------------------------------------------------------------

;       ==================================================================
;                                Sub HOUSEKEEPING
;       ==================================================================;
; Insert any housekeeping function here. Currently there is none.

HOUSEKEEPING:
                        ; Simply return to main loop
        RTS
;       ------------------------------------------------------------------
        END
; ========================================================================
; ========================================================================
```

DIGITAL PID CONTROLLER

A PID controller contains terms that require integration and differentiation of error. In order to execute the PID algorithm in an efficient manner, the integration and differential processes have to be reduced to simple algebraic manipulations (multiplication/division and addition/subtraction). It is best to take another look at the analog control algorithm.

Analog PID Algorithm

The analog PID control algorithm is given as

$$v_o = K_P e + K_i \int e\,dt + K_d \frac{de}{dt}$$

In an analog world, this algorithm is implemented (or solved) by the use of analog computation blocks, such as multipliers, dividers, integrators, and differentiators. These analog blocks are easily constructed using operational amplifiers. The proportional coefficient K_P, integral coefficient K_I, and derivative coefficient K_D are set with the help of variable resistors in op-amp circuits. The tuning (gain adjustment) of an analog controller is carried out through adjustment of these variable resistors.

Digital PID Algorithm

A digital control system is not a continuous system but a discrete (sampled-data) control system (Figure 12.7). In a digital system, measurement of error is available only at regular sampling intervals. Sampling intervals are the instants when samples (A/D measurements) are taken. Once new error data is available, another output can be calculated.

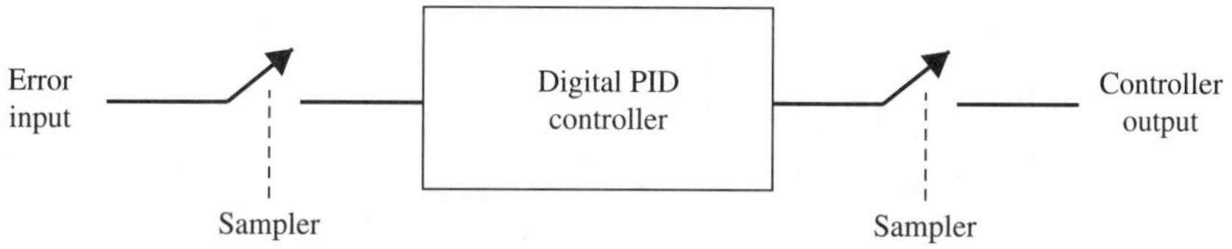

Figure 12.7 Digital PID controller.

The error data can be thought of as a set of pulses, which appear at fixed (sampling) intervals—hence the name pulse train. So, the error data appear as a pulse train spaced apart by one sampling interval, T. The controller calculates a new output only when new error data becomes available; hence, output from the controller also appears as a set of pulses (pulse train) spaced apart by one sampling interval.

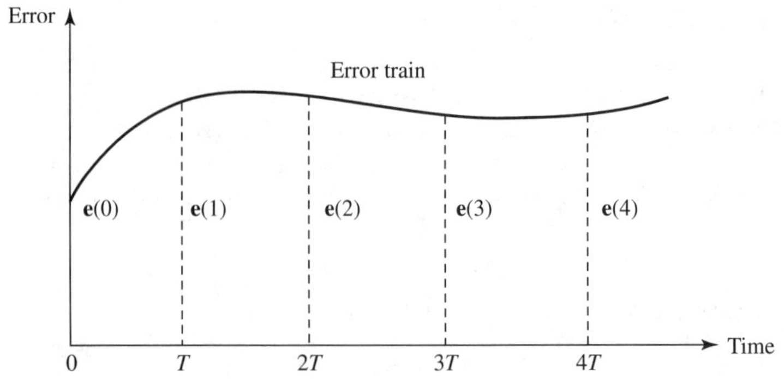

$\mathbf{e}(0)$ = error at time $t = 0$ (zeroth sample)
$\mathbf{e}(1)$ = error at time $t = T$ (first sample)
$\mathbf{e}(2)$ = error at time $t = 2T$ (second sample)
$\mathbf{e}(3)$ = error at time $t = 3T$ (third sample)
$\mathbf{e}(4)$ = error at time $t = 4T$ (fourth sample)

Figure 12.8 Error pulse train.

Conventionally, a pulse train is represented by a bold character containing an n in parentheses. The character is chosen so that it provides some meaningful information about the signal being used. When indicating the pulse value corresponding to a specific time interval, the letter n is replaced by the sample number. For example, $\mathbf{e}(n)$ can be used to represent the error train, and $\mathbf{e}(0)$ represents the zeroth sample.
Figures 12.8 and 12.9 show an error pulse train and output pulse train, respectively.

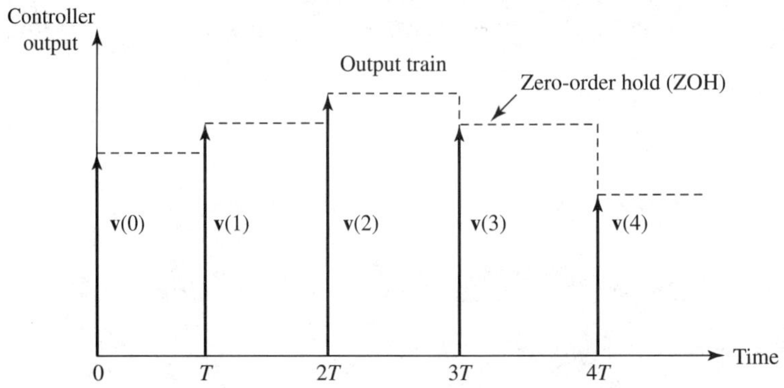

$\mathbf{v}(0)$ = controller output at time $t = 0$ (zeroth output to ZOH)
$\mathbf{v}(1)$ = controller output at time $t = T$ (first output to ZOH)
$\mathbf{v}(2)$ = controller output at time $t = 2T$ (second output to ZOH)
$\mathbf{v}(3)$ = controller output at time $t = 3T$ (third output to ZOH)
$\mathbf{v}(4)$ = controller output at time $t = 4T$ (fourth output to ZOH)

Figure 12.9 Output pulse train.

Zero-Order Hold (ZOH)

The controller output is connected to the process only for an instant, when the sampler switch is closed. After this brief instant the controller is disconnected from the process, and the connection is not made again until the next sampling instant. Output applied in this fashion will not be able to properly control the process. A device is needed to capture this output and retain it until the next sampling instant. A ZOH (Figure 12.10) is such a device. It is a sample-and-hold amplifier, which accepts the controller output at the sampling instant and holds this value until it is updated at the next sampling instant. As soon as the controller has completed the next round of calculations, it generates a new output and sends it to the ZOH. The ZOH accepts the new controller output and again retains it for the duration of the next sample period. As far as the process is concerned, an output is always available from the controller, and the control loop can function normally. Without a ZOH, controller output would be reduced to sheer impulses and would be of no practical use.

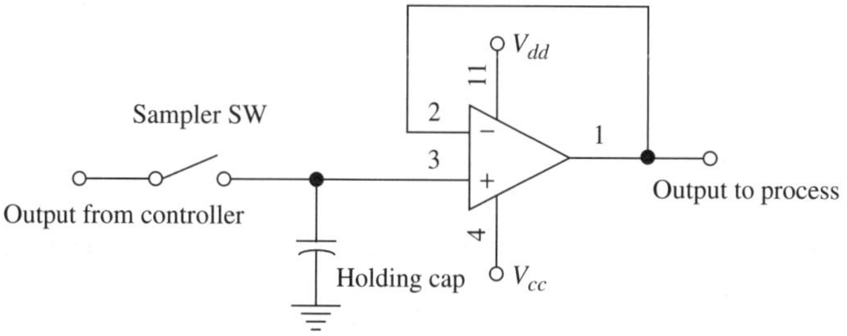

Figure 12.10 ZOH.

Proportional Control Algorithm Proportional output is simply the product of current (instantaneous) error and the proportional coefficient.

$$\text{current controller output} = \text{current error} \times \text{proportional coefficient}$$

$$v_o = K_P e \qquad \text{(analog)}$$

$$V(n) = K_P E(n) \qquad \text{(digital)}$$

Integral Control Algorithm In integral mode, controller output depends on the integral coefficient (integral time) and the past error history.

$$v_o = K_i \int e dt$$

The term $\int e\, dt$ is nothing but the area under the error-time curve (Figure 12.11). In the digital domain, a continuous error curve is not available. Error information is available

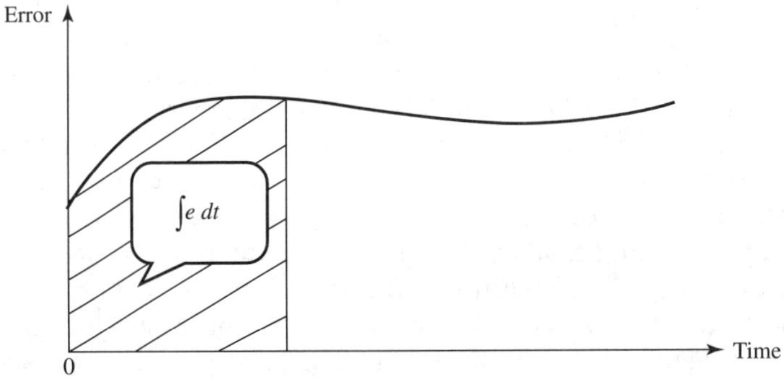

Figure 12.11 Integral output.

only at the sampling instants, and it is retained (stored) by the digital controller in its internal memory.

Time $t = 0$: At time $t = 0$, there is a finite error $E(0)$, but it is just a vertical line, so $\int e\,dt$, the area under the error curve, is zero (Figure 12.12).

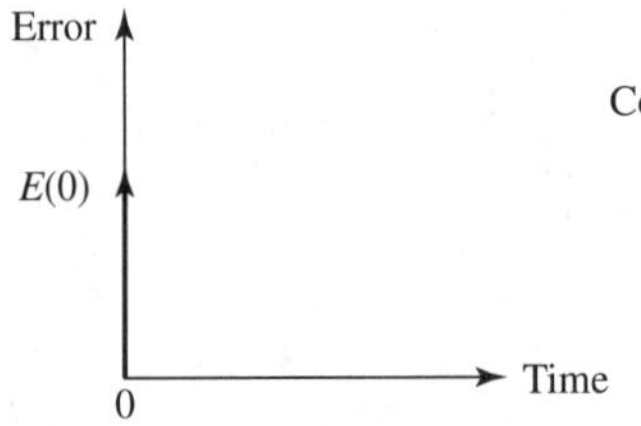

Figure 12.12 Integral output ($t = 0$).

Time $t = T$: In the real world, an error curve is a continuous function and is defined for every single point between $t = 0$ and $t = T$. But, to the digital controller, error information is available only at two discrete points (0 and T). In order to get an estimate of the area under the error curve, it can be assumed that error has been changing linearly from $t = 0$ to $t = T$. This leads to a trapezoidal approximation of error, whose area can be easily calculated.

$$\int e dt = \text{area under error curve from } t = 0 \text{ to } t = T$$

$$\int e dt = \left[\frac{E(0) + E(1)}{2} T\right]$$

$$\text{controller output } V(1) = \frac{K_I T}{2}[E(0) + E(1)]$$

See Figure 12.13.

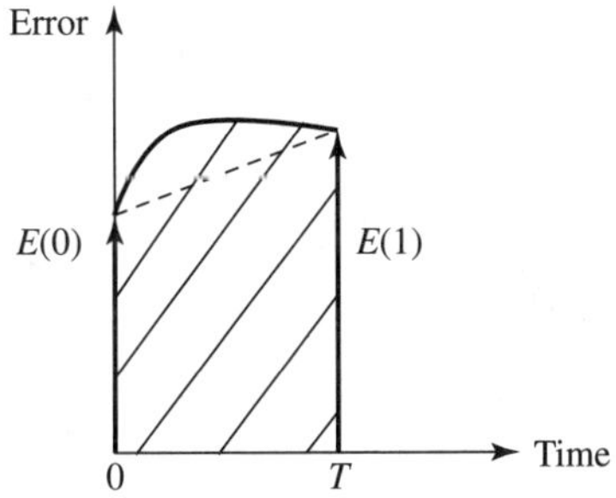

Figure 12.13 Integral output ($t = T$).

Time $t = 2T$: Similarly, it can be assumed that error is changing linearly from $t = T$ to $t = 2T$, and the total area under the error curve from $t = 0$ to $t = 2T$ can be determined (Figure 12.14).

$$\int e dt = \text{area under error curve from } t = 0 \text{ to } t = 2T$$

$$\int e dt = \left[\frac{E(0) + E(1)}{2} T\right] + \left[\frac{E(1) + E(2)}{2} T\right]$$

$$\int e dt = \frac{T}{2}[E(0) + 2E(1) + E(2)]$$

$$\text{controller output } V(1) = \frac{K_I T}{2}[E(0) + 2E(1) + E(2)]$$

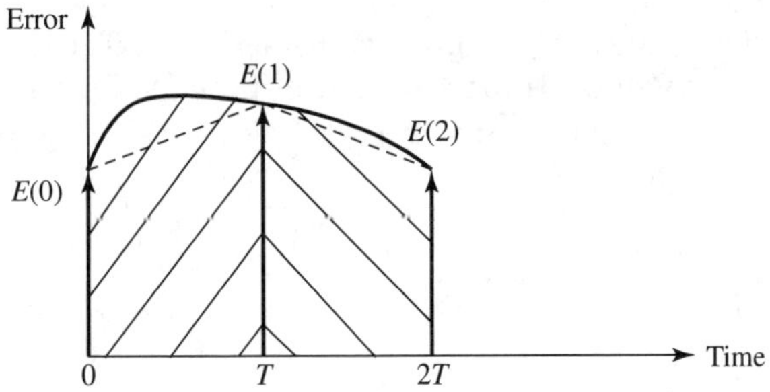

Figure 12.14 Integral output ($t = 2T$).

Time $t = nT$: Following the same train of thought, the area under the error curve at time $t = nT$ can be calculated:

$$\int e dt = \frac{T}{2}[E(0) + 2E(1) + 2E(1) + 2E(2) + \cdots + 2E(n-1) + E(n)]$$

$$\int e dt = T\left[\frac{E(0)}{2} + E(1) + E(1) + E(2) + \cdots + E(n-1) + \frac{E(n)}{2}\right]$$

An interesting observation can be made here. In the final expression for area under the error curve, every single error term appears as a whole, except the first and the last terms. This stems from the fact that over a large period of time, the trapezoidal rule essentially approaches the rectangular rule for calculation of area (Figure 12.15).

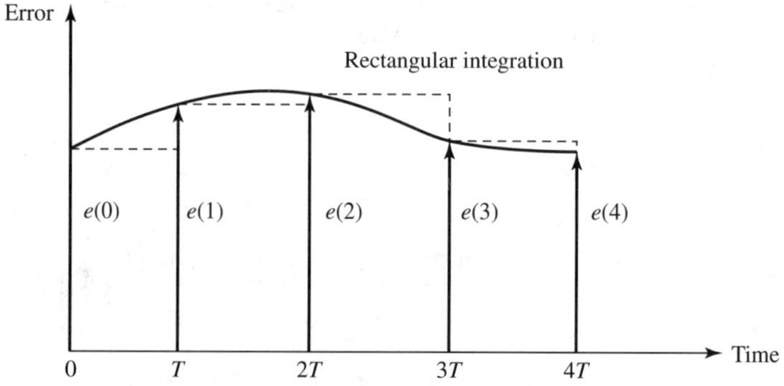

Figure 12.15 Rectangular integration.

In a rectangular integration rule, the error is assumed to remain constant at the previous value and change only at the sampling intervals. Differences between the trapezoidal calculation and the rectangular calculation diminishes as the number of samples increases. This is a fairly valid assumption in real life; when a controller is once put into service, it will continue to operate (practically) forever. So, simplifying the expression for area under the error curve, the integral at time $t = nT \int_{t=0}^{t=nT} e\,dt$ is given as

$$\int e dt = T[E(0) + E(1) + E(2) + \cdots + E(n-2) + E(n-1) + E(n)]$$

$$\int e dt = T \sum_{i=0}^{i=nT} E(i)$$

Correspondingly, controller output for the integral mode is

$$V(n) = K_I T \sum_{i=0}^{i=nT} E(i)$$

Derivative Control Algorithm In this operating mode, controller output depends on the slope of the error curve at that time (Figure 12.16).

$$v_o = K_D \frac{de}{dt}$$

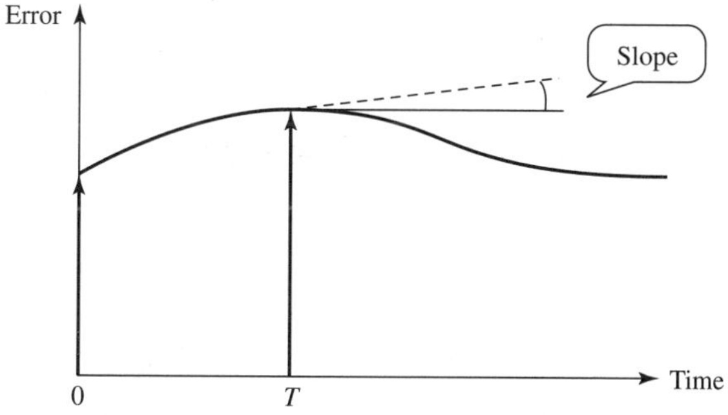

Figure 12.16 Derivative output.

The instantaneous slope of the error curve provides information on the *tendency* of error. Thus, a corrective action can be taken even before error has reached that level. This seems to provide a sort of intelligence to the controller—hence the name *anticipatory control*. Because, in a sampled data system, error information is available only at discrete intervals, the slope can only be estimated based upon these terminal values.

$$\frac{de}{dt} = \frac{E(n+1) - E(n)}{T}$$

Unfortunately, the data value $E(n+1)$ is a future value and is not available at this instant ($t = nT$). So, the next-best estimate of slope is made by using the present error, $E(n)$, and the previous error, $E(n-1)$. Controller output $V(n)$ can thus be calculated:

$$V(n) = K_D \left[\frac{E(n) - E(n-1)}{T} \right]$$

$$V(n) = \frac{K_D}{T} [E(n) - E(n-1)]$$

IDEAL DIGITAL PID ALGORITHM

Combining all three modes of control, the controller output is

$$V(n) = K_P E(n) + K_I T \sum_{i=0}^{i=nT} E(i) + \frac{K_D}{T}[E(n) - E(n-1)]$$

This algorithm is also known as the *position algorithm*. It is considered to be a safe algorithm where a temporary loss of control—e.g., controller malfunction or loss of communication—can be tolerated. In such a case, after control is restored, the control system will act to recover and eventually reach the desired state.

Simplifying the preceding equation,

$$V(n) = K_P + S_n + \frac{K_D}{T}[E(n) - E(n-1)]$$

$$V(n) = A_1 E(n) + S_n + A_3[E(n) - E(n-1)]$$

where

$$S_n = A_2 E(n) + S_{n-1}$$

$$S_{n-1} = K_I T \sum_{i=0}^{i=(n-1)T} E(i)$$

$$A_1 = K_P$$

$$A_2 = K_I T$$

$$A_3 = \frac{K_D}{T}$$

Once the coefficients K_P, K_I, and K_D and the sampling interval T_S are set, the values of A_1, A_2, and A_3 become constant. This formulation requires fewer calculations and is great for a moderate-sized microprocessor. In a very small system, the values of A_1, A_2, and A_3 can be calculated externally and substituted in the code. This will tend to reduce the truncation error, which is a problem in systems without the capability for floating-point calculations.

Skeleton 'BASIC' language routines for implementing a discrete PID controller are shown next.

Discrete PID Algorithm—BASIC Language

Here only the main PID control loop has been considered. For a functional controller, the routines for Read_SP(), Read_MV(), and Write_Output() need to be added. The implementation of these routines depends upon the hardware used. Some controllers have built-in A/D and D/A, whereas in others, these may be located remotely and communicating to the controller through a serial link.

```
'E0  = Current error
'E1  = Previous error
'E2  = Error two sample intervals before
'S0  = Current integral output
'S1  = Previous integral output
'V0  = Current controller output
'V1  = Previous controller output

Init_Controller()             'Get coefficients & calculate A1, A2, and A3

While (NOT Stop())
       IF (Loop_Time) THEN
              Read_SP()
              Read_MV()
              Run_Loop()
              Write_Output()
       ENDIF

       IF (Change_Coefficient_Flag) THEN
               Read_Coefficients()
       ENDIF

       Do_Housekeeping()
WEND


Sub Run_Loop()

       S0 = A2 E0 + S1                      ' Calculate current integral output
       V0 = A1 E0 + S0 + A3 (E0 - E1)       ' Controller output
       S1 = S0                              ' Shift data for next time
       E1 = E0

       RETURN
End Sub


Sub Init_Controller()

       T = 0.1          'Change Sample interval to suit your cycle time
       E0 = 0
       E1 = 0
       S0 = 0
       S1 = 0

       Read_Coefficients ()
       ' Do other initialization
       RETURN
End Sub
```

```
Sub Read_Coefficients()

        PRINT " Proportional Coefficient Kp :" Kp
        PRINT " Enter Y to change ";
        INPUT ans$
        IF (UCASE$(ans$) = "Y") THEN
              INPUT "Enter new value of Kp :"Kp
        ENDIF

        PRINT " Integral Coefficient Ki : " Ki
        PRINT " Enter Y to change ";
        INPUT ans$
        IF (UCASE$(ans$) = "Y") THEN
              INPUT "Enter new value of Ki : "Ki
        ENDIF

        PRINT " Derivative Coefficient Kd : " Kd
        PRINT " Enter Y to change ";
        INPUT ans$
        IF (UCASE$(ans$) = "Y") THEN
              INPUT "Enter new value of Kd : "Kd
        ENDIF

        ' Calculate A1, A2, and A3
        A1 = Kp
        A2 = Ki x T
        A3 = Kd / T

        RETURN

End Sub
```

Manual Control Mode

An ideal discrete PID control algorithm does not allow for a manual control mode. In a number of applications, a plant is started in *manual control mode* and switched to automatic control later (Figure 12.17). During this manual run, proportional, integral, and derivative coefficients are adjusted, and the plant is brought to the desired speed.

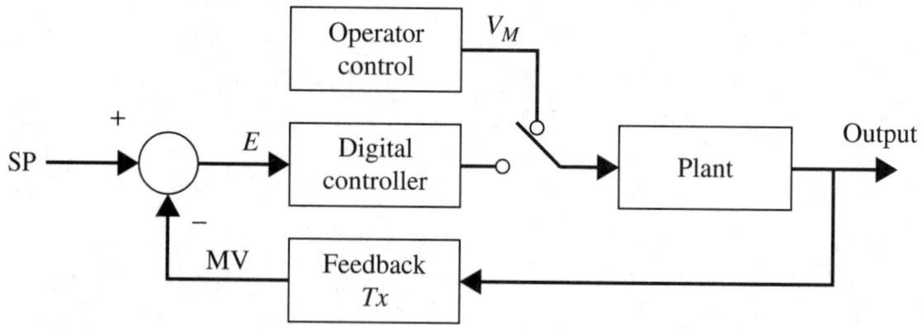

Figure 12.17 Manual control mode.

Once the coefficients have proven to be satisfactory, the plant is switched over to computer control. The ideal PID control algorithm can be modified by adding an additional term, V_M, to the basic equation.

$$V(n) = K_P E(n) + K_I T \sum_{i=0}^{i=nT} E(i) + \frac{K_D}{T}[E(n) - E(n-1)] + V_M$$

where V_M is the controller output under manual control.

During the manual mode, the plant FCE (final control element) does not accept the output from the PID controller but is controlled by the operator. Thus, in this mode

$$V(n) = V_M$$

The operator continues to modify V_M while observing the error (error = SP – MV) and controller output. When the error is very small (approximately zero), the controller can be switched to operate in *AUTO (automatic mode)*. If the error at the time of the switchover is a very small quantity, controller output will not suddenly change (jump) to a very different value, and the transition from *MANUAL* to *AUTO* will be relatively *bumpless*. But if there is a significant error during the time of switchover, there may be a considerable change (bump) in controller output. It should be noted that this manual control at startup is akin to the offset value introduced in a proportional controller (refer to Chapter 9).

Velocity Algorithm

True bumpless transfer can be achieved by using another type of PID algorithm. Controller output at the nth sample interval is given as

$$V(n) = K_P E(n) + K_I T \sum_{i=0}^{i=nT} E(i) + \frac{K_D}{T}[E(n) - E(n-1)]$$

Controller output at one sample time before ($t = (n-1)T$) is given as

$$V(n-1) = K_P E(n-1) + K_I T \sum_{i=0}^{i=(n-1)T} E(i) + \frac{K_D}{T}[E(n-1) - E(n-2)]$$

Subtracting controller output at $t = (n-1)T$ from the output at $t = nT$,

$$V(n) - V(n-1) = K_P[E(n) - E(n-1)] + K_I T E(n) + \frac{K_D}{T}[E(n) - 2E(n-1) + E(n-2)]$$

$$V(n) = V(n-1) + \left[K_P + K_I T + \frac{K_D}{T}\right]E(n) - \left[K_P + 2\frac{K_D}{T}\right]E(n-1) + \frac{K_D}{T}E(n-2)$$

$$V(n) = V(n-1) + A_1 E(n) + A_2 E(n-1) + A_3 E(n-2)$$

where
$$A_1 = \left[K_P + K_I T + \frac{K_D}{T}\right]$$
$$A_2 = -\left[K_P + 2\frac{K_D}{T}\right]$$
$$A_3 = \frac{K_D}{T}$$
$$V(n) = \text{current controller output}$$
$$V(n-1) = \text{previous controller output}$$

This is a fairly simple equation and can be easily implemented. In the manual mode of operation, the operator directly controls the controller output, $V(n)$. When the plant is at the desired operating point (small error), the control mode can be switched to AUTO. As soon as the controller is switched to AUTO, the last manual output is treated as the previous output, and the controller generates a new output by slightly modifying the previous output term.

Improved Derivative Term

Calculations for the derivative term can be affected by measurement noise (error). An improved four-point averaging technique [3] provides a better result. Here, the slope of the error curve at time $t = nT$ is estimated as

$$\frac{de}{dt} = \frac{1}{6T}[e(n) + 3e(n-1) - 3e(n-2) - e(n-3)]$$

Substituting this value of slope into controller output at $t = n$T,

$$V(n) = K_P E(n) + S_n + \frac{K_D}{6T}[E(n) + 3E(n-1) - 3E(n-2) - E(n-3)]$$
$$V(n) = A_1 E(n) + A_2 E(n-1) + A_3 E(n-2) + A_4 E(n-3) + S_n$$

where
$$A_1 = K_P + \frac{K_D}{6T}$$
$$A_2 = \frac{K_D}{2T}$$
$$A_3 = -\frac{K_D}{2T}$$
$$A_4 = -\frac{K_D}{6T}$$

MICROCONTROLLER IMPLEMENTATION OF A PID CONTROLLER

A detailed project design is presented here that covers a complete implementation of a digital PID controler.

DC Motor Control

Requirements: A DC motor is required to be run at a controlled velocity under varying load. A tachometer coupled to the DC motor provides the velocity feedback (Figure 12.18).

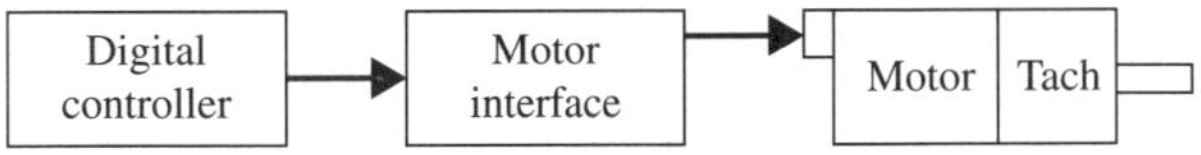

Figure 12.18 Motor control system.

The development of a digital controller can be divided into two major sections:

1. Controller hardware
2. Controller software

Controller Hardware Controller hardware consists essentially of two major components:

- Microcontroller unit
- Interface unit

Today's designer is faced with a very large number of microcontrollers to choose from. These microcontrollers range from basic 8-bit units to powerful 32-bit units. Although 4-bit microcontrollers are still being used in some simple applications, these do not lend themselves to moderate to fast industrial control. The development of a microcontroller unit is somewhat involved and does not warrant the time or expense of designing and developing a special purpose unit. For this project a commercially available microcontroller unit PIC Module 2 [14] has been chosen, incorporating Microchip Technology's PIC16C773 uC. The motor interface has been developed especially for this control project.

PIC Module 2: The PIC Module 2 is a small controller board that contains all the features necessary for implementing a dedicated and high-performance digital controller. The salient features are as follows:

- 8-bit PIC16C773 microcontroller
  - Six channels of fast 12-bit A/D conversion
  - PWM output
  - Serial communications
  - IIC/SPI interface
- RS-232 communication
- SPI EEPROM for coefficient storage
- On-board 5-V regulation

All interface signals are brought out to an edge connector, which can then be connected to any additional signal-conditioning circuits. Figure 12.19 shows the schematic of the controller board.

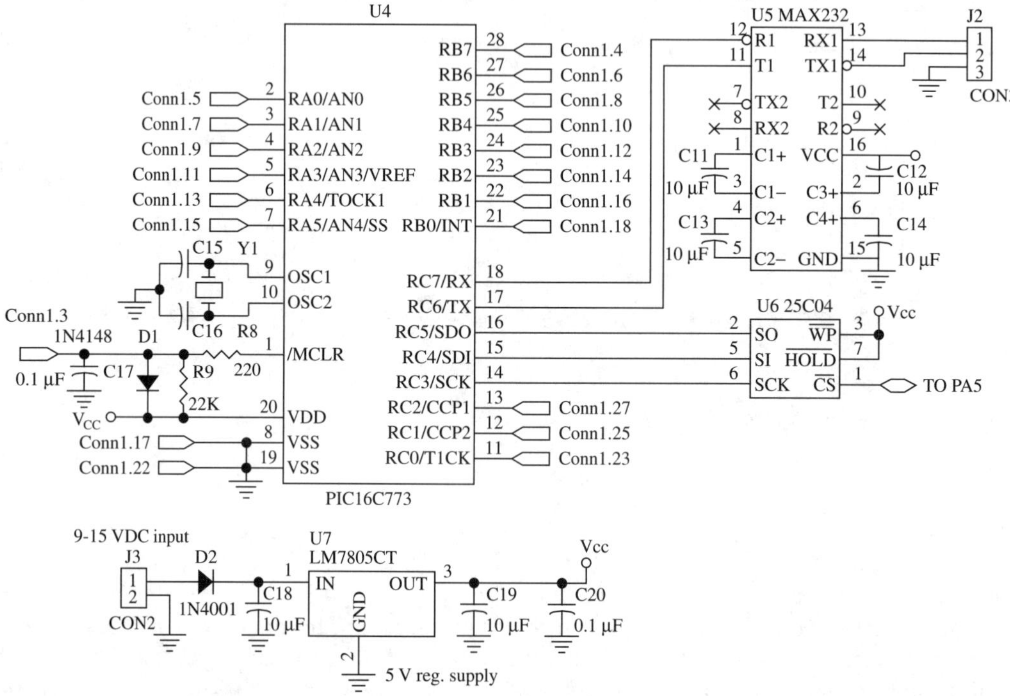

Figure 12.19 Microcontroller board.

Motor Interface: The motor interface schematic diagram is presented in Figure 12.20. The motor is driven by an H-bridge driver LMD 18201 from National Semiconductor [11]. It is a dedicated motor driver chip with up to 3 A of continuous

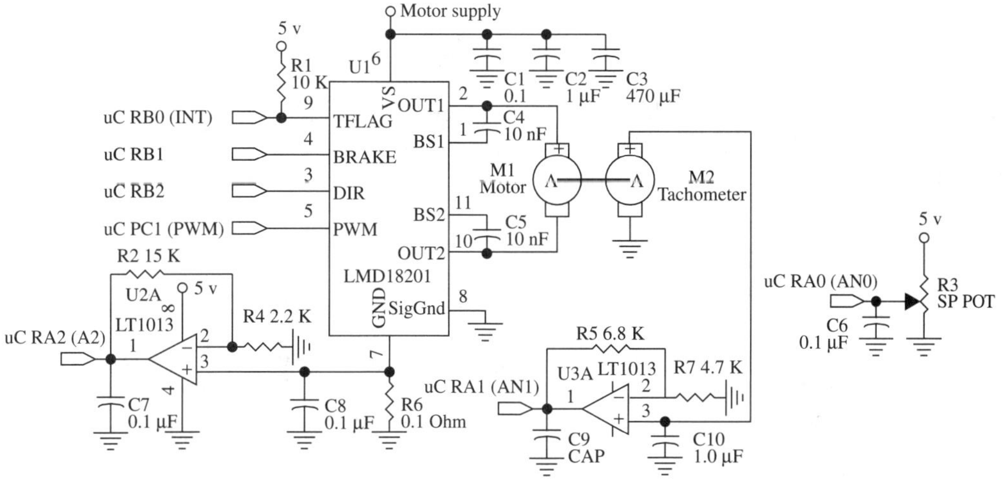

Figure 12.20 Motor interface.

current at voltages reaching 55 V DC. The speed feedback is provided by a tachometer coupled to a motor. Motor current is monitored through a sense resistor located in the ground path of the H-bridge. The current flowing through this sense resistor develops a voltage proportional to the motor current. This voltage is filtered through a low-pass RC filter and amplified before being fed to one of the A/D inputs of PIC uC.

The H-bridge is controlled by a PWM signal originating from the microcontroller. Directional control input controls the motor direction. In this project, only a unidirectional motor movement is considered. Bidirectional control will require additional hardware to condition the bipolar tachometer output (A/D input is only unidirectional, as is the case for almost all uC-based devices). A potentiometer is used to enter the SP (desired) velocity information into the digital controller.

Controller Software Assembly language is probably the best choice in terms of operating speed and exact control of every single bit of microcontroller peripheral I/O (input and output). MPLAB [12] provides an excellent platform for simulating, testing, and debugging the assembler code. Unfortunately, the assembly language programming is also very unforgiving and demands every bit of understanding and attention from the programmer. In order to assist the control system designer/developer, high-level language tools have proliferated recently. The motor control software was developed using CCS C Compiler [15], which allows a PIC microcontroller to be programmed in C language. It has built-in functions to directly control microcontroller peripherals.

CCS C Compiler A C compiler hides the somewhat cryptic instruction set of a microcontroller from the programmer, allowing the development to be carried out in almost universal C code. CCS C compiler goes a little further than that. It contains several prebuilt functions dedicated to PIC microcontroller hardware. Thus, the user is sheltered from the details of properly setting up some of the on-board peripherals. It has a built-in interface for MPLAB, allowing the developed program to be tested and simulated under MPLAB. The data variables can be declared as bit, byte, integer, or floating-point types.

Control Software Description

The control program consists of two major parts.

System Initialization On coming out of power-on reset, the microcontroller peripherals are configured. This includes setting up the A/D converter, PWM modulation circuit, timer and port I/O. The user is then prompted to enter the PID control coefficients.

Endless Control Loop Once the coefficients are entered, the program enters an indefinite loop. At the beginning of this loop, SP and MV are read through the A/D converter, and the value of error is computed. This error signal, along with previous error values, is used in PID algorithm to calculate controller output. This output is checked for out-of-bound conditions. A PWM signal is then generated in accordance with this output, which in turn controls the motor velocity. The program loops once every millisecond (sample time = 1 ms). The control loop only takes about 500 μs to execute, and a balance of an additional 500 μs (idle time) is available for other tasks. A commercial PID controller will have some additional requirements, such as looking for upper and lower operating limits, communication with the supervisory computer, etc., and these tasks can be executed during this idle time period. A fully commented listing (Program Listing 12.2) of the program is included.

Program Listing 12.2

```
////////////////////////////////////////////////////////////////////////
/
//                          PID1.C
///
//
///
////////////////////////////////////////////////////////////////////////
/

// S.Gupta
// Mar 01,00
// Time: 13:10

// Digital Implementation of conventional PID algorithm
// PIC16C773 microcontroller with 8 MHz crystal frequency
```

```
#include <16C773.H>

#BYTE OPTION = 0x81
#bit TOIF = 0x0B.2

#use Delay(Clock=8000000)

#use RS232(Baud=9600, Xmit=PIN_C6, Rcv=PIN_C7, brgh1ok)

#define BRAKE PIN_B1        // Output to apply brake to motor
#define DIRECTION PIN_B2    // Output to control motor direction
#define OVERHEATED PIN_B0   // Input H-bridge temperature flag
#define ALARM PIN_B4        // Output to buzzer/alarm

// *****************************************************************
//    Global Variables
// *****************************************************************
// All the control coefficients are entered as multiples of
// desired values. This provides a hidden decimal point
// to improve calculations

   long int KP;         // Entered as  Kp * 8
   long int KI;         // KI is entered as Ki*TS
                        // KI = (Ki * 0.001024) * 256
   long int KD;         // Entered as (Kd/TS) * 256
                        // KD = (Kd /0.001024) * 256
   long int TS;         // Cycle time
   long int V0;         // Current output
   long int V1;         // Last output
   long int ER0;        // Current error
   long int ER1;        // Previous error
   long int ER2;        // Previous error
   int SP_VEL;          // Desired velocity
   int MV_VEL;          // Measured velocity
   byte duty;
// ***************************************************************

initialize()
{
   // Initialize all variables

   byte tmp;
   // Arbitrary assignment of controller coefficients
   KP = 2;        // Proportional Coefficient
   KI = 1;        // Integral Coefficient
   KD = 1;        // Derivative Coefficient
   V0 = 0;        // Current Controller output
   V1 = 0;        // Previous Controller output
   ER0 = 0;       // Current error
   ER1 = 1;       // Last error
   ER2 = 0;       // Second last error
   SP_VEL = 0;    // Current desired velocity
```

```
    MV_VEL = 0;      // Current motor velocity
    TS = 977;        // Sample time 977 ms
    duty = 0;        // Duty cycle

    // Set up peripheral functions
    // set motor direction, and no brake
    output_bit(DIRECTION,1);        // Set direction as forward
    output_bit(BRAKE,0);            // No brakes
    output_bit(ALARM,0);            // Reset alarm
                                    // A buzzer/light can be connected to
                                    // this output

    set_pwm1_duty(duty);    // Set up 0 duty cycle at the start
    setup_ccp1(CCP_PWM);    // Configure CCP1 as a PWM
    setup_timer_2(T2_DIV_BY_4, 255, 1);
    set_pwm1_duty(duty);            // This sets duty cycle of 0

    // Set up A/D
    // First two channels are used to read SP & MV
    setup_port_a  (ALL_ANALOG);
    setup_adc  (ADC_CLOCK_INTERNAL);

    // Set up timer0 for Sample Clock
    // Timer0 overflow every 1 ms Xtal = 8 MHz

    tmp = OPTION;          // Read option register
    tmp = tmp & 0xC0;      // Clear PSA bit. Assign Prescaler to Timer0
    tmp = tmp | 0x02;      // Assign Prescaler (001) to div by 8
                           // Timer0 will overflo every msec.
    #asm
       clrwdt              // Start watch dog timer
    #endasm
    OPTION = tmp;         // Set option register
}

long int input_num()
    {
    // There is no built in routine to read an integer number through
    // serial port.

    // Read an integer through serial port
    // Routine is terminated
         // on receiving a non decimal char
         // on receiving 4 chars
    // Routine returns a long integer (16 bit value)

    short exit_flag;  // Local flag to indicate a non decimal char
    byte b;
    byte d;           // Number of digits.
    long int n;
```

```
   exit_flag = 0;     // reset the flag
   d = 0;
   b = 0;
   n = 0;

   do
      {
      b = getch();    // Read a char
      if ((b >='0') && (b <= '9'))
         {
         b = b - '0';
         n = n * 10;
         n = n + b;
         d = d + 1;      // Increment digit counter
         if (d > 3)      // Exit if four digits received
            exit_flag = 1;
         }
      else            // Non decimal char received
         exit_flag = 1;
      }
   while (!exit_flag);  // repeat if flag is not set
   return n;
   }


read_coefficients()
   {
   // Accept the control coefficient through serial port
   // and set the global variables

   printf("\nEnter KP: ");
   KP = input_num();

   printf("\nEnter KI: ");
   KI = input_num();

   printf("\nEnter KD: ");
   KD = input_num();

   //  Display the coefficients to user
   printf("\n KP: lu",KP);
   printf("\n KI: lu",KI);
   printf("\n KD: lu",KD);
   }


main() {
   byte tick ;
   long int ad_result;
   long int tmp_x, tmp_y;

   long int K1;      // First Coefficient
```

```
long int K2;      // Second Coeffcient
long int K3;      // Third Coefficient

initialize();     // Initialize variables & setup peripherals

read_coefficients();     // Read the control coefficients
                         // Coefficients are only read once. Reset
                         // uC in order to enter new values
                         // Code can be modified to enter the values
                         // on the fly and store these in EEPROM

While(1)                 // Begining of infinite loop
   {
   while(! TOIF);        // wait till TOIF flag is zero
                         // It will be set to 1 on Timer0 overflow.
   TOIF = 0;             // Reset the TOIF flag
   restart_wdt();        // Reset watch dog timer

   set_adc_channel(0);  // Read SP_Vel
   ad_result = Read_ADC();
   SP_VEL = ad_result;

   set_adc_channel(1);  // Read MV_Vel
   ad_result = read_adc();
   MV_VEL = ad_result;

   ER0 = SP_VEL- MV_VEL;       // Current error

   // Calculate output
   // Velocity algorithm used
   // V0 = V1 + KP*(ER0-ER1) + KI*ER0*TS + KD*(ER0-2*ER1+ER2)/TS;
   // V0 = V1 + K1*(ER0-ER1)/8 + K2*ER0/256 + K3*(ER0-2*ER1+ER2)/256;

   tmp_x = (ER0 - ER1);     // Proportional
   tmp_x = KP * tmp_x ;
   tmp_x = tmp_x >> 3;      // Divide by 8
   V0 = V1 + tmp_x;

   tmp_x = K2* ER0;         // Integral


   tmp_x = tmp_x >> 8;      // Divide by 256
   V0 = V0 + tmp_x;

   tmp_x = ER0 + ER2;
   tmp_y = 2 * ER1 ;
   tmp_x = tmp_x - tmp_y;
   tmp_y = tmp_x * K3;
   tmp_y = tmp_y >> 8; // Divide by 256


   V0 = V0 + tmp_y;
```

```
        // Check the output for bounds
        if (V0 < 0)
           V0 = 0;
        if (V0 > 255)
           V0 = 255;
        duty = V0;

        // Shift Variables
        V1 = V0;
        ER2 = ER1;
        ER1 = ER0;

        set_pwm1_duty(duty); // Set output duty cycle

        // Housekeeping
        // Check temperature flag and if overheated sound a buzzer
        // Reset buzzer if temperature flag resets
        if(input(OVERHEATED))
           output_bit(ALARM,1);
        else
          output_bit(ALARM,0);
        // Additional tasks
        // Enter Tasks here
        // ...
        }
}
```

PROBLEMS

1. A closed-loop control system uses an analog PID controller, as shown here. The controller needs replacement, and it is decided to replace it with a digital PID controller. Can a typical commercial digital controller be used as a drop-in replacement of an analog unit?

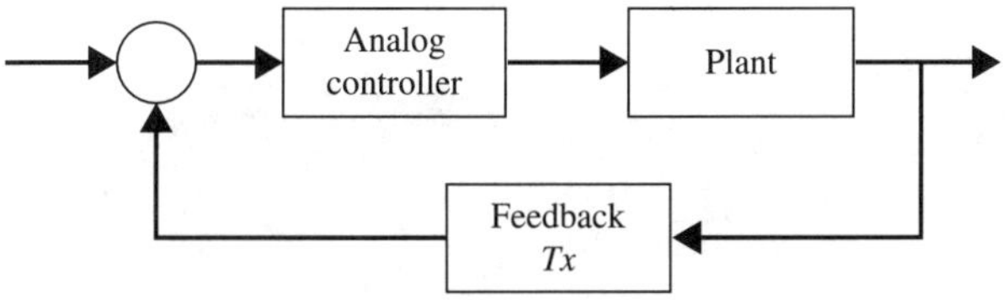

2. Redraw the block diagram with a digital controller replacing the analog unit.
3. Is it possible to set up any computer to be used as a drop-in replacement for an analog controller?

4. An A/D converter is required to be able to measure to better than 0.25% of its FS (full-scale) range. Which of the following converters are suitable?
   a. 8-bit
   b. 10-bit
   c. 12-bit
5. A flow control valve is being used as an FCE in a digital level control system. The valve requires an input of 0 to 10 V to operate over its FS range of 0 to 100 L/min. An 8-bit D/A converter controls the valve opening. What is the smallest change in flow rate that can be caused by the digital controller? Assume a linear input/output relationship for the flow control valve.
6. Review the theory behind a two-position control system and the microprocessor implementation of it. Modify the system (software and hardware) to implement a floating control mode, where the output drops to 50% of controller output while the error is within the dead band.

   Can a scheme be devised so that a power transistor is still employed as the final control element and 50% output is generated without using any additional components?
7. What is the advantage of the velocity algorithm over the position algorithm?
8. Do the names *velocity* and *position* in the velocity and position algorithm indicate that these are designed only for use in velocity and position control systems, respectively?
9. Write a BASIC language program to implement a PD controller.

13

FUZZY CONTROLLER

INTRODUCTION

Both modern and classical control system analysis and design techniques require that a plant mathematical model be formed. These techniques further require a good understanding of control system theory. It is not possible to integrate the experience and knowledge gained in the operation of a similar plant into control algorithm. But the real world is full of examples of successful control systems where a mathematical model is never formed or is difficult to obtain. A common example is the control of an automobile. The driver can anticipate the control requirements and execute them successfully (*most of the time*). Fuzzy control, which has its roots in fuzzy set development proposed by Professor Zadeh, allows the experience and knowledge gained in previous systems to be applied for the control of processes. Some may even refer to this method as *common-sense control*. In this chapter the concepts of fuzzy control are introduced. Terms commonly referred to in industry in relation to fuzzy logic and fuzzy control are described. Worked examples are included to show the application of fuzzy control for industrial applications.

CONTINUUM WORLD

The following real-world properties are continuous in nature and do not change in discrete steps:

- Changes in atmospheric temperature and pressure occur in a continuous form and not in steps.
- Earth's atmosphere varies continuously from sea level to the boundaries of space. It is not possible to pinpoint a distance from Earth as the (sudden) end of atmosphere and beginning of space.
- Although it is well understood that there is a range of people from poor to rich, there is no magical number (wealth) below which everyone can be considered to be poor.
- The velocity of a bullet, even when fired from a high-powered rifle, changes in a continuous fashion from zero to the maximum. The maximum amount of acceleration, although occurring during the initial fraction of a second, is still a continuous change.
- In the decimal number system, a number can always be found that is closer to a given number than the one just defined.

EXAMPLE 13.1

Find all the decimal numbers between 2 and 3.

Solution

An infinite number of decimal numbers lie between 2 and 3.

EXAMPLE 13.2

Find a number just greater than a given number, such as 4.0.

Solution

5 is a solution to this problem.

4.1 is a better solution.

4.01 is still better.

4.001

⋮

4.000000001, and so on

In short, it is not possible to specify a number as the one just greater than 4. Once a number is defined, it is possible to find another number closer to 4 than the defined number.

Conventional Logic

In spite of the continuum nature of the world, it is a common practice to try to express quantities using discrete logic. Classifications are set up using discrete and well-defined boundaries. Atmospheric temperature is routinely expressed in integer degrees and sometimes to the first decimal place, such as 68°F or 20°C. When temperature increases, it may be expressed as 20.1°C. The temperature does not suddenly jump from 20.0 to 20.1°C, but it takes a time interval (however small), during which the temperature increases in a continuous fashion.

In order to classify the height of a person, three (or more) categories are commonly used, *short, medium, and tall* (Figure 13.1). But this classification is by no means absolute. When considering the height of people, it is possible to find a person shorter or taller than the one in question. Also, the definition of short and tall varies tremendously based upon the geological location. Again, based upon these discrete boundaries, a person with a height of 155.01 cm will be considered to be of medium height and a person with a height of 150 cm will be considered short. In reality, this differentiation does not seem to make a lot of sense.

Figure 13.1 Discrete boundaries for height classification.

Conventional logic deals with sets (groups) that are mutually exclusive. If an element belongs to set *A*, it does not belong to set *B*. Conventional computer logic follows *binary logic*, which works with only two mutually exclusive states, 0 and 1. If the state of an object is defined as 1, then it is not equal to zero. A bulb is either ON or OFF. It is not possible to define the state of a bulb which is partially (dimmed) ON.

FUZZY LOGIC AND FUZZY LOGIC TERMINOLOGY

Fuzzy logic is an extension of conventional (binary) logic. It allows an object to occupy more than one set (state, loosely speaking) at any time. An object can (theoretically speaking) belong to an infinite number of sets at any given time. In fuzzy logic, a bulb (controlled with a dimmer switch) can occupy a state that may be partially ON as well as partially OFF. It is made possible through the concept of *degree of membership*.

Degree of Membership, μ

The **degree of membership** is the degree to which a value (crisp variable) belongs to a fuzzy set (variable, characteristic, or specification). It is expressed either as a fractional value ranging from 0.0 to 1.0 or as a percentage ranging from 0% to 100%.

See Table 13.1 for a sample of degree of membership.

Table 13.1 Degree of Membership

Degree of Membership	***Inference***
0.0 (0%)	The (crisp) variable does not belong to the set.
0.0 < to < 1.0 (0% < to < 100%)	The (crisp) variable partially belongs to the set.
1.0 (100%)	The (crisp) variable fully belongs to the set.

It is possible to have membership in multiple sets, but for the sake of simplicity, the sets are chosen so that at any given time, a value occupies only two sets at most. Figure 13.2 shows the state of a light bulb, ranging from totally OFF (dark) to fully ON (bright). The state of the bulb can be defined by two properties (sets) known as ON and OFF. It can further be stated that at any given time, the bulb occupies both the sets. When the bulb is fully ON, it has 100% membership in the ON set and 0% membership in the OFF set. The total of both memberships is always equal to 100%. Thus a

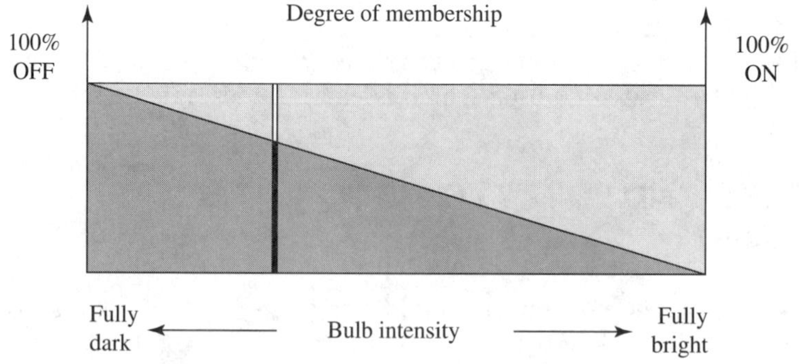

Figure 13.2 Degree of membership.

partially lit bulb (marked by a vertical bar) occupies both sets and has approximately 70% membership in the OFF set and 30% membership in the ON set.

The method in Figure 13.2 for indicating the degree of membership is somewhat awkward and does not lend itself for inclusion in multiple sets. Figure 13.3 shows the same diagram redrawn in a different style. OFF membership has a value of 100% at the extreme left and 0% at the extreme right. A sloping line is drawn joining 0% on the right to 100% on the left. Similarly, a sloping line is drawn representing the ON set. The state of a partially lit bulb is again located on the horizontal axis, and a vertical line drawn from it intersects the two sloping lines. The vertical distance (of the intercept) indicates the degree of membership for each set.

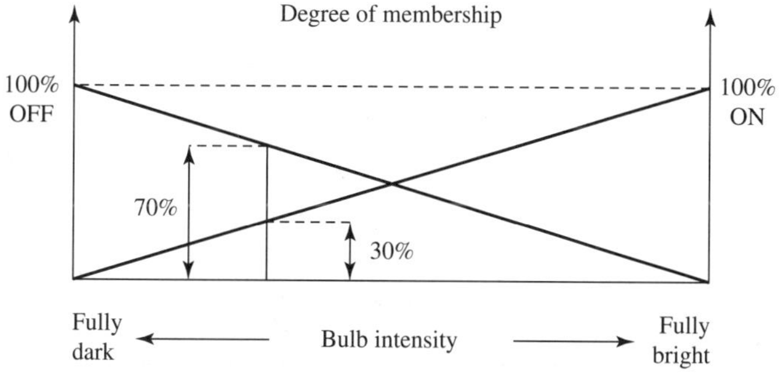

Figure 13.3 Degree of membership: alternative method.

This form of representation also permits the degree of membership to be calculated, without having to resort to graphical representation. Having done that, a computer can be utilized to calculate the membership, and a computer-based fuzzy controller is well on its way.

EXAMPLE 13.3

Draw a diagram to represent the membership of water temperature per the following logic:

hot = water temperature > 60°C
warm = water temperature < 40°C
water temperature range = 30°C to 80°C

Solution

Water temperature ranges from 30°C to 80°C. There are two sets, the HOT set and COLD set. Any time water temperature drops below 40°C, it is considered to be fully warm (100% membership). Similarly, water temperature greater than 60°C qualifies for fully hot (100% membership). Thus the overlap between HOT and WARM lasts only for a temperature ranging from 40°C to 60°C. Once outside this temperature range, the temperature value belongs fully to only one of the two sets. This is represented in Figure 13.4.

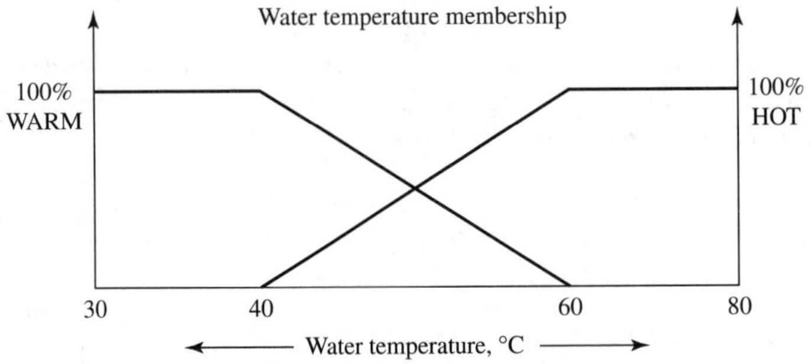

Figure 13.4 Water temperature for Example 13.3.

EXAMPLE 13.4

Based upon Example 13.4, determine the degree of membership for the following temperature values:

a. 35°C
b. 70°C
c. 45°C

Solution

a. A vertical line drawn from the 35°C point on the x-axis intersects only the line representing the WARM set. The point of intersection is at the horizontal section (100%). Thus,

$$\text{degree of membership in WARM set} = 100\%$$
$$\text{degree of membership in HOT set} = 0\%$$

b. Similarly, a vertical line drawn from the 70°C point intersects only the line representing the HOT set. The point of intersection is at the horizontal section (100%). Thus,

$$\text{degree of membership in WARM set} = 0\%$$
$$\text{degree of membership in HOT set} = 100\%$$

c. A vertical line drawn from the 45°C point intersects both the lines representing the WARM and HOT sets. Thus, this temperature value belongs to both the sets. The degree of membership can be determined graphically, but a mathematical approach is taken here. Considering WARM, the sloping line runs from 40°C (100%) to 60°C (0%); thus, the slope of this line is

$$\begin{aligned} \text{slope} &= \frac{100 - 0}{40 - 60}\%/^\circ\text{C} \\ &= -5\%/^\circ\text{C} \end{aligned}$$

Thus,

$$\text{degree of membership} = 100\% + \delta t \cdot \text{slope}$$

where

$$\begin{aligned} \delta t &= \text{current temperature} - \text{temperature corresponding to 100\% point} \\ &= 45^\circ\text{C} - 40^\circ\text{C} \\ &= 5^\circ\text{C} \end{aligned}$$

Thus,

$$\begin{aligned} \text{degree of membership in WARM set} &= 100\% + 5^\circ\text{C} \cdot (-5\%/^\circ\text{C}) \\ &= 100\% - 25\% \\ &= 75\% \end{aligned}$$

Similarly considering the HOT set, the sloping line runs from 40°C (0%) to 60°C (100%); thus, the slope of this line is

$$\begin{aligned} \text{slope} &= \frac{0 - 100}{40 - 60}\%/^\circ\text{C} \\ &= 5\%/^\circ\text{C} \end{aligned}$$

Thus,

$$\text{degree of membership} = 0\% + \delta t \cdot \text{slope}$$

where

$$\begin{aligned} \delta t &= \text{current temperature} - \text{temperature corresponding to 0\% point} \\ &= 45^\circ\text{C} - 40^\circ\text{C} \\ &= 5^\circ\text{C} \end{aligned}$$

Thus,

$$\text{degree of membership in the HOT set} = 0\% + 5^\circ\text{C} \cdot (5\%/^\circ\text{C})$$
$$= 25\%$$

The last result could have been obtained from the fact that the total degree of membership in all sets is always equal to 100%.

Thus,

$$\text{degree of membership in the HOT set} = 100\% - \text{degree of membership in the WARM set}$$
$$= 100\% - 75\%$$
$$= 25\%$$

EXAMPLE 13.5

Draw a diagram showing the membership for water temperature for the following conditions:

$$\begin{aligned} \text{hot} &= \text{temperature} > 60^\circ\text{C} \\ \text{warm} &= 30^\circ\text{C} < \text{temperature} < 60^\circ\text{C} \\ \text{cold} &= \text{temperature} < 30^\circ\text{C} \\ \text{temperature range} &= 10^\circ\text{C–}80^\circ\text{C} \end{aligned}$$

Solution

There are three sets. In order to keep the graphic representation simple, a temperature value can belong to only two sets at the most. A temperature below 30°C has 100% membership in the COLD set; a temperature greater than 60°C has 100% membership in the HOT set; and a temperature ranging from 30°C to 60°C will be shared by the WARM set and one of the other two sets. Two questions need to be answered before a diagram can be drawn.

1. What is the upper limit of the COLD temperature set?
2. What is the lower limit of the HOT temperature set?

There are two possible solutions here.

1. *The same temperature value satisfies both limits.* Any temperature limit in between 30°C and 60°C can be used, but a common approach is to take the middle of this

Table 13.2 Temperature Logic for Example 13.5

Temperature (°C)	Degree of Membership (%)		
	COLD	WARM	HOT
≤30	100	0	0
30 < *t* ≤ 45	100 to 0	0 to 100	0
45 < *t* ≤ 60	0	100 to 0	0 to 100
t > 60	0	0	100

temperature range. Thus, a value of 45°C will be used as the limit. This leads to the assignment in Table 13.2.

Again, the sum of degree of membership in all sets for any temperature value is always equal to 100%. The results are shown graphically in Figure 13.5. The middle set is shaped as a triangle to accommodate the degree of membership in the adjoining sets.

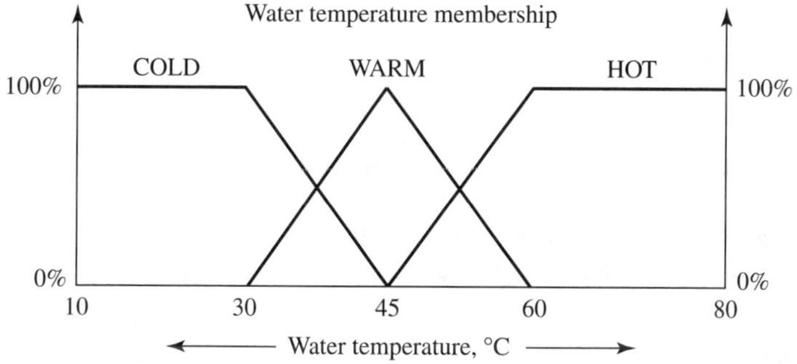

Figure 13.5 Water temperature for Example 13.5.

2. *Separate temperature limits are needed.* If the two limits are different, then the membership function curves (lines in this case) do not meet at a common temperature value. This scenario is useful when water at a certain range will be considered to completely (100%) satisfy the requirement for the middle set (WARM). The actual temperature range will depend upon the application requirements, but let this be from 40°C to 50°C. Thus, between a temperature range of 40°C to 50°C, the degree of membership for the other two sets will be zero. Table 13.3 summarizes the distribution.

Table 13.3 Separate Temperature Limits for Example 13.5

Temperature (°C)	Degree of Membership (%)		
	COLD	WARM	HOT
≤ 30	100	0	0
$30 < t \leq 40$	100 to 0	0 to 100	0
$40 < t < 50$	0	100	0
$50 \leq t \leq 60$	0	100 to 0	0 to 100
$t > 60$	0	0	100

The completed diagram is shown in Figure 13.6. It should be noted that the shape of the middle set (WARM) has changed from a triangle to a trapezoid. The shapes of the other two sets remain the same, but the slopes of lines and the limits have changed.

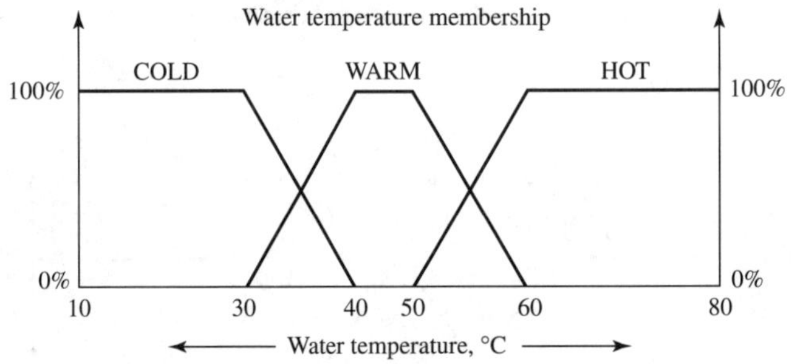

Figure 13.6 Separate temperature limits for Example 13.5.

Crisp Value

A **crisp value** is a value that can be specified numerically, such as 12.5, 36, or 60. A temperature value expressed in degrees Celsius or degrees Fahrenheit is a crisp value.

Crisp Variable

A **crisp variable** is a physical variable that can be measured through instruments and can be assigned a numerical (crisp) value.

EXAMPLE 13.6

The following are a few examples of crisp variables:

a. A temperature of 25°C
b. A pressure reading of 103.1 KPa
c. An output voltage of 4.75 V
d. A variable with a numerical value

Fuzzy Value (Linguistic Term)

A **fuzzy value** is a value that cannot be expressed using a numerical value but is expressed in linguistic (language) terms, i.e., very tall, tall, average, short, very short, etc.

These are the terms commonly used in day-to-day conversation, i.e., the weather is clear, the day is nice, it is hot, etc.

Fuzzy (Linguistic) Variable

A **fuzzy variable** is variable that cannot be specified as a numerical (crisp) value. A fuzzy variable can only be described linguistically.

Note: In this text, the term *fuzzy variable* is used to describe all the linguistic terms.

EXAMPLE 13.7

The following are fuzzy variables:

a. The temperature is *hot*.
b. The voltage output is *low*.
c. The humidity is *tolerable*.

Fuzzification

Fuzzification is the process of taking the numerical (crisp) value of a (crisp) variable and relating it to a fuzzy set through a membership function.

Membership Function

A **membership function** is normally expressed graphically and tends to illustrate how completely a crisp variable belongs to a fuzzy set.

Trapezoids and triangles are the two most popular membership functions. The four commonly used membership functions are shown in Figure 13.7. The ideal membership function is a bell-shaped curve with a Gaussian distribution. A Gaussian function is difficult to implement on a simple 8-bit microcontroller and requires a significant execution time. The trapezoid function (C) comes fairly close to a bell-shaped curve and is a lot easier to implement and faster to execute.

Singleton

A **singleton** is the single crisp value being used to represent a set.

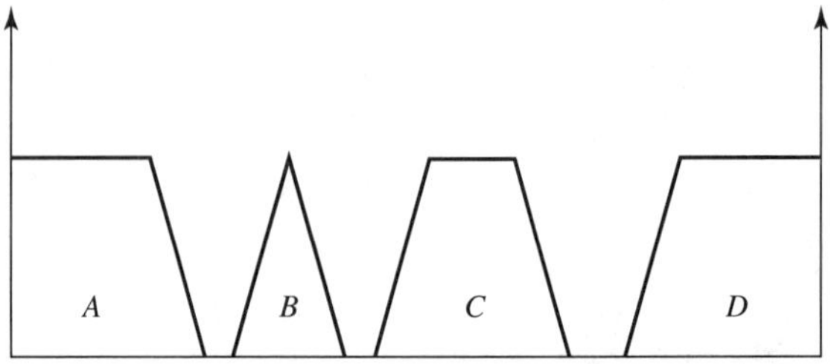

Figure 13.7 Common membership functions.

A fuzzy value cannot be used to control a physical apparatus. A control element requires a numerical (crisp) input in order to generate a physical output. A 90-V DC motor requires 45 V (50% of rated voltage) and not *medium voltage* applied to it when commanded to rotate at medium speed. Here 45 (V) represents the singleton corresponding to medium set. It is customary to use singletons in output calculations.

Defuzzification

Defuzzification is the process of taking a fuzzy value and converting it into a numerical (crisp) value (*quantifications*).

Defuzzification is required for generating a real-world output. People routinely use (fuzzy) instructions such as *close the door a little* or *open the tap some more* and can intuitively transform these fuzzy instructions for the control of physical devices. Unfortunately, such instructions are difficult to implement on a digital computer. A computer requires that the term *a little* be spelled out either as the voltage to be applied to the door motor or as the required degree of rotation of the door (in absolute or relative terms), and only then it can carry out the task.

Single Rule When only a single rule is applicable, the calculation of output is straightforward. Output is a product of the rule grade and the crisp value (singleton) of resultant action.

$$\text{output (crisp)} = \text{rule grade} \cdot \text{action singleton}$$

Multiple Rules When several rules with varying grades of membership are applicable, the output is based upon the weighted sum of products of all grades and singletons. For a case where N rules are applicable, the output is given by the following expression:

$$\text{output (crisp)} = \frac{\sum_{I=1}^{N} (\text{rule grade} \cdot \text{action singleton})_I}{\sum_{I=1}^{N} (\text{rule grade})_I}$$

EXAMPLE 13.8

Determine the output voltage for a DC motor control system from the following data and given set of rules.

Rules

1. If velocity is *LOW*, then voltage is *HIGH*.
2. If velocity is *OK*, then voltage is *MEDIUM*.
3. If velocity is *HIGH*, then voltage is *LOW*.

Output Singletons

1. *HIGH* voltage = 90 V DC
2. *MEDIUM* voltage = 45 V DC
3. *LOW* voltage = 15 V DC

***Rule Grades* (Calculated)**

1. Grade for rule 1 = 0.0
2. Grade for rule 2 = 0.6
3. Grade for rule 3 = 0.4

Solution

Only rules 2 and 3 are applicable. Rule 1 can still be utilized, but its contribution to the output will be zero.

$$\text{output (crisp)} = \frac{\sum_{I=1}^{N} (\text{rule grade} \cdot \text{action singleton})_I}{\sum_{I=1}^{N} (\text{rule grade})_I}$$

$$\text{output} = \frac{(\text{grade} \cdot \text{action singleton})_{\text{rule 2}} + (\text{grade} \cdot \text{action singleton})_{\text{rule 3}}}{\text{grade}_{\text{rule 2}} + \text{grade}_{\text{rule 3}}}$$

$$\text{output} = \frac{(0.6 \cdot 45\ \text{V})_{\text{rulo 2}} + (0.4 \cdot 100\ \text{V})_{\text{rule 3}}}{0.6_{\text{rule 2}} + 0.4_{\text{rule 2}}}$$

$$\text{output} = \frac{27\ \text{V} + 36\ \text{V}}{0.6 + 0.4}$$

$$= 63\ \text{V}$$

EXAMPLE 13.9

The data given in Table 13.4 summarizes the results of rule evaluation for a DC motor position control system. Determine the output voltage.

Table 13.4 Table of Rules for Example 13.9

Rule	Rule Grade	Action Singleton (V)
1	0	0
2	0	10
3	0.25	20
4	0.667	40
5	0.25	90
6	0.333	60

Solution

$$\text{output (crisp)} = \frac{\sum_{I=1}^{N} (\text{rule grade} \cdot \text{action singleton})_I}{\sum_{I=1}^{N} (\text{rule grade})_I}$$

$$\text{output} = \frac{(0.0 \cdot 0) + (0.0 \cdot 10) + (0.25 \cdot 20) + (0.667 \cdot 40) + (0.25 \cdot 90) + (0.333 \cdot 60)}{(0.0 + 0.0 + 0.25 + 0.667 + 0.25 + 0.333)} \text{V}$$

$$= 49.44 \text{ V}$$

The fuzzy controller will generate an output of 49.44 V to control the motor.

Rule (Knowledge-Based Rule)

A **knowledge-based rule** is the knowledge or the experience expressed as a rule. A rule typically takes the form "*If* ___, *then* ___.

EXAMPLE 13.10

The following represent some of the rules that may be adopted for highway driving.

a. If the *speed* (crisp variable—input) is *high* (fuzzy variable—input), then *release pressure on accelerator* (*fuzzy* variable—output).

b. If it's *raining* (crisp variable—input) *lightly* (fuzzy variable—input), then *slow down a little* (*fuzzy* variable—output).
c. If it's *raining* (crisp variable—input) *hard* (fuzzy variable—input), then *slow down significantly* (*fuzzy* variable—output).

The last two rules pose an interesting challenge on a computer control system. For a human driver, the evaluation of the strength of rain is very simple, although it may vary from person to person. A human operator does not need to take any measurement of the rate of waterfall to determine if it is raining lightly or hard. A simple glance through the windshield allows a quick and reliable judgment to be made. When using a computer control system, evaluation of this rule becomes a bit involved. Either a rain gauge has to be used (an expensive and slow process) or a visibility (optical) sensor needs to be incorporated. This sensor will measure (quantify) the rainfall and feed this information to the digital controller. The controller will determine the degree of membership for the measured quantity of rainfall in various fuzzy sets, i.e., raining *very lightly, lightly,* and *hard*.

Rules can involve single or multiple input variables. In practice, an attempt is made to limit the number of input variables to two or three, making it easier to graphically view the relationship between input variables and the fuzzy controller output. A rule incorporating multiple inputs would require the use of the terms ***AND*** and ***OR*** to combine the parts of the rule.

EXAMPLE 13.11

A residential temperature control system using an electric heater may incorporate the following rules encompassing two variables: room temperature and the rate of change of temperature

1. If the temperature is *HOT* and is rising *FAST*, then reduce *heater* power *A LOT*.
2. If the temperature is *HOT* and is *STEADY*, then reduce the heater power *A LITTLE*.
3. If the temperature is *COMFORTABLE* and *NOT CHANGING*, then change the heater power *BY ZERO* (no change).
4. If the temperature is *LOW* and *NOT CHANGING*, then increase the heater power *A LITTLE*.
5. If the temperature is *LOW* and is dropping *FAST*, then increase the heater power *A LOT*.

Rule Grade (Fuzzy Inferencing)

The **rule grade** is the degree to which the current input conditions satisfy a rule. This process is also known as *fuzzy inferencing*.

Single Input Variable For a simple rule containing only one input variable, the rule grade is the same as the degree of membership in a fuzzy set.

rule grade = degree of membership

EXAMPLE 13.12

Determine the rule grade for a hot water heater using the given set of rules and the membership diagram in Figure 13.8 for two values of temperature.

a. Measured temperature = 40°C
b. Measured temperature = 54°C

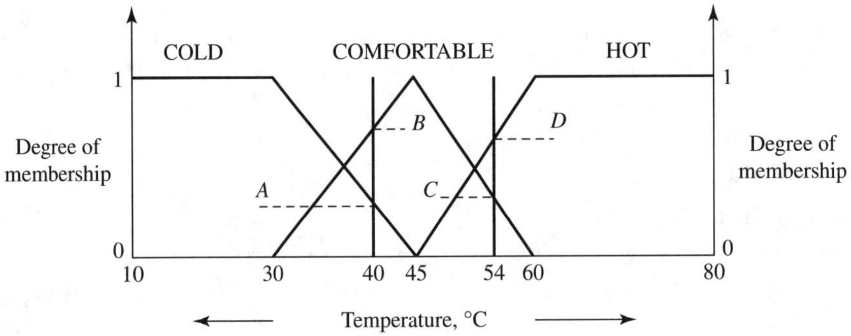

Figure 13.8 Membership diagram for Example 13.12.

1. If the temperature is *HOT*, then set the heat to *LOW*.
2. If the temperature is *COMFORTABLE*, then set the heat to *MEDIUM*.
3. If the temperature is *LOW*, then set the heat to *HIGH*.

Solution

The two measured temperature values of 40°C and 54°C are shown superimposed on the diagram.

a. The vertical line drawn at 40°C intersects two fuzzy sets: COLD and COMFORTABLE. The degree of membership in the COLD set (for 40°C), marked as *A*, is

$$\frac{45-40}{45-30} = 0.333 \quad (33.3\%)$$

Thus the rule grade for the first rule is 0.333.

Similarly, the degree of membership in the COMFORTABLE set (for 40°C), marked as B, is

$$\frac{40-30}{45-30} \text{ (for symmetrical triangle)} = 0.667 \quad (66.7\%)$$

Thus the rule grade for the second rule is 0.667. The rule grade for the third rule is 0.0.

b. The vertical line drawn at 54°C intersects two fuzzy sets: COMFORTABLE and HOT. The degree of membership in the COMFORTABLE set (for 54°C), marked as C, is

$$\frac{60-54}{60-45} = 0.4 \quad (44\%)$$

Thus the rule grade for the second rule is 0.4. Similarly, the degree of membership in the HOT set (for 54°C) is

$$\frac{54-45}{60-45} = 0.6 \quad (60\%)$$

Thus the rule grade for the third rule is 0.6. The rule grade for the first rule is 0.0.

Multiple Input Variables For a rule containing two or more input variables, the rule grade is the same as the degree of membership, which is the smallest of all the fuzzy sets relevant to the rule.

rule grade = smallest of all the degrees of membership

EXAMPLE 13.13

Using the data in Example 13.12, also take into consideration the fluid flow rate. The following revised set of rules control the amount of power sent to the heater.

1. If the temperature is *HOT* **AND** the flow rate is *MEDIUM*, then *set the heat* to *MEDIUM LOW*.
2. If the temperature is *HOT* **AND** the flow rate is *LOW*, then *set the heat* to *LOW*.

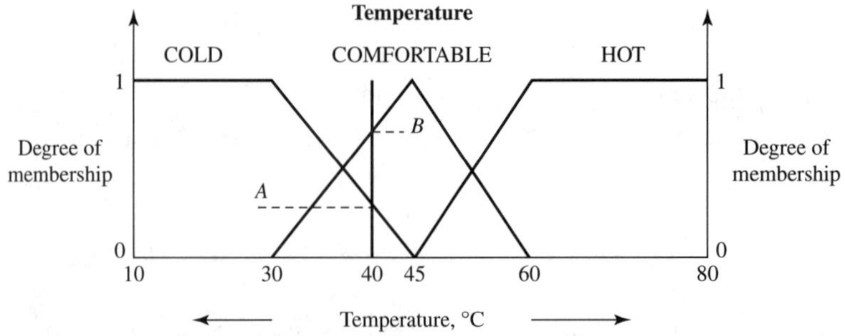

Figure 13.9 Temperature membership diagram for Example 13.13.

3. If the temperature is *COMFORTABLE* **AND** the flow rate is *MEDIUM*, then set the heat to *MEDIUM*.
4. If the temperature is *COMFORTABLE* **AND** the flow rate is LOW, then set the heat to *MEDIUM LOW*.
5. If the temperature is *LOW* **AND** the flow rate is *MEDIUM*, then set the heat to *HIGH*.
6. If the temperature is *LOW* **AND** the flow rate is *LOW*, then set the heat to *MEDIUM*.

Determine the rule grade for the given set of rules and Figures 13.9 and 13.10. The measured temperature is 40°C and flow rate is 1.5 L/min.

Solution

Two input variables and two degrees of membership diagrams are provided. The crisp (numerical) value of 40°C, representing the first variable temperature, is superimposed on the first diagram. Similarly, the crisp (numerical) value of 1.5 L/min, representing the second variable flow rate, is superimposed on the second diagram.

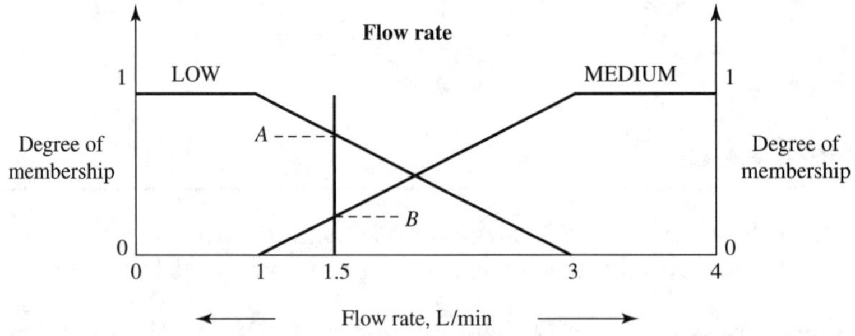

Figure 13.10 Flow rate membership diagram for Example 13.13.

1. *Input variable 1—temperature:* This was calculated in Example 13.12. The degree of membership for these sets is as shown in Table 13.5.

Table 13.5 Fuzzification of Temperature Data

Fuzzy Set	***Degree of Membership***
COLD	0.333
COMFORTABLE	0.667
HOT	0.0

2. *Input variable 2—flow rate:* There are only two fuzzy sets for flow rate. A vertical line drawn at 1.5 L/min intersects both fuzzy sets LOW and MEDIUM.
   The degree of membership in the LOW set (for a flow rate of 1.5 L/min) is

$$\frac{3 - 1.5}{3 - 1.0} = 0.75 \quad (75\%)$$

Similarly, the degree of membership in the MEDIUM set is

$$\frac{1.5 - 1.0}{3.0 - 1.0} = 0.25 \quad (25\%)$$

See Table 13.6.

Table 13.6 Fuzzification of Flow Rate Data

Fuzzy Set	***Degree of Membership***
LOW	0.75
MEDIUM	0.25

Rule evaluation: Once the degree of membership for each fuzzy set has been determined, the rules that are applicable are identified from the complete list of rules. For the measured value of 1.5 L/min, the second variable (flow rate) has a certain degree of membership in both fuzzy sets. But corresponding to a temperature of 40°C, the first variable (temperature) has membership in only two sets, COLD and COMFORTABLE. It has zero membership in the third fuzzy set, HOT. Thus, any rule that involves the HOT temperature will not be applicable. Thus rules 1 and 2 are not applicable, but all the other rules will be. *The rules that are not applicable are assigned a grade of* 0.

The applicable rules are listed again:

3. If the temperature is *COMFORTABLE* **AND** the flow rate is *MEDIUM*, then set the heat to *MEDIUM*.
4. If the temperature is *COMFORTABLE* **AND** the flow rate is *LOW*, then set the heat to *MEDIUM LOW*.
5. If the temperature is *LOW* **AND** the flow rate is *MEDIUM*, then set the heat to *HIGH*.
6. If the temperature is *LOW* **AND** the flow rate is *LOW*, then set the heat to *MEDIUM*.

Grades for the applicable rules (3, 4, 5, and 6) now need to be evaluated one by one.

Grade for rule 3: There are two fuzzy sets specified in this rule, COMFORTABLE and MEDIUM. The degree of membership for each of these sets is as shown.

Fuzzy Set	*Degree of Membership*	*Inference (Conclusion)*
COMFORTABLE (temperature)	0.667	
MEDIUM (flow rate)	0.25	0.25 (lower of the two)

From the two degrees of membership, the lower of the two values, 0.25, is selected as the grade for this rule. In other words, it can be stated that rule 3 is applicable to a grade of 0.25 (25%).

Grade for rule 4: Again, there are two fuzzy sets specified in this rule, COMFORTABLE and LOW. The degree of membership for each of these sets is as shown.

Fuzzy Set	*Degree of Membership*	*Inference (Conclusion)*
COMFORTABLE (temperature)	0.667	0.667 (lower of the two)
LOW (flow rate)	0.75	

From the two degrees of membership, the lower of the two values, 0.667, is selected as the grade for this rule.

Rules 5 and 6 can be similarly evaluated.

Grade for rule 5:

Fuzzy Set	*Degree of Membership*	*Inference (Conclusion)*
LOW (temperature)	0.333	
MEDIUM (flow rate)	0.25	0.25 (lower of the two)

From the two degrees of membership, the lower of the two values, 0.25, is selected as the grade for this rule.

Grade for rule 6:

Fuzzy Set	***Degree of Membership***	***Inference (Conclusion)***
LOW (temperature)	0.333	0.333 (lower of the two)
LOW (flow rate)	0.75	

From the two degrees of membership, the lower of the two values, 0.333, is selected as the grade for this rule.

The following table summarizes the grade for all the rules.

Summary of Rule Grades

Rule No.	***Rule Grade***	***Resultant Output (Fuzzy)***
1	0	MEDIUM LOW
2	0	LOW
3	0.25	MEDIUM LOW
4	0.667	MEDIUM LOW
5	0.25	HIGH
6	0.333	MEDIUM

FUZZY CONTROL SYSTEM

A fuzzy control system deals with real-world (crisp) inputs and generates a real-world (crisp) output. But unlike a PID controller, where the output is based upon the result of the PID equation, in a fuzzy controller the output is dependent upon fuzzy logic, which is primarily the experience and/or knowledge gained with the process.

A fuzzy control system is a knowledge-based system. The driver of a car does not have to do any calculations for maneuvering the car through a winding hillside road. All the driver (control) actions are based upon the past experience and knowledge of the driver. A complete fuzzy control system design requires four steps.

1. Problem identification
2. Conceptual design
3. Computer simulation
4. Computer implementation

Problem Identification

Problem identification should always be investigated seriously, but unfortunately such investigation is rarely done. Whenever a problem is discovered, someone usually comes up with a perceived solution (right or wrong) without much consideration. The following example illustrates this point.

EXAMPLE 13.14

A small industrial extrusion press was producing poor-quality extruded forms. A faulty hydraulic pressure control system was suspected and as a solution, a new controller was substituted. The problem did not correct itself, and it was later discovered that the temperature control system was at fault. Once corrected, the production resumed to normal.

Conceptual Design

Conceptual design is the heart of control system design. The complete control system is designed and tested in principle and without any serious concern about the sensors and actuators used for input and output generation. If a control system is conceptually sound, it can be implemented on a computer. This process can be divided into the following steps:

1. Identify the input variables.
2. Identify the output variables.
3. Set up the fuzzy logic rules.
4. Set the method of calculating the output.

The process can perhaps be best understood through an example.

EXAMPLE 13.15

Develop a fuzzy control system for a position control system. The system has the following components:

DC motor:

$$\text{operating voltage range} = \pm 12 \text{ V DC}$$

$$\text{operating speed} = 0 \text{ to } 3000 \text{ rpm} \quad \text{(bidirectional)}$$

Displacement (position) sensor:

linear optical encoder = 50 pulses/cm

DC tachogenerator:

output – 2 V DC/1000 rpm (bidirectional)

operating distance = 0 to 50 cm (measured from one end of slide)

Solution

The fuzzy controller design is carried out following the design steps as outlined.

Identification of input variables: Two measured variables are available:

Position

Motor velocity

The slide can be commanded to move in a forward or reverse direction from the current position. Thus, the position data can be processed to determine the relative position, which may be a positive or negative quantity.

relative position (Rel_Pos) = target position – current position

The range of relative position will be ±50 cm. The slide may be at one end of the movement, and it may be commanded to go to the other end. The entire operating range can be broken down into at least three segments, but to provide better control, it will be divided into five segments. Table 13.7 lists the fuzzy sets related to relative position measurement. See Figure 13.11.

Table 13.7 Fuzzy Sets of Position Variable (Rel_Pos)

Fuzzy Set (Label)	*Description*	*Range (cm)*	*Membership Function*
Pos_Far (E)	Large position difference in positive direction	2 to 10 10 to 50	S-curve
Pos_Near (D)	Small position difference in positive direction	0.1 to 2 2 to 10	Triangular
Near (C)	Very small position difference	−1 to −0.1 −0.1 to 0.1 0.1 to 1	Trapezoidal
Neg_Near (B)	Small position difference in negative direction	−0.1 to −2 −2 to 10	Triangular
Neg_Far (A)	Large position difference in negative direction	−2 to −10 −10 to −50	Z-curve

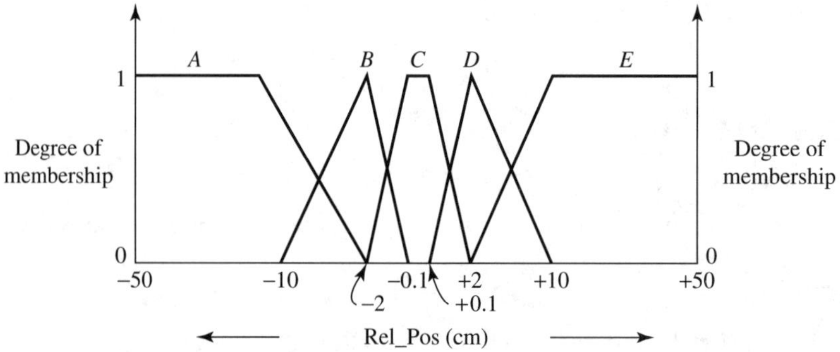

Figure 13.11 Membership in Rel_Pos for Example 13.15.

Slide (motor) velocity is a bidirectional quantity as well, with the maximum output corresponding to the maximum operating speed of 3000 rpm.

The range of the tachogenerator output **Mot_Vel** is ±6 V DC.

Because this variable is a bidirectional quantity, a minimum of five segments is recommended. Table 13.8 describes the fuzzy sets related to slide velocity. See Figure 13.12.

Identification of output variables: The only mechanism for controlling the motor position is the voltage applied to the motor. Thus, there is only one output variable, *motor voltage*. The operating range of motor voltage is ±12 V DC. In order to provide a smoother variation in motor control voltage, it may be desirable to divide the operating range (±12 V DC) into multiple segments (fuzzy sets). A large number

Table 13.8 Fuzzy Sets of the Velocity Variable (Mot_Vel)

Fuzzy Set (Label)	***Description***	***Range (rpm)***	***Membership Function***
Pos_Large	Large velocity in positive direction	100 to 1000	S-curve
(E)		1000 to 3000	
Pos_Small	Small velocity in positive direction	10 to 100	Triangular
(D)		100 to 1000	
Small	Very small velocity	−100 to −10	Trapezoidal
(C)		−10 to +10	
		+10 to +100	
Neg_Small	Small velocity in negative direction	−10 to −100	Trangular
(B)		−100 to −1000	
Neg_Large	Large velocity in negative direction	−100 to −1000	Z-curve
(A)		−1000 to −3000	

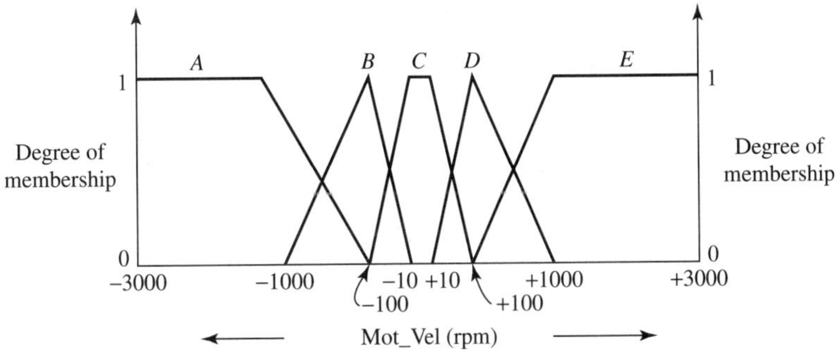

Figure 13.12 Membership in Mot_Vel for Example 13.15.

of segments will result in a greater number of calculations and a slower execution of the control system. Five to seven segments in this case will provide a reasonable compromise. The fuzzy sets for motor control voltage are shown in Table 13.9.

Figure 13.13 shows the fuzzy sets related to control output. In some cases, these output fuzzy sets may be defined in detail, but it is quite satisfactory to use crisp values (singletons) to denote each fuzzy set.

Fuzzy logic rules: Based upon the past experience with this type (position control) of control system, a set of rules can now be developed. These rules will vary from individual to individual and are strongly influenced by personal experiences. A tentative list of rules is shown below. There are five fuzzy sets in each input variable, allowing for a total of 25 rules.

Table 13.9 Fuzzy Sets for Motor Voltage Output (Mot_Volt)

Fuzzy Set (Label)	***Description***	***Action Singleton (volt)***
Pos_Max (E)	Large positive voltage	+12
Pos_Min (D)	Small positive voltage	+5
Min (C)	Zero voltage	0
Neg_Min (B)	Small negative voltage	−5
Neg_Max (A)	Large negative voltage	−12

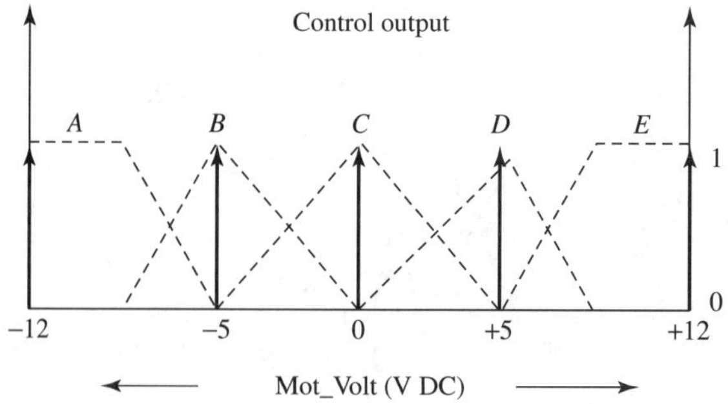

Figure 13.13 Output singletons for Example 13.15.

Group 1

1. If Rel_Pos is Pos_Far and velocity is Pos_Large, then Mot_Volt is Pos_Max. The rule simply states that if there is a large distance to cover in the positive direction and the motor is moving in the right direction, then apply the maximum voltage to continue to move the motor in that direction.
2. If Rel_Pos is Pos_Far and velocity is Pos_Small, then Mot_Volt is Pos_Max.
3. If Rel_Pos is Pos_Far and velocity is small, then Mot_Volt is Pos_Max.
4. If Rel_Pos is Pos_Far and velocity is Neg_Small, then Mot_Volt is Pos_Min.

Even though it is desirable to move the motor fast in the opposite direction, it is better to slow down the motor and then reverse the direction of rotation. Otherwise, a large current surge may appear and it may also result in a jerky movement.

5. If Rel_Pos is Pos_Far and velocity is Neg_Large, then Mot_Volt is Min. Again, slow down the motor and avoid a large in-rush current.

Group 2

6. If Rel_Pos is Pos_Near and velocity is Pos_Large, then Mot_Volt is Pos_Min.
7. If Rel_Pos is Pos_Near and velocity is Pos_Small, then Mot_Volt is Min.
8. If Rel_Pos is Pos_Near and velocity is small, then Mot_Volt is Pos_Min.
9. If Rel_Pos is Pos_Near and velocity is Neg_Small, then Mot_Volt is Pos_Min.
10. If Rel_Pos is Pos_Near and velocity is Neg_Large, then Mot_Volt is Pos_Min.

Group 3

11. If Rel_Pos is Near and Velocity is Pos_Large, then Mot_Volt is Neg_Min. Apply negative voltage, or else the motor will overshoot the target.

12. If Rel_Pos is near and velocity is Pos_Small, then Mot_Volt is Min.
13. If Rel_Pos is near and velocity is small, then Mot_Volt is Min.
14. If Rel_Pos is near and velocity is Neg_Small, then Mot_Volt is Min.
15. If Rel_Pos is near and velocity is Neg_Large, then Mot_Volt is Pos_Min.

Group 4

16. If Rel_Pos is Neg_Near and velocity is Pos_Small, then Mot_Volt is Neg_Min.
17. If Rel_Pos is Neg_Near and velocity is Pos_Large, then Mot_Volt is Neg_Min.
18. If Rel_Pos is Neg_Near and velocity is small, then Mot_Volt is Min.
19. If Rel_Pos is Neg_Near and velocity is Neg_Small, then Mot_Volt is Neg_Min.
20. If Rel_Pos is Neg_Near and velocity is Neg_Large, then Mot_Volt is Pos_Min.

Group 5

21. If Rel_Pos is Neg_Far and velocity is Pos_Large, then Mot_Volt is Min.
22. If Rel_Pos is Neg_Far and velocity is Pos_Small, then Mot_Volt is Neg_Min.
23. If Rel_Pos is Neg_Far and velocity is small, then Mot_Volt is Neg_Max.
24. If Rel_Pos is Neg_Far and velocity is Neg_Small, then Mot_Volt is Neg_Max.
25. If Rel_Pos is Neg_Far and velocity is Neg_large, then Mot_Volt is Neg_Max.

From the sake of simplicity in implementation and execution speed, it is preferable to reduce the number of applicable rules. On review, some of the rules may be eliminated. The set of rules can be better represented through a table or a rule matrix. The rule matrix for the given set of rules is shown in Table 13.10.

Table 13.10 Rule Matrix for Motor Voltage Output

Mot_Vel / Rel_Pos	Pos_Large (E)	Pos_Small (D)	Small (C)	Neg_Small (B)	Neg_Large (A)
Pos_Far (*E*)	Pos_Max	Pos_Max	Pos_Max	Pos_Min	Min
Pos_Near (*D*)	Pos_Min	Min	Pos_Min	Pos_Min	Pos_Min
Near (*C*)	Neg_Min	Min	Min	Min	Pos_Min
Neg_Near (*B*)	Neg_Min	Neg_Min	Neg_Min	Min	Pos_Min
Neg_Far (*A*)	Min	Neg_Min	Neg_Max	Neg_Max	Neg_Max

This concludes the preliminary design of the fuzzy controller. The design can be tested for various input conditions through simulation and/or field testing.

EXAMPLE 13.16

For the fuzzy controller given in Example 13.15, determine the (motor) output voltage for the following operating conditions.

$$\begin{aligned}
&\text{tachogenerator output voltage} &&= 1.5\text{ V}\\
&\text{current pulse accumulator count} &&= 500\text{ pulses}\\
&\text{target position} &&= 15\text{ cm} \quad \text{(absolute position from one end of the table)}\\
&\text{output} &&= \frac{2\text{ V DC}}{1000\text{ rpm} \quad \text{(bidirectional)}}
\end{aligned}$$

Solution

The first step in determining the output is to convert the sensor inputs into desired input variables, as indicated in Example 13.15.

Relative position: The relative position data need to be calculated in terms of distance (in cm).

$$\begin{aligned}
\text{current distance (from starting point)} &= \frac{\text{current pulse count}}{\text{encoder resolution}}\\
&= \frac{500\text{ pulses}}{50\text{ pulses/cm}}\\
&= 10\text{ cm}\\
\text{Rel_Pos} &= \text{target position} - \text{current position}\\
&= 15\text{ cm} - 10\text{ cm}\\
&= 5\text{ cm}
\end{aligned}$$

Motor velocity: Motor velocity is calculated from the tachogenerator output voltage.

$$\begin{aligned}
\text{Mot_Vel} &= \frac{\text{current tach output}}{\text{V/rpm}}\\
&= \frac{1.5\text{ V}}{2\text{ V}/1000\text{ rpm}}\\
&= 750\text{ rpm} \qquad \text{(positive direction)}
\end{aligned}$$

The next step is determining the fuzzy outputs corresponding to each input (fuzzification). This step will determine how well an input condition (crisp input) meets the relevant fuzzy sets.

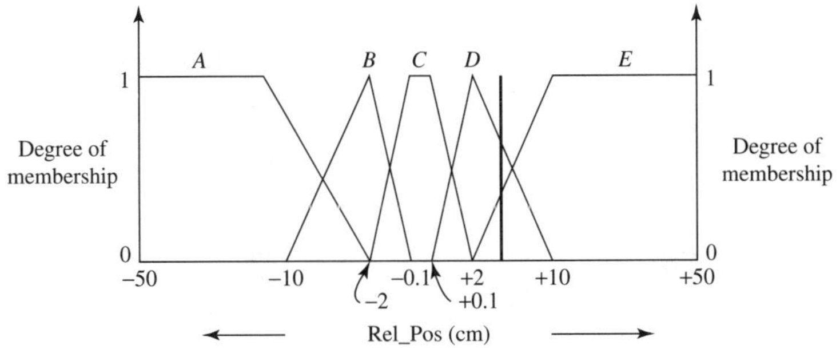

Figure 13.14 Membership in Rel_Pos for Example 13.16.

Fuzzification: There are two input variables, and fuzzification is required for each of the two variables. The value of variable Rel_Pos (5 cm) is superimposed on the fuzzy set diagram and is shown in Figure 13.14. From the diagram it is evident that a value of 5 cm satisfies two sets, D and E. The membership of the relative distance variable (Rel_Pos = 5 cm) can be either read from the diagram or calculated.

Using the calculation, the membership in set D is

$$\frac{10-5}{10-2} = 0.625$$

The degree of membership for a 5-cm value in set E can either be calculated or determined from the fact that the total sum of all memberships is equal to 1.00. Thus, degree of membership in set E is $1.00 - 0.625 = 0.375$.

Similarly, the value of the second variable, Mot_Vel (750 rpm), is superimposed on its fuzzy set diagram (Figure 13.15). This value of Mot_Vel belongs to two sets, D and E. The degree of membership in set D is

$$\frac{1000-750}{1000-100} = 0.278$$

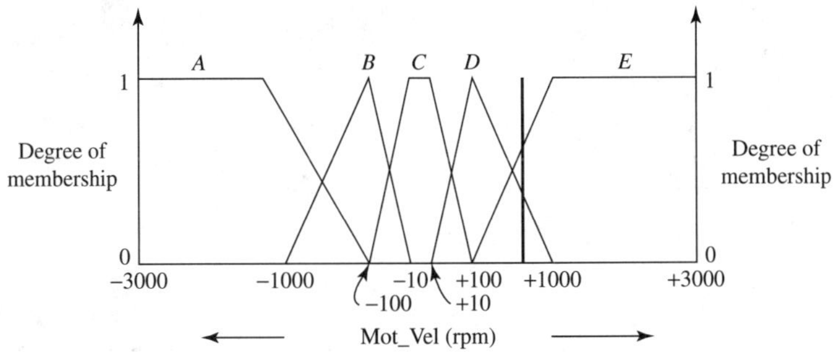

Figure 13.15 Membership in Mot_Vel for Example 13.16.

The degree of membership in set E is $1.0 - 0.278 = 0.722$.

Rule evaluation: Only four rules are applicable (highlighted in the rule matrix table, Table 13.11).

Table 13.11 Rule Matrix for Motor Voltage Output

Mot_Vel / Rel_Pos	Pos_Large (E)	Pos_Small (D)	Small (C)	Neg_Small (B)	Neg_Large (A)
Pos_Far (E)	Pos_Max	Pos_Max	Pos_Max	Pos_Min	Min
Pos_Near (D)	Pos_Min	Min	Pos_Min	Pos_Min	Pos_Min
Near (C)	Neg_Min	Min	Min	Min	Pos_Min
Neg_Near (B)	Neg_Min	Neg_Min	Neg_Min	Min	Pos_Min
Neg_Far (A)	Min	Neg_Min	Neg_Max	Neg_Max	Neg_Max

1. If Rel_Pos is Pos_Far (0.625) and velocity is Pos_Large (0.722), then Mot_Volt is Pos_Max (+12 V).
2. If Rel_Pos is Pos_Far (0.625) and velocity is Pos_Small (0.278), then Mot_Volt is Pos_Max (+12 V).
3. If Rel_Pos is Pos_Near (0.375) and velocity is Pos_Large (0.722), then Mot_Volt is Pos_Min (+5 V).
4. If Rel_Pos is Pos_Near (0.375) and velocity is Pos_Small (0.278), then Mot_Volt is Min (0 V).

The grade of each rule is taken as the smallest of the two degrees of membership in that rule. Thus the rule grades are

Rule 1 Minimum of 0.625 and 0.722 = 0.625

Rule 2 Minimum of 0.625 and 0.278 = 0.278

Rule 3 Minimum of 0.375 and 0.722 = 0.375

Rule 4 Minimum of 0.375 and 0.278 = 0.375

Crisp output:

$$\text{output (crisp)} = \frac{\sum_{l=1}^{N} (\text{rule grade} \times \text{action singleton})_l}{\sum_{l=1}^{N} (\text{rule grade})_l}$$

$$= \frac{0.625 \times 12\ \text{V} + 0.278 \times 12\ \text{V} + 0.375 \times 5\ \text{V} + 0.375 \times 0\ \text{V}}{0.625 + 0.278 + 0.375 + 0.375}$$

$$= 7.69\ \text{V}$$

Thus, based on the current input conditions, the fuzzy controller will generate an output of +7.69 V. This is a relatively large voltage, and it will cause the slide to move fast toward the target. With a change in slide position and motor velocity, the input to the fuzzy controller will change, resulting in a revised output. This process will continue until the slide reaches the target position.

EXAMPLE 13.17

The slide in the position control system in Example 13.16 is stationary at a distance of 30 cm from the start (left end). What will be the new controller output if it is commanded to go to a target position of 28.5 cm (from the starting position)?

Solution

Because the slide is stationary, the motor velocity is zero. Zero velocity falls only in fuzzy set C (MIN) and has a membership of 1.0 (100%). The slide is required to move to the left (negative direction).

$$\begin{aligned} \text{Rel\_Pos} &= \text{target position} - \text{current position} \\ &= 28.5 \text{ cm} - 30 \text{ cm} \\ &= -1.5 \text{ cm} \end{aligned}$$

Superimposing the relative position information on the fuzzy diagram, the membership in the fuzzy sets can be determined (Figure 13.16).

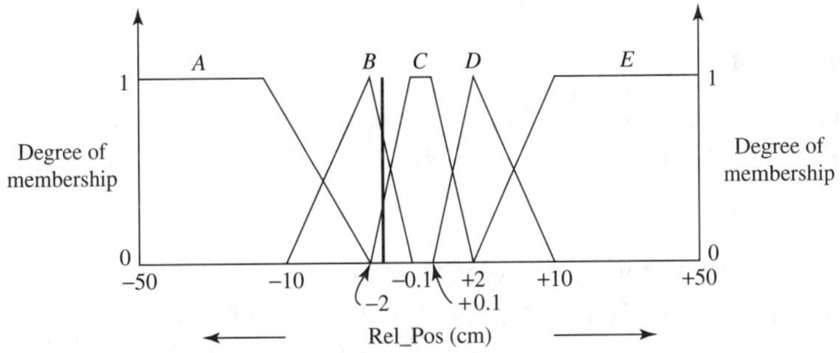

Figure 13.16 Membership function of Rel_Pos for Example 13.17.

From the diagram it is evident that −1.5 cm satisfies two fuzzy sets, B (Neg_Near) and C (Near).

Using calculations, the membership in set B is

$$\frac{0.1 - 1.5}{0.1 - 2} = 0.737$$

Similarly, the membership in set C is

$$1.0 - 0.737 = 0.263.$$

For those input conditions two rules are applicable.

1. If Rel_Pos is Near (0.263) and velocity is small (1.00), then Mot_Volt is Min (0 V)

$$\begin{aligned}\text{rule grade} &= \text{minimum of } (0.263) \text{ and } (1.00)\\ &= 0.263\end{aligned}$$

2. If Rel_Pos is Neg_Near (0.737) and velocity is small (1.00), then Mot_Volt is Min (0 V).

$$\begin{aligned}\text{rule grade} &= \text{minimum of } (0.737) \text{ and } (1.00)\\ &= 0.737\\ \text{output voltage} &= \frac{0.263 \times 0\text{ V} + 0.737 \times 0\text{ V}}{0.263 + 0.737}\\ &= 0\text{ V}\end{aligned}$$

The output of the fuzzy controller is 0 V. Since the motor is already stationary, the application of 0 V to the motor is not going to cause it to move. This indicates a fault in rule formulation and needs to be corrected. Two rules, at a minimum, need to be revised. It may also be possible to divide the controller output into significantly more sets (singletons), rather than just 5.

Current Rules

If Rel_Pos is Pos_Near and velocity is Pos_Small, then Mot_Volt is Min.

If Rel_Pos is Neg_Near and velocity is Neg_Small, then Mot_Volt is Min.

Revised Rules

If Rel_Pos is Pos_Near and velocity is Pos_Small, then Mot_Volt is Pos_Min.

If Rel_Pos is Neg_Near and velocity is Neg_Small, then Mot_Volt is Neg_Min.

The revised rule matrix is shown in Table 13.12.
The rule grades do not change. The rules are repeated with revised outputs.

1. If Rel_Pos is Near (0.263) and velocity is small (1.00), then Mot_Volt is Min (0 V).

$$\begin{aligned}\text{rule grade} &= \text{minimum of } (0.263) \text{ and } (1.00)\\ &= 0.263\end{aligned}$$

Table 13.12 Rule Matrix for Motor Voltage Output

Rel_Pos \ Mot_Vel	Pos_Large (E)	Pos_Small (D)	Small (C)	Neg_Small (B)	Neg_Large (A)
Pos_Far (E)	Pos_Max	Pos_Max	Pos_Max	Pos_Min	Min
Pos_Near (D)	Pos_Min	Pos_Min	Pos_Min	Pos_Min	Pos_Min
Near (C)	Neg_Min	Min	Min	Min	Pos_Min
Neg_Near (B)	Neg_Min	Neg_Min	Neg_Min	Neg_Min	Pos_Min
Neg_Far (A)	Min	Neg_Min	Neg_Max	Neg_Max	Neg_Max

2. If Rel_Pos is Neg_Near (0.737) and velocity is small (1.00), then Mot_Volt is Neg_Min (−5 V).

$$\text{rule grade} = \text{minimum of } (0.737) \text{ and } (1.00)$$
$$= 0.737$$
$$\text{output voltage} = \frac{0.263 \times 0\text{ V} + 0.737 \times (-5\text{ V})}{0.263 + 0.737}$$
$$= -3.685\text{ V}$$

This negative voltage will cause the motor to start accelerating toward the left. As soon as there is movement, motor velocity changes, and so does the relative position. New values of these inputs will cause a revised controller output, and this process will continue until the target is reached. At this point the controller output would be zero and the slide should be stationary.

Note: It should be emphasized that a fuzzy control system is not a miracle system. Its success in an application depends upon the experience of the designer and careful consideration of all the possible operating conditions. There are several commercial software programs that can assist in the design of fuzzy control systems and provide computer simulation of the control system operation. A final test is obviously a rigorous field test of the newly designed system.

PROBLEMS

1. How does a fuzzy controller differ from a conventional analog controller, such as a proportional controller?
2. Can the operation of a fuzzy controller be considered similar to an on-off controller?
3. What is a fuzzy set? Give an example.
4. How many fuzzy sets are generally used to represent an input measurement? Is it better to use a larger or smaller number of fuzzy sets? Explain your answer.

5. Explain the meaning of the term *fuzzify*, in the context of fuzzy control.
6. What is a singleton? Is it used to represent an input or an output quantity, and why?
7. Explain the following terms:
   a. Degree of membership
   b. Grade of a rule
8. Draw a fuzzy diagram to represent the motor speed measured through a tachometer.

$$\text{tachometer output} = 10\ \text{V}/1000\ \text{rpm}$$

The following fuzzy sets are used to represent the motor speed.

	Fuzzy Set	Range (rpm)
1	Very low (VL)	0–300
2	Low (LO)	200–500
3	Mid (MD)	300–700
4	High (HH)	500–800
5	Very high (VH)	700–1000

9. Using the data in Example 8, determine the degree of membership in each fuzzy set for the following tachometer outputs.
   a. 0.5 V
   b. 1.25 V
   c. 6.75 V
   d. 4.75 V
10. The rule matrix for a fuzzy controller is shown here.

Rule	1A	2A	3A
1B	H	H	M
2B	H	M	M
3B	M	L	L

The singletons for outputs are H = 12, M = 6, L = 1. Determine the controller output for the following degrees of membership.
   a. 1A = 0, 1B = 0.1, 1C = 0.9; 2A = 1.0, 2B = 0, 2C = 0
   b. 1A = 0.75, 1B = 0.25, 1C = 0; 2A = 0, 2B = 0.3, 2C = 0.7
11. A residential environmental controller is to be designed to control the temperature and humidity. Humidity is controlled by controlling the speed of a drum motor in the humidifying tank. Heat is provided by an electric heater, where the SCR firing angle is varied continuously from 0° to 180°. There are three

possible speed settings for the drum motor: off, midspeed, and full speed. Explain if it is possible to use a conventional analog PID controller to achieve this task.

12. Develop a fuzzy controller for control of a home-heating system. Heat is supplied through a gas-heated furnace. The furnace gas blower and gas control valve are interlocked, and a single output controls both the units. Furthermore, both the units operate in fully on or fully off state.

13. Develop a fuzzy control strategy for speed control of a small kitchen blender. The desired speed of operation is entered through a potentiometer, providing a voltage reading from 0 to 12 V. The blender motor is a DC motor operating over a power supply range of 0–12 V DC. It should be noted that blender speed changes with the type of ingredients used as well as with the quantity in the bowl. The actual blender motor speed is directly proportional to back emf. Back emf can be determined from the following equation.

$$\text{back emf} = \text{motor terminal voltage} - IR_A$$

Assume that a circuit is available that directly provides a voltage output equal to the instantaneous back emf voltage.

14. A fuzzy controller has been configured for the following input and output fuzzy diagrams.

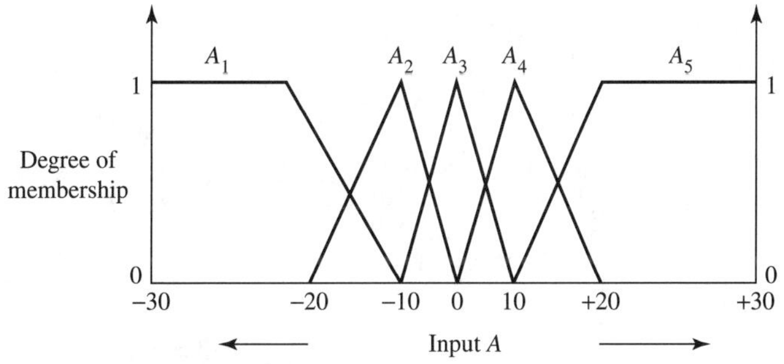

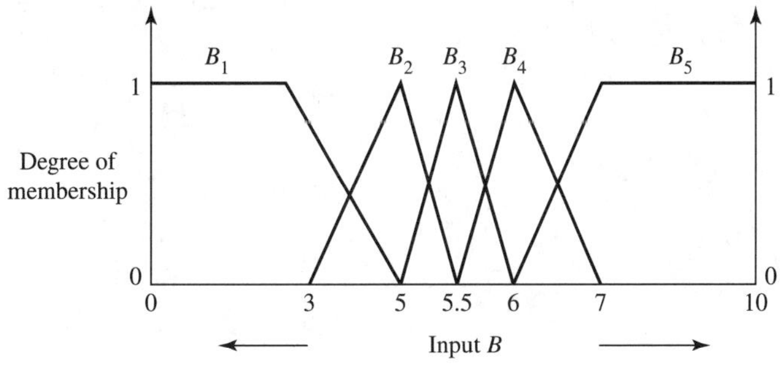

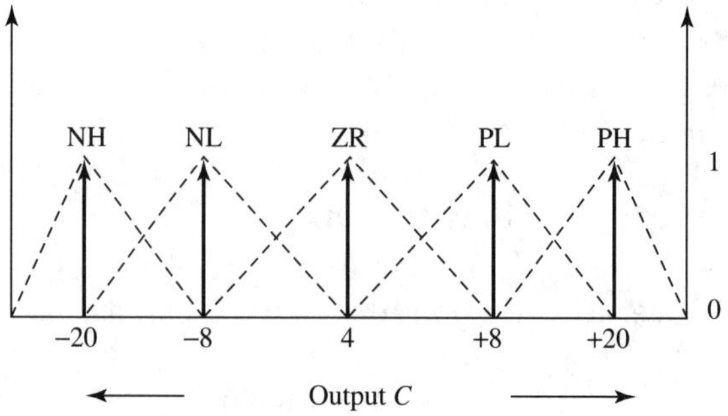

The rule matrix is as follows.

Rule	B_1	B_2	B_3	B_4	B_5
A_1	NH	NH	NH	NH	NH
A_2	NL	NL	NL	NL	NH
A_3	NL	ZR	ZR	ZR	PL
A_4	PL	PL	PL	PH	PH
A_5	PL	PH	PH	PH	PH

Calculate the crisp output for the following input conditions:

a. Input $A = +1.6$, input $B = 2.1$

b. Input $A = -4.5$, input $B = 6.8$

c. Input $A = 2.5$, input $B = 3.75$

14

CONTROLLER TUNING AND SYSTEM DESIGN

INTRODUCTION

Closed-loop control system operation is dictated by its components, and the disturbances to which it is subjected. The disturbances are controlled by external events and cannot be predicted in advance (except in some statistically controlled processes). The controller is the only component of a control loop whose behavior can be modified by changing its operating parameters, such as the proportional, integral, and derivative constants. Control system design covers the selection of controller and its parameters to meet the system operating requirements. The modification of controller parameters is known as *controller tuning*, and it affects the operation of the entire control loop. Controller tuning is carried out to obtain a desired performance from the control loop. Thus, before any such modification is carried out, it is essential that the objectives for control system operation be clearly defined.

Controller tuning (or retuning) is necessitated by one of the following situations:

- *New plant*: The controller coefficients are undefined and need to be set before the plant is started. During start-up, the coefficients may be adjusted based upon previous analysis or the start-up data.
- *Old plant, new requirements:* In an existing plant, a change in the operating conditions may be warranted. This may be caused by market conditions or improvement in quality control. An adjustment of the controller may be needed.
- *Old plant, component aging*: When a plant has been operating over a long period of time, the components and subsystems start to show the effect of wear and tear. Flow rating of control valves change, spring stiffness changes, resistors and capacitors

change in value, gear trains develop significant play and backlash, etc. All these affect the overall operation of the control system. Also, if an aging component is replaced by a new one, its characteristics are somewhat different and may also dictate readjustment of controller coefficients.

PERFORMANCE CRITERIA

Because the primary reason for controller tuning is to ensure that the process be operated in a desired manner, a question can be raised.

What is the desired performance criterion (objective or specification)?

In order to provide a fair assessment of control system performance, some quantitative performance criteria need to be established. Performance criteria should be well defined and relate to physical variables so that these can be measured and quantified. In some systems, the criterion may be as simple as the *peak overshoot, settling time*, or *residual error*.

It is easy to become overenthusiastic and start defining a number of criteria. Unfortunately, it is not possible to simultaneously meet a number of criteria. It is best to identify the primary criterion and list the secondary ones. Sometimes it is not possible to single out a criterion. In this case, performance is based upon a suitable combination of several factors of interest.

Some of commonly used criteria are as follows.

Time-Domain Specifications

Residual error
Settling time
Speed of response
Damping ratio
Peak overshoot
Relative stability
Integral of absolute error

Frequency-Domain Specifications

Gain margin
Phase margin
System bandwidth

Controller Tuning

Once performance criteria have been established, controller tuning can be undertaken. The approach taken for controller parameter adjustments depends upon the availability (or lack of it) of a suitable plant mathematical model (transfer function). There are two distinct possibilities:

1. *Known plant model:* This is usually the case for a new design. For a small existing plant, a mathematical model can be determined if sufficient information is available about the characteristics (mathematical model) of each and every component and their interactions with each other.
2. *Unknown plant model:* An unknown model appears to be more of a rule than an exception in an industrial setting. Mathematical models of older and existing plants are generally not available. Even if a plant mathematical model is available, with age plant characteristics may deviate significantly from the original model, rendering the original model more or less useless. For tuning of such plants, there are two possible options.
   a. *Experimental determination of plant model*: It is possible to carry out tests on a plant to determine its transfer function (mathematical model). These tests can range from a simple step test to a more complex frequency response test. Once experimental data have been obtained, analysis can be performed on the collected data and a suitable plant model can be extracted.
   b. *Experimental determination of controller setting*: Sometimes, it is not feasible to carry out a comprehensive test on a plant, or plant authorities may not permit such a test. At other times, the nature of the plant may be such that a meaningful model determination may require a costly and lengthy test. In all such cases, it is still possible to find a suitable controller setting, provided simple experimental test procedures are allowed.

Controller Tuning: Known Plant Model

If the plant model is known, suitable controller parameters can be determined through analysis and/or simulations. The effect of controller tuning can be observed by changing the coefficients in the mathematical equation representing the controller and repeating the analysis. This process is continued until a satisfactory performance is obtained. The analysis can be manually carried out, or a computer program such as MATLAB can be gainfully employed. Several powerful and process-specific software packages are also available and are sometimes developed in-house to facilitate an almost lifelike simulation.

EXAMPLE 14.1

A proportional controller is used to control the speed of a DC motor in a small rolling mill. The mill velocity is monitored through a tachometer, and the velocity reference signal is obtained through a potentiometer. The closed-loop diagram is shown in Figure 14.1. The following transfer functions are known:

$$\text{plant (DC motor + rolling mill)} = \frac{2.25}{s(03s + 1)}$$

$$\text{tachometer} = 0.2$$

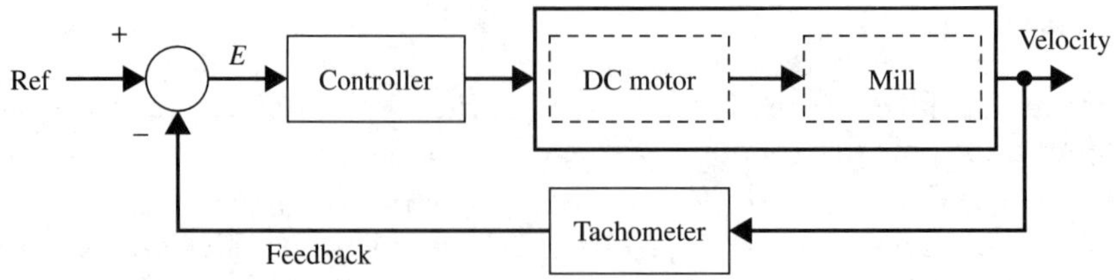

Figure 14.1 Control system for Example 14.1.

Determine the value of the proportionality constant (K_p) that will provide a fast rise time without any overshoot.

Solution

$$\text{closed transfer function} = \frac{G(s)}{1 + GH(s)}$$

$$= \frac{K_P \dfrac{2.25}{s(0.3s+1)}}{1 + K_P \dfrac{2.25}{s(0.3s+1)} \times 0.2}$$

$$= \frac{2.25K_P}{s(0.3s+1) + 0.45K_P}$$

$$= \frac{2.25K_P}{0.3s^2 + s + 0.45K_P}$$

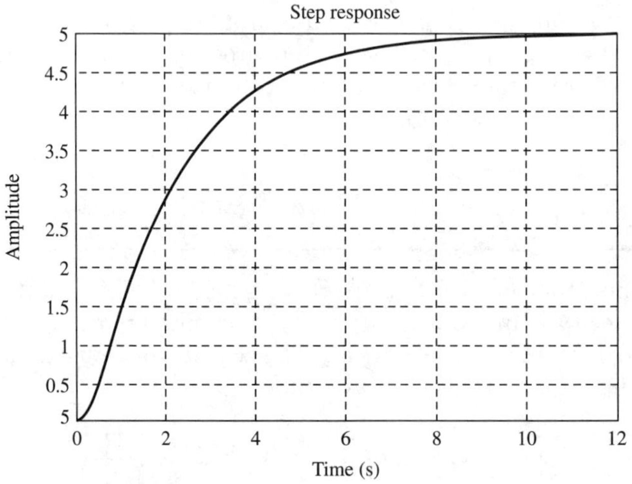

Figure 14.2 Step response for $K_P = 1$.

The effect of the proportionality constant K_P can be observed by substituting different values of K_P in this transfer function.

Case I: $K_P = 1$:

$$\text{CLTF} = \frac{2.25}{0.3s^2 + s + 0.45}$$

The step response of this result is shown in Figure 14.2. The output rises exponentially and settles for a final value of 5 units. There is no overshoot in the output. This (from CLTF) is a second-order system, so the damping ratio is (equal to or) greater than 1. The output reaches the (99% of) final value in just about 10 s.

Case II: $K_P = 10$: Increasing the proportional gain K_P to 10, the transfer function becomes

$$\text{CLTF} = \frac{22.5}{0.3s^2 + s + 4.5}$$

The new step response is shown in Figure 14.3.

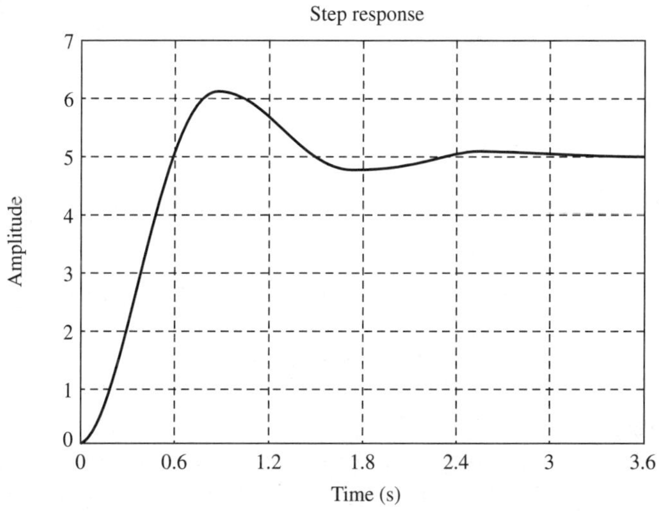

Figure 14.3 Step response for $K_P = 10$.

It can be clearly seen that increasing the proportional gain to 10 results in an overshoot. On the other hand, output rises much faster and settles to final value within 4 s. Even though in some processes it may be considered to be a significant improvement in output response, it is not acceptable here. It was clearly stated that an overshoot was considered unacceptable.

The optimum value of the proportionality constant K_P lies somewhere in between 1 and 10. Output response can be checked for different values of K_P. Unfortunately, this type of guesswork may turn out to be a lengthy process and may still not reveal the optimum value of K_P. Because it is a simple transfer function (second-order), mathematical analysis can easily be carried out, revealing the optimum value of K_P.

Roots of the characteristic equation: For an overdamped response, all the roots should be real, with no imaginary component (Chapter 5). The general characteristic equation for this closed-loop control system can be obtained from the expression for CLTF.

The characteristic equation is $1 + GH(s) = 0$, or

$$0.3s^2 + s + 0.45K_P = 0$$

The roots of the equation are

$$s = \frac{-1 \pm \sqrt{1^2 - 4 \times 0.3 \times 0.45K_P}}{2 \times 0.3}$$

$$s = \frac{-1 \pm \sqrt{1 - 0.54K_P}}{0.6}$$

In order to have no imaginary component, the radicand (quantity under the radical sign) should not be negative. Therefore,

$$\sqrt{1 - 0.54K_P} \geq 0$$

$$1 - 0.54K_P \geq 0$$

$$1 \geq 0.54K_P$$

$$0.54K_P \leq 1$$

$$K_P \leq \frac{1}{0.54}$$

$$K_P \leq 1.85$$

Thus, the proportionality constant K_P should be less than 1.85. Because an increase in the value of K_P is accompanied by an improvement in the output response time, the largest possible value of K_P should be selected. Therefore, the optimum value of K_P is 1.85.

Check: Substituting the calculated value of K_P,

$$\text{CLTF} = \frac{2.25 \times 1.85}{0.3s^2 + s + 0.45 \times 1.85}$$

$$= \frac{4.1625}{0.3s^2 + s + 0.8325}$$

The step response of the compensated system is shown in Figure 14.4.

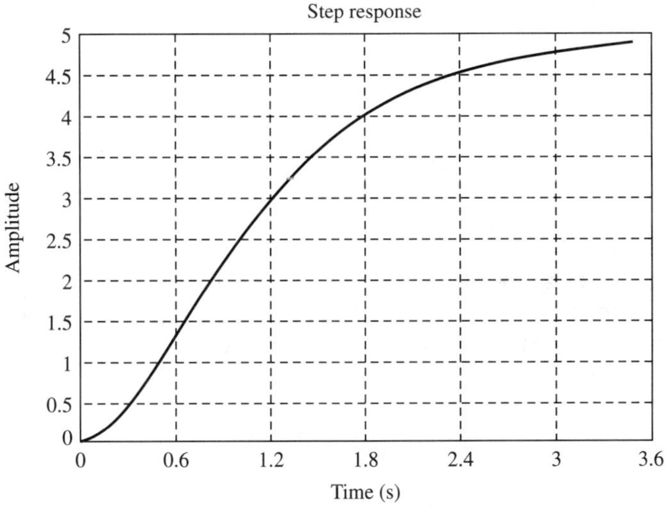

Figure 14.4 Compensated step response.

It can be seen that there is no overshoot in the output response. The system also settles much faster than the one with a proportionality constant of 1.

EXAMPLE 14.2

A PI controller is incorporated in the closed-loop control system in Figure 14.5. Determine suitable controller settings so that the following operational specifications may be achieved:

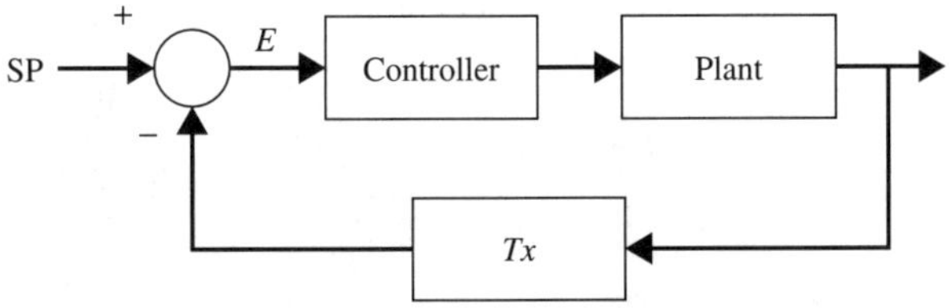

Figure 14.5 System block diagram for Example 14.2.

1. Gain margin of 10 dB
2. Minimum phase margin of 30°, although the desirable value is 45°.
3. Settling time <5 s

The plant and transducer transfer functions are

$$\text{plant} = \frac{1.25}{(s+1)(s+4)}$$

$$\text{transducer} = 0.5$$

Solution

The first two specifications relate to gain and phase margin, and these are established through the Bode plot of the open-loop transfer function. The third specification relates to the control system step (time) response. The closed-loop transfer function is required to determine the step response.

$$\text{PI controller transfer function} = \frac{K_P s + K_I}{s}$$

$$\begin{aligned}
\text{open-loop transfer function (OLTF)} &= GH(s) \\
&= \frac{K_P s + K_I}{s}\frac{1.25}{(s+1)(s+4)}0.5 \\
&= \frac{0.625(K_P s + K_I)}{s(s+1)(s+4)} \\
&= \frac{0.625K_P s + 0.625K_I}{s^3 + 5s^2 + 4s}
\end{aligned}$$

$$\begin{aligned}
\text{closed-loop transfer function (CLTF)} &= \frac{G(s)}{1 + GH(s)} \\
&= \frac{\dfrac{K_P s + K_I}{s}\dfrac{1.25}{(s+1)(s+4)}}{1 + \dfrac{K_P s + K_I}{s}\dfrac{1.25}{(s+1)(s+4)}0.5} \\
&= \frac{\dfrac{1.25K_P s + 1.25K_I}{s(s+1)(s+4)}}{1 + \dfrac{0.625K_P s + 0.625K_I}{s(s+1)(s+4)}} \\
&= \frac{1.25K_P s + 1.25K_I}{s(s+1)(s+4) + 0.625K_P s + 0.625K_I} \\
&= \frac{1.25K_P s + 1.25K_I}{s^3 + 5s^2 + (4 + 0.625K_P)s + 0.625K_I}
\end{aligned}$$

In a PI controller, its two coefficients, proportional gain K_P and integral gain K_I, are varied. Several combinations of K_P and K_I may yield a satisfactory solution. A number of different values for the gain coefficients will be tried, and their impact on gain, phase margin, and time response will be observed. A suitable combination of K_P and K_I can then be selected.

There is no hard-and-fast rule for selecting the starting values of gain coefficients. In most of the systems, proportional gain is much larger than integral gain, although it is by no means universal in nature. An initial value of 1 can be tried for proportional gain, and a relatively small value (say 0.01) for integral gain.

Case I $K_P = 1$ and $K_I = 0.01$: Substituting the values of controller gain in the expression for the open-loop transfer function gives

$$\text{OLTF} = \frac{0.625s + 0.625 \times 0.01}{s^3 + 5s^2 + 4s}$$
$$= \frac{0.625s + 0.00625}{s^3 + 5s^2 + 4s}$$

The resultant frequency response plot is shown in Figure 14.6.

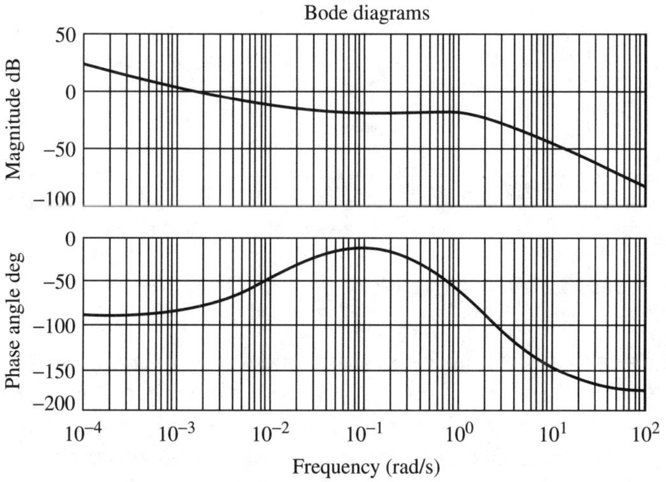

Figure 14.6 Frequency response for Example 14.2.

From the frequency response plot, the following can be estimated:

- Gain crossover frequency $\approx$ 0.001 rad/s
- Phase crossover frequency $\approx$ 100 rad/s
- Gain margin $\approx$ 80 dB
- Phase margin $\approx 100^\circ$

Such large margins obviously make the system very stable but it also results in a sluggish operation.

Similarly, substituting the value of the gain coefficients in the expression for the closed-loop transfer function gives

$$\text{CLTF} = \frac{1.25s + 0.0125}{s^3 + 5s^2 + 4.625s + 0.00625}$$

The time response of this system when subjected to a unit step input is shown in Figure 14.7.

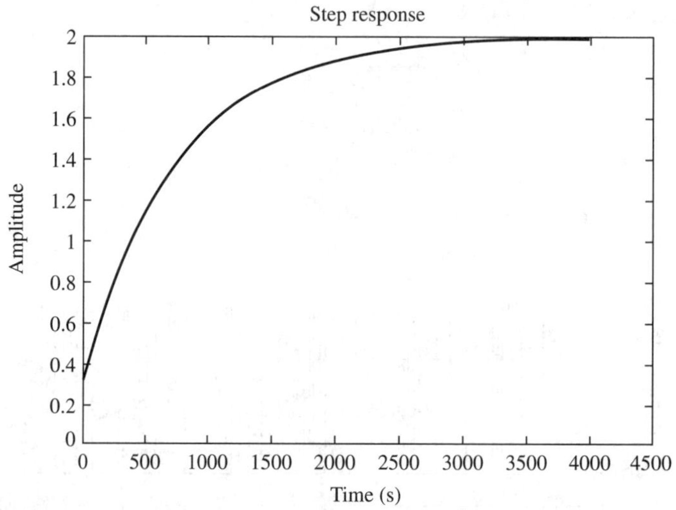

Figure 14.7 Step response for $K_P = 1$ and $K_I = 0.01$.

The step response indicates that the system takes a long time (>4000 s) to reach the final value. Obviously, this is not acceptable and the controller gain values need to be increased. Because there can be an infinite number of combinations of K_P and K_I, a systematic approach is necessary. First, the effect of varying the proportional gain will be observed, followed by the effect of integral gain.

The effect of increasing K_P on system time response can be seen from the step response plot (Figure 14.8).

Case II $K_P = 10$ and $K_I = 0.01$: There is a slight improvement in output response as proportional gain is increased to 10. Although it is not as apparent from the plot (because of the limited span of 0–5 s), a large amount of time (>1000 s) is taken for the output to rise to the final value. Again, it is much slower than the required settling time of 5 s.

Case III $K_P = 100$ and $K_I = 0.01$: There is a remarkable improvement in output response as K_P is increased to 100. Output settles well within the required time of 5 s, although a small overshoot and some oscillations occur. Before proceeding any

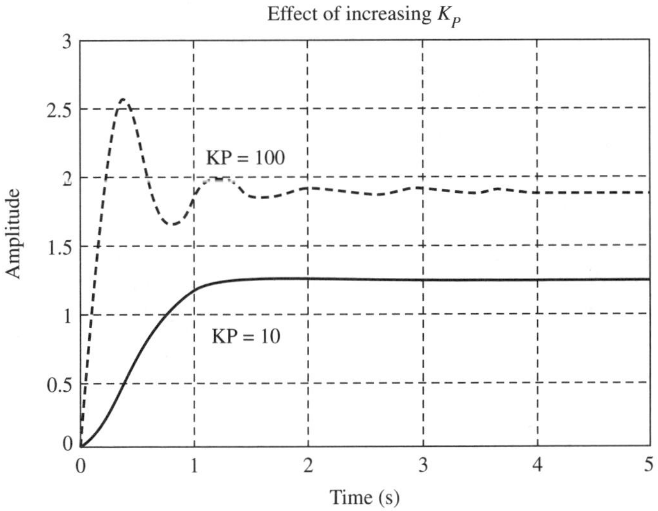

Figure 14.8 Effect of increasing K_p on step response.

further, gain and phase margin need to be verified. The frequency response plot for K_P of 100 ($K_I = 0.01$) is shown in Figure 14.9. The frequency response plot indicates a very low value of gain and phase margin. This indicates that proportional gain is too high and needs to be reduced. At the same time, the integral gain can be varied.

Case IV $K_P = 10$ and $K_I = 1$: Proportional gain is now reduced to 10 and integral gain is increased to 1. The step response (Figure 14.10) of the control system shows an

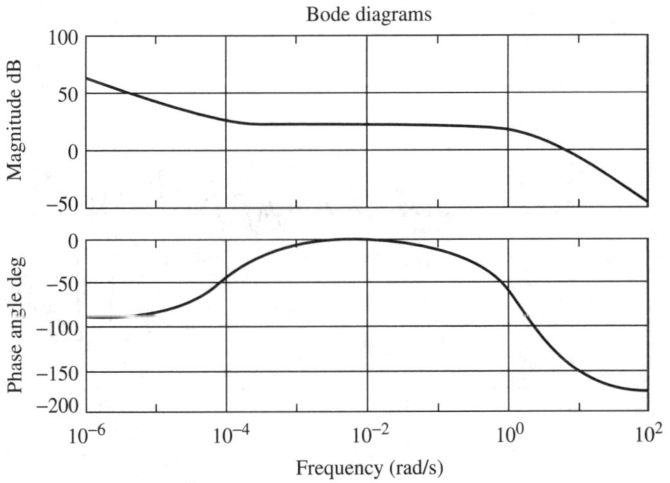

Figure 14.9 Frequency response for $K_p = 100$ and $K_I = 0.01$.

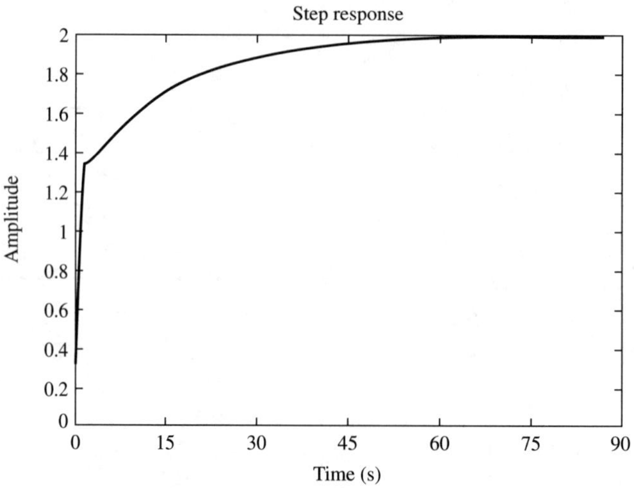

Figure 14.10 Step response for $K_p = 10$ and $K_I = 1$.

improvement in the operating speed, but it still does not quite meet the settling-time requirement.

Case V $K_P = 10$ and $K_I = 100$: The proportional gain is kept at 10 and the integral gain is increased to a value of 100. The resultant step response (Figure 14.11) shows very remarkable results. The system output goes into serious oscillations and eventually crashes. This combination of controller gain has rendered the system *unstable*.

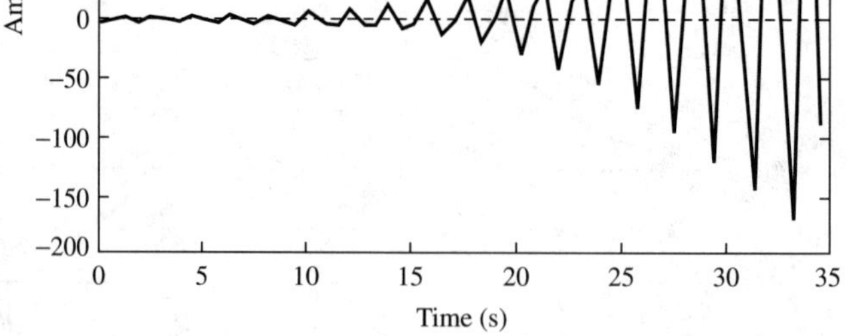

Figure 14.11 Oscillations due to increase in K_I.

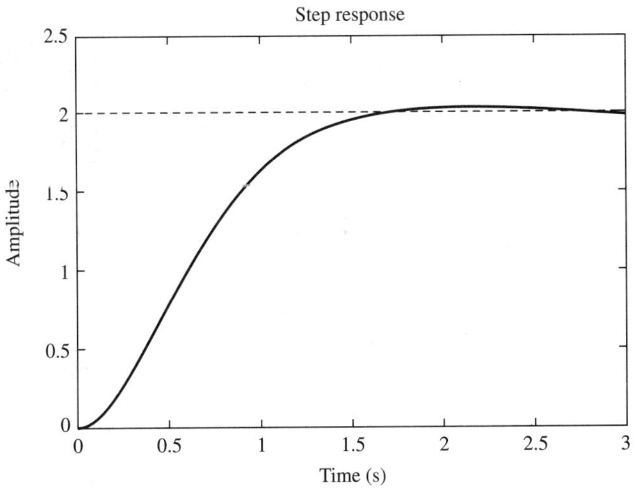

Figure 14.12 Step response for $K_p = 10$ and $K_I = 10$.

Obviously, the integral gain is too high and needs to be reduced. This is typical characteristic of the integral mode. A large value of integral gain generally results in oscillations and tends to make the system unstable.

Case VI $K_P = 10$ and $K_I = 10$: Keeping the proportional gain to 10 and reducing the integral gain to 10 reveals the step response shown in Figure 14.12. It exhibits a very fast settling time and very small overshoot. Because there was no specification on the amount of overshoot, it will be considered acceptable. But before these gain settings can be considered acceptable, the values of phase and gain margin also need to be verified.

The frequency plot (Figure 14.13) shows that the phase angle never drops below $-180°$. It means that the system does not become unstable at high frequency. The following data can be estimated from the frequency response plot:

- Gain crossover frequency $\approx$ 1.5 rad/s.
- Phase crossover does not occur.
- Phase margin $\approx 60°$.
- The gain (margin) is more than 60 dB below the 0-dB line at the frequency where the phase angle approaches the $-180°$ line.

All criteria for control system performance have been met, so these gain settings can be considered acceptable. It should be pointed out, however, that these gain values are by no means the optimum. Improved settings can be obtained by continued estimation, by computer simulation, or through field tests.

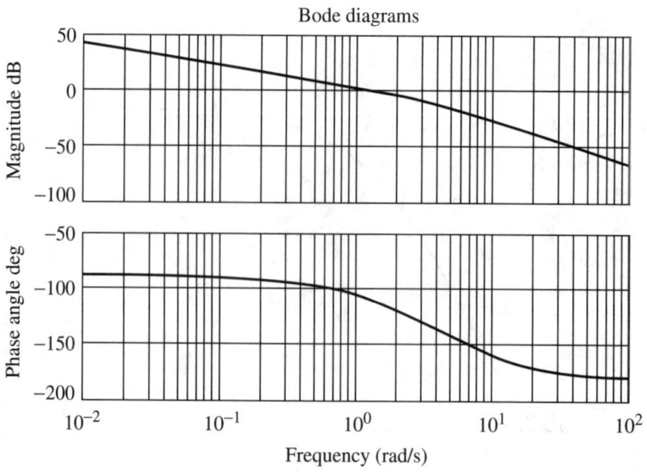

Figure 14.13 Frequency response plot for $K_p = 10$ and $K_I = 10$.

PLANT MODEL FROM EXPERIMENTAL RESPONSE TESTING

Even a small industrial plant contains several first-, second-, and third-order elements, and a much-higher-order mathematical model may be required for precise representation. Fortunately, for all practical purposes it is generally sufficient to use a first- or second-order model. Much more precise models, however, are needed for systems requiring very tight control, such as high-speed servomotor control systems, aircraft flight control systems, and systems requiring energy-mass balance. There are two types of tests that are commonly performed for plant model identification:

1. Step response test
2. Frequency response test

STEP RESPONSE TEST

The step response test (Figure 14.14) is also known as the *process reaction curve test.* It provides a quick and simple approach for development of a plant dynamic response model. Unfortunately, the model thus obtained is subject to interpretation errors. This test is best suited for extracting a simple first- or second-order model.

Chart recorders were previously used for such testing, but recently, high-speed data-acquisition systems have gained significant ground. Data-acquisition systems have a significant advantage in that the data is inherently in digital format and can be directly fed to an analytical program or printed in numerical or graphical form. Computer-based data-acquisition systems generally do not perform well when a process with very small

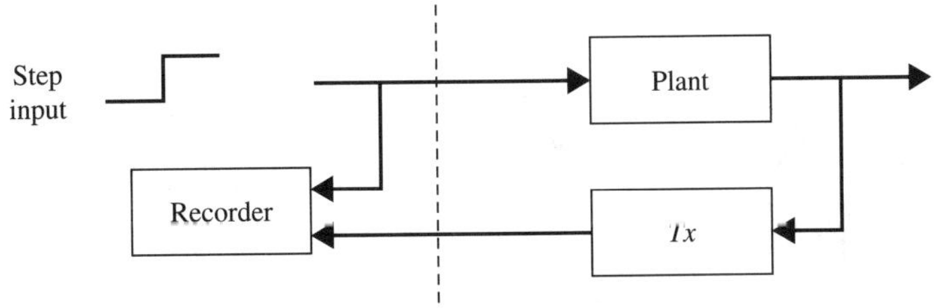

Figure 14.14 Step response test.

time constant, such as a servomotor motor control system, is tested. High-speed digital storage scopes are preferred in all such applications. The following test procedure is used for step response testing

Test Procedure

- Attach a recording device to continuously monitor the step input and the output from the feedback transducer.
- Switch the control system from automatic to manual mode.
- Manually adjust the set point to operate at 50% of range (if permitted).
- Wait until the transients in the process output die down and a steady-state output is obtained.
- Apply a step change of 5% to 10% (or larger if the transducer output is noisy).
- Continue recording until a steady-state output is reached.
- Analyze the recorded observations for extracting a suitable plant model.

Once experimental data is obtained, a model can be devised either through graphical analysis or through computation.

Graphical Analysis

Since the early days of chart recorders, step response data has been made available in graphical form. Data-acquisition units have replaced or at least supplemented most chart recorders. High-resolution A/D converters used in these recording units allow test data to be recorded with high precision; when fed to a computer, the numerical data can be viewed as a real-time graph (*process reaction curve*). See Figure 14.15.

The graphical method is satisfactory for identifying a *first-order lag model with dead time*. Even though it is possible to extract a *second-order model*, the results are not very accurate and are subject to the user's interpretation. Here, only the first-order model is discussed.

Graphical Analysis for First-Order plus Dead-Time Model

- Obtain the process reaction curve (step response curve).
- Draw a tangent at the point of the steepest slope (ascent).

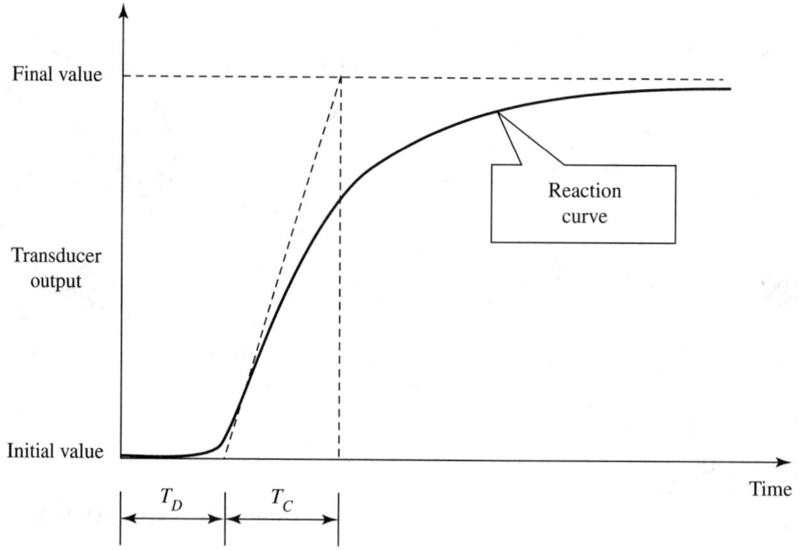

Figure 14.15 Process reaction curve.

- Extend the line (tangent) down to the initial value.
- Extend the line (tangent) up to the final value.
- *Process dead time* T_D, is the horizontal distance (time) from the point of initial upset (step input) to the point where the tangent meets the initial value.
- *Time* T_C is the horizontal distance (time) between the points, where the tangent intersects the initial and final (horizontal) output lines.
- *Time* $T_{63\%}$ is the horizontal distance (time) from start to the time when the process reaches 63.2% of the final value.

$$T_{63\%} = T_D + \tau$$

- *The process time constant* τ is calculated from the preceding time values.
- *Process gain* K is the ratio of change in the process output to the applied change in process input:

$$K = \frac{\text{change in process output}}{\text{change in process input}}$$

Once the calculations for dead time and process time constant are finished,

$$\text{TF} = \frac{Ke^{-sT_D}}{\tau s + 1}$$

EXAMPLE 14.3

The reaction curve obtained from the step response test on a process is shown in Figure 14.15. The following values were extracted from the graph:

$$
\begin{aligned}
\text{initial process input} &= 5\% \\
\text{final process input} &= 10\% \\
\text{initial process output} &= 45\% \\
\text{final process output} &= 55\% \\
\text{time at which tangent meets the initial value} &= 2.5\text{ s} \\
\text{time at which tangent meets the final value} &= 7.5\text{ s} \\
\text{time for process to reach 63.2\% of final value} &= 6.0\text{ s}
\end{aligned}
$$

Determine the process transfer function.

Solution

The data given allows a model to be determined based on first-order lag and dead time.

$$
\begin{aligned}
\text{process gain } K &= \frac{\text{change in process output}}{\text{change in process input}} \\
&= \frac{(55-45)\%}{(10-5)\%} \\
&= 2 \\
\text{dead time } T_D &= 2.5\text{ s}
\end{aligned}
$$

The time difference (points where tangent meets initial and final values) is

$$
\begin{aligned}
T_C &= 7.5 - 2.5\text{ s} \\
&= 5.0\text{ s} \\
T_{63\%} &= T_D + \tau \\
&= 6\text{ s}
\end{aligned}
$$

Therefore,

$$
\begin{aligned}
\tau &= T_{63\%} - T_D \\
&= 6 - 2.5\text{ s} \\
&= 3.5\text{ s}
\end{aligned}
$$

Thus the process model is

$$\mathrm{TF} = \frac{Ke^{-sT_D}}{\tau s + 1}$$

$$= \frac{2e^{-2.5\,s}}{3.5s + 1}$$

Computational Method

The graphical method requires a tangent to be drawn at the point of steepest slope. Based upon individual judgment, slightly different values of dead time and system time may be obtained. As a result, the model obtained may vary from person to person. The computational method eliminates the errors caused by false interpretation and gives a more accurate model. This method is not limited to a first-order model; second- and higher-order plant models can also be estimated. Here, for the sake of simplicity, only a first-order model with dead time and unity gain is considered.

First-Order plus Dead-Time Model (Unity Gain) A first-order system is described by the following time equation:

$$v_o = v_i\left(1 - e^{-\frac{t}{\tau}}\right)$$

where τ = system time constant. Considering a unit step input, the equation simplifies to

$$V_{\text{out}} = (1 - e^{-t/\tau})$$

When coupled with a dead-time element T_D, the equation becomes

$$V_{\text{out}} = \left(1 - e^{-(t-T_D)/\tau}\right) \qquad (\text{valid for time } t \geq T_D)$$

As can be seen, there are two unknown variables, τ and T_D, and in order to determine these values, two points on the process reaction curve are needed. Suitable values will be the times when process output reaches $\frac{1}{3}$ and $\frac{2}{3}$ of the final value.

Time t_{13}, when output reaches $\frac{1}{3}$ the final output, is given as

$$\frac{1}{3} = \left(1 - e^{-(t_{13}-T_D)/\tau}\right)$$

$$\frac{1}{3} - 1 = -e^{-(t_{13}-T_D)/\tau}$$

$$e^{-(t_{13}-T_D)/\tau} = \frac{2}{3}$$

Taking the natural logarithm of both sides,

$$\ln\left(e^{-(t_{13}-T_D)/\tau}\right) = \ln\left(\frac{2}{3}\right)$$

$$-\left(\frac{t_{13}-T_D}{\tau}\right) = -0.405465$$

$$-(t_{13}-T_D) = -0.405465\tau$$

$$0.405465\tau = t_{13}-T_D$$

Solving for the time constant τ gives

$$\tau = \frac{t_{13}-T_D}{0.405465}$$

Similarly, considering the time t_{23}, when the output has reached $\frac{2}{3}$ of the final value,

$$\frac{2}{3} = \left(1 - e^{-(t_{23}-T_D)/\tau}\right)$$

Simplifying and solving for the time constant τ gives

$$\frac{2}{3} - 1 = -e^{-(t_{23}-T_D)/\tau}$$

$$\ln\left(e^{-(t_{23}-T_D)/\tau}\right) = \ln\left(\frac{1}{3}\right)$$

$$-\left(\frac{t_{23}-T_D}{\tau}\right) = -1.098612$$

$$t_{23}-T_D = 1.098612\tau$$

$$\tau = \frac{t_{23}-T_D}{1.098612}$$

From these two equations, the expressions for the system time constant, τ, and dead time, T_D, can be derived.

System time constant:

$$\tau = 1.443(t_{23}-t_{13})$$

Dead time:

$$T_D = 1.585t_{13} - 0.585t_{23}$$

EXAMPLE 14.4

The following data were obtained from a step response test on a process.

$$\text{time to reach } \frac{1}{3} \text{ of final value} = 25.5 \text{ s}$$

$$\text{time to reach } \frac{2}{3} \text{ of final value} = 57.5 \text{ s}$$

Considering a first-order lag model with dead time, determine the transfer function of the process. The system steady-state gain is unity.

Solution

From the given data,

$$t_{13} = 25.5 \text{ s}$$
$$t_{23} = 57.5 \text{ s}$$

Substituting the values in the expression for the time constant gives

$$\begin{aligned}\tau &= 1.443(t_{23} - t_{13}) \\ &= 1.443(57.5 - 25.5) \\ &= 46.2 \text{ s}\end{aligned}$$

Similarly, the dead time is

$$\begin{aligned}T_D &= 1.585t_{13} - 0.585t_{23} \\ &= 1.585 \times 25.5 - 0.585 \times 57.5 \\ &= 6.78 \text{ s}\end{aligned}$$

Check: Checking for the time taken to reach $\frac{1}{3}$ of the final value,

$$\begin{aligned}\frac{1}{3} &= \left(1 - e^{-(t_{13}-T_D)/\tau}\right) \\ \frac{1}{3} &= \left(1 - e^{-(t_{13}-6.78)/46.2}\right) \\ e^{-(t_{13}-6.78)/46.2} &= \frac{2}{3}\end{aligned}$$

Taking the (natural) logarithm of both sides,

$$\ln\left(e^{-(t_{13}-6.78)/46.2}\right) = \ln\left(\frac{2}{3}\right)$$

$$-\left(\frac{t_{13}-6.78}{46.2}\right) = -0.405465$$

$$t_{13} - 6.78 = 0.405465 \times 46.2$$

Simplifying, $T_{13} = 25.51$ s, which checks.

Frequency Response Test

A frequency response test (Figure 14.16) provides considerably more information about the plant under test. It can indicate the presence of integral, derivative, first-order lead or lag, and second-order components. In practical terms, it is more difficult than a step response test. Some of the following problems may be encountered in conducting this test.

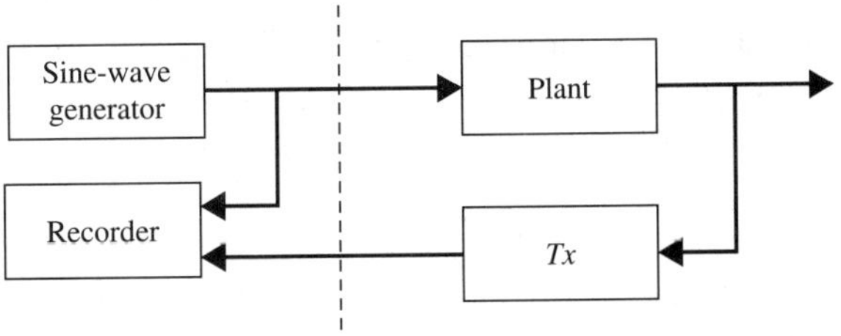

Figure 14.16 Frequency response test.

- Some plants cannot be put in continuous oscillation without a serious risk of injury or damage.
- A plant with a large time constant will require a considerable time and effort in carrying out this test.
- The test is susceptible to signal noise.
- Unwarranted variation in plant load and/or environment may occur, resulting in incorrect data.
- It is difficult to determine the optimum signal amplitude that will not result in excessive oscillation, yet will yield meaningful measurements.
- It may be difficult to develop a sinusoidal variation in the process input. This difficulty is encountered when a plant input is nonelectrical in nature, such as pneumatic, thermal, or mechanical.

The following procedure is used for frequency response testing.

Test Procedure

- Take precautions to ensure that the plant does not suffer from load disturbances during the test.
- Estimate the frequency range over which the test needs to be performed.
- Connect a sine-wave generator, which will be able to modulate the process input over the required frequency range.
- Attach a recording device to monitor the process input and the output from the feedback transducer.
- Allow the plant output to stabilize.
- Switch the control system from automatic to manual mode.
- Set the sine-wave generator for the lowest test frequency.
- Adjust the signal amplitude to a suitable value.
- Wait until the transients in process output die down and constant oscillations are observed in plant output.
- Record the signal amplitude ratio as well as the signal phase shift.
- Repeat the preceding steps until the entire frequency range of interest is covered.
- Plot the recorded observations on semilog paper (Bode plot).
- Analyze the Bode plot to extract a suitable plant model. (Use the standard frequency response characteristics of common components.)

EXAMPLE 14.5

The Bode plot obtained from the frequency response test on a simple plant is shown in Figure 14.17. Estimate a model for the plant from the graph.

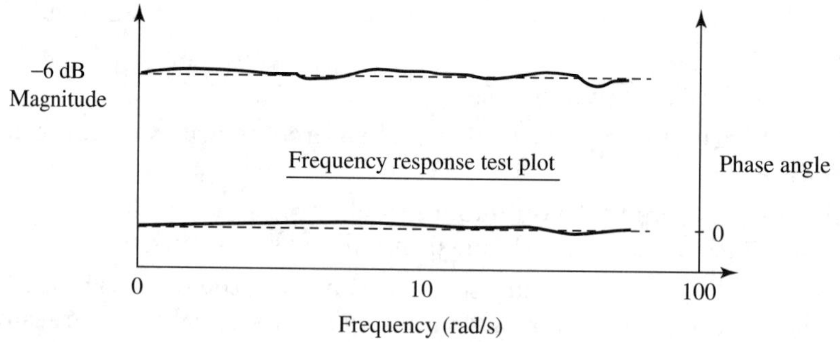

Figure 14.17 Process reaction curve for Example 14.5.

Solution

The magnitude plot represents almost a straight line (shown as dotted). Similarly, the phase-angle plot can also be estimated to be a straight line at a 0° angle. Both these characteristics are typical of a constant block.

$$\text{process gain} = -6\ \text{dB}$$
$$= 10^{(-6/20)}$$
$$= 0.5$$

Thus,

$$\text{process model (TF)} = 0.5$$

Note: In the following examples the frequency response plots shown do not contain any noise data. The noise signals have been completely filtered out, and these curves represent ideal plant components. This has been intentionally done to eliminate any confusion that may be caused by actual noisy recording. However, careful attention should be paid in the field, where test data reflect the overall effect of several time constants present in the system along with signal noise. Only some of these time constants are of significance, and the effect of signal noise should be taken into consideration.

EXAMPLE 14.6

Estimate a plant model from the frequency response in Figure 14.18 obtained through a frequency response test.

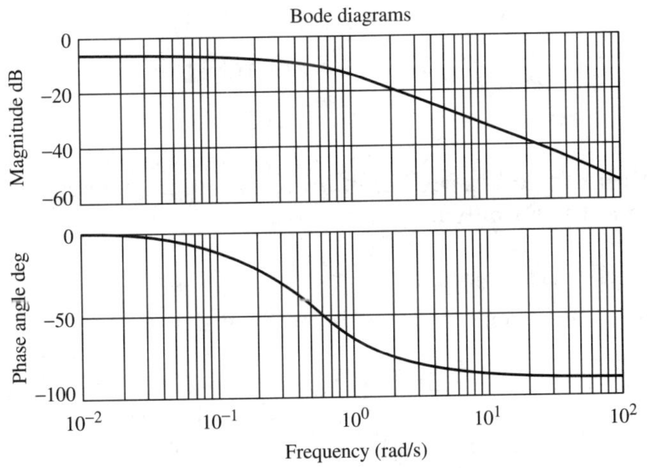

Figure 14.18 Frequency response curve for Example 14.6.

Solution

From the observation of the frequency response plot, it appears that the plant consists of a constant plus a lag block. It appears that two asymptotes can be drawn on the magnitude plot; a horizontal line covering the low-frequency section and a sloping line covering the higher-frequency section. The Bode plot is reproduced in Figure 14.19 superimposed with the two asymptotes.

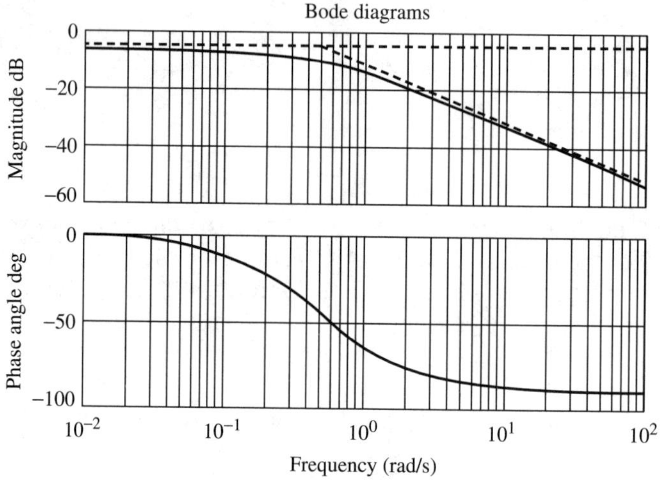

Figure 14.19 Asymptotic approximation of frequency response data.

Identification of a constant block: The horizontal asymptote has a magnitude of about -5 dB. Thus, the gain for the constant block can be calculated.

$$\begin{aligned}\text{process gain } K &= -5 \text{ dB}\\ &= 10^{(-5/20)}\\ &= 0.562\end{aligned}$$

Identification of the lag block: In order to completely describe a lag block, two parameters need to be estimated:

- Order of the block (first or second)
- Time constant (or two time constants, if second-order)

The order of the lag system is the first parameter to be determined. There are a couple of factors that aid in this determination. The phase angle varies from a maximum of 0° to a minimum of $-90°$. This is consistent with a first-order lag process. When considering the magnitude plot, it can be seen that the asymptote drops by 20 dB for a decade increase in frequency. This confirms a first-order lag block.

The next factor to be determined is the time constant. The time constant is related to break frequency:

$$\text{time constant } \tau\,(\text{s}) = \frac{1}{\text{break frequency in rad/s}}$$

The time constant can thus be calculated if break frequency can be determined. For a first-order lag process, there are two clues:

1. Break frequency is the frequency at which the two asymptotes meet. From the graph, it is estimated to be 0.5 rad/s.

$$\text{time constant } \tau = \frac{1}{0.5 \text{ rad/s}}$$
$$= 2 \text{ s}$$

2. The phase angle reaches -45° at the break frequency. From the graph, it can be seen that the phase angle starts at about 0°, drops to -90°, and reaches -45° at a frequency of 0.5 rad/s. This verifies the time constant.

   Based upon this information,

$$\text{plant model (transfer function)} = \frac{K}{(\tau s + 1)}$$
$$= \frac{0.562}{(2s + 1)}$$

EXAMPLE 14.7

Estimate a plant model from the Bode plot in Figure 14.20 obtained through a frequency response test.

Solution

Careful examination of magnitude and phase-angle plots indicate that the plant consists of a minimum of two blocks: a constant block and a lag block. The phase-angle plot further reveals that the phase angle starts from a maximum of 0° and drops to -180° at high frequency. This indicates the presence of either a second-order lag block or a combination of two first-order blocks. Because there is no noticeable resonant effect in the magnitude plot, it can be safely assumed that even if a second-order lag block exists, it is an overdamped type. Furthermore, it is known that an overdamped second-order lag block can be isolated into two first-order lag blocks.

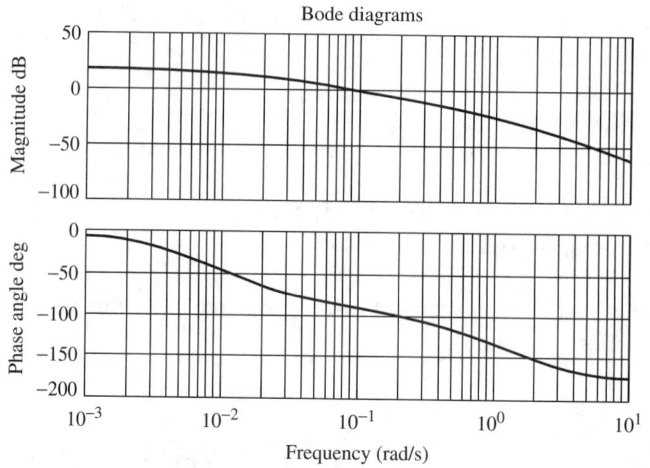

Figure 14.20 Frequency response curve for Example 14.7.

The plant can be considered to consist of the following:

- A constant block
- A first-order lag block with time constant τ_1
- A first-order lag block with time constant τ_2

Asymptotes are drawn to the magnitude plot. There are three asymptotes. One horizontal line represents the constant block. A sloping line represents the first lag

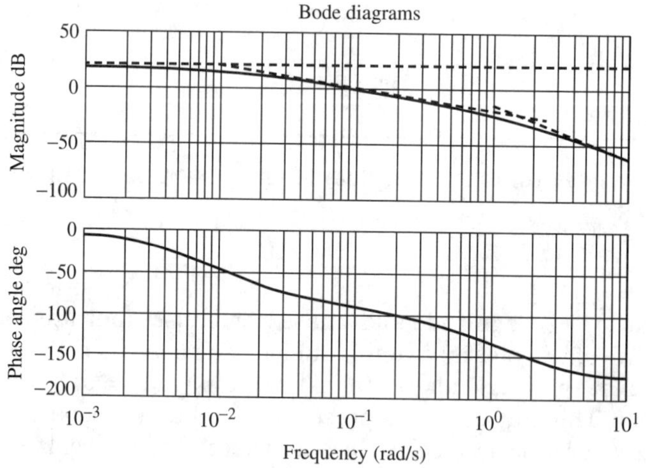

Figure 14.21 Asymptotic approximation for Example 14.7.

block and the second sloping line results from the combination of two first-order lag blocks. The superimposed Bode plot is shown in Figure 14.21.

Constant block:

$$\begin{aligned}\text{magnitude of horizontal asymptote} &= 20\text{ dB}\\ \text{plant gain} &= 10^{20/20}\\ &= 10\end{aligned}$$

Time constant τ_1: The first sloping line meets the horizontal asymptote at a frequency of 0.01 rad/s. Checking with the phase-angle plot confirms that at this frequency the phase angle has reached an angle of $-45°$. Thus, the break frequency for the first lag block is 0.01 rad/s.

Thus,

$$\begin{aligned}\text{time constant } \tau_1 &= \frac{1}{\text{break frequency in rad/s}}\\ &= \frac{1}{0.01\text{ rad/s}}\\ &= 100\text{ s}\end{aligned}$$

Time constant τ_2: The second sloping line meets the first sloping line at a frequency of 1 rad/s. Checking with the phase-angle plot confirms that at this frequency the phase angle has reached an angle of $-135°$. This phase angle is the result of $-90°$ reached by the first lag block and an additional drop of $-45°$ because of the second lag block (τ_2).

Note: The phase angle of $-135°$ is reached only because the two time constants are significantly different. If the two constants were identical, the phase angle would be only $-90°$ at the common break frequency. Caution should be exercised when interpreting the test plots.

Thus, the break frequency for the second first-order lag block is 1 rad/s and

$$\begin{aligned}\text{time constant } \tau_2 &= \frac{1}{\text{break frequency in rad/s}}\\ &= \frac{1}{1\text{ rad/s}}\\ &= 1\text{ s}\\ \text{plant model (TF)} &= \frac{K}{(\tau_1 s+1)(\tau_2 s+1)}\\ &= \frac{10}{(100s+1)(s+1)}\end{aligned}$$

EXPERIMENTAL DETERMINATION OF CONTROLLER SETTING

Experimental determination of the controller setting can yield very satisfactory results, even if there is no information about the plant transfer function. This approach is generally utilized to verify results, even when the plant model is known. These procedures aim to find the (more or less) optimum tuning coefficients for a three-term (PID) controller. These approaches are based upon experience gathered by control engineers in operation and control of various plants over the years. There are two main approaches for determining tuning coefficients, and both are based upon the pioneering work of two control practitioners, J. G. Ziegler and N. B. Nichols.

Both methods strive to provide a slightly underdamped response with a decay ratio of 1:4. These settings tend to provide a reasonably fast response with small settling time. These settings obviously are not suitable for control systems, where an underdamped response cannot be tolerated, e.g., cutting-tool movement in milling machines or pen movement in plotters. But they are very suitable for process control as well as some positioning systems.

Continuous Cycling Method

In this method, the plant is operated in automatic mode with integral and derivative modes disabled. The method relies on finding the value of the proportional gain that will result in a continuous oscillation of output. This really places the plant on the verge of marginal stability and may require careful consideration for some plants. The test procedure is as follows:

1. Switch the plant to operate in automatic mode.
2. Set the integral and derivative gains to zero.
3. Set the proportional gain K_U to a small value.
4. Wait until the plant operation stabilizes.
5. Introduce a small change in set point (5% to 10%).
6. Observe the plant output response. For a small value of K_U, the output should settle down, resulting in a damped response. If this is not the case, then the value of K_U is too high and should be reduced significantly. Steps 4 to 6 should be repeated.
7. Return the set point to its original value.
8. Wait until the output stabilizes.
9. Increase the value of K_U.
10. Introduce a small change in set point (5% to 10%).
11. Observe the plant output response. The output response should be less damped, resulting in greater overshoot and longer settling time.
12. Continue to increase the value of K_U (repeat from step 9) until sustained oscillations are obtained at the output. If the oscillations are increasing in

amplitude, then continue to reduce K_U until the oscillations are of constant amplitude.

13. Record the value of the proportional gain, K_U, and the time period of oscillation, T_U. This value of K_U is known as the *ultimate gain*, and the T_U is known as the *ultimate time period*.
14. Return the plant to safe operating conditions.
15. Use Table 14.1 to view the recommended controller setting.

Table 14.1 Recommended Setting—Continuous Cycling

Controller Type	K_P	K_I	K_D
P	$0.5\ K_U$	—	—
PI	$0.45\ K_U$	$\frac{1.2}{T_U}$	—
PID	$0.6\ K_U$	$\frac{2.0}{T_U}$	$\frac{T_U}{8}$

Reaction Curve Method

The reaction curve method, which is used primarily for developing a model of an open-loop system (refer to earlier discussions), can also be used to directly estimate the controller setting. The process reaction curve is reproduced here (Figure 14.22).

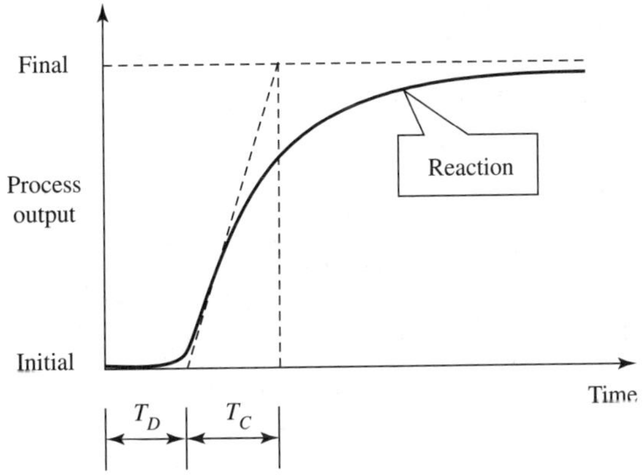

Figure 14.22 Process reaction curve (reproduced).

Based upon the data extracted from the process reaction curve, the controller setting can be determined (Table 14.2). This method is far easier and safer than the continuous cycle method.

Table 14.2 Recommended Setting—Process Reaction Curve

Controller Type	K_P	K_I	K_D
P	$\frac{\tau}{KT_D}$	—	—
PI	$\frac{\tau}{KT_D}$	$\frac{0.3}{T_D}$	—
PID	$\frac{1.25\tau}{KT_D}$	$\frac{0.4}{T_D}$	$\frac{0.4}{T_D}$

EXAMPLE 14.8

The following results were obtained from a process reaction curve on a process.

$$\text{applied process input} = 10\%$$
$$\text{final process output} = 15\%$$
$$\text{time } T_D = 5\text{ s}$$
$$\text{time } T_C = 22.5\text{ s}$$
$$\text{time } T_{63\%} = 18\text{ s}$$

Determine the tuning coefficients for a PI controller.

Solution

The controller constants K_P and K_I can be determined from the reaction curve data and the entry in Table 14.2 for the PI controller.

Proportional constant K_P:

$$K_P = \frac{\tau}{KT_D}$$

$$\text{process time constant } \tau = T_{63\%} - T_D$$
$$= 18 - 5$$
$$= 13\text{ s}$$

$$\text{process gain } K = \frac{\text{change in process output}}{\text{change in process input}}$$
$$= \frac{(55 - 45)\%}{(10 - 5)\%}$$
$$= 2$$

Substituting the values gives

$$K_P = \frac{13}{1.5 \times 5} = 1.7$$

Integral constant K_I:

$$K_I = \frac{0.3}{T_D} = \frac{0.3}{5} = 0.06 \quad \text{(reset/s)}$$

Hence, the coefficients are $K_P = 1.7$ and $K_I = 0.06$.

CASE STUDY: MACHINE POSITIONAL CONTROL SYSTEM

A servomotor is used to position a light metallic plate in a drilling machine Figure 14.23. The machine drills a number of holes in a straight line. The number of holes and the spacing between the holes can vary from part to part. In order to improve the part production, a computer is interfaced to the machine. The computer sends the reference command to the motor controller through a D/A converter. The motor controller is an analog unit capable of full PID mode of operation. The equipment details follow.

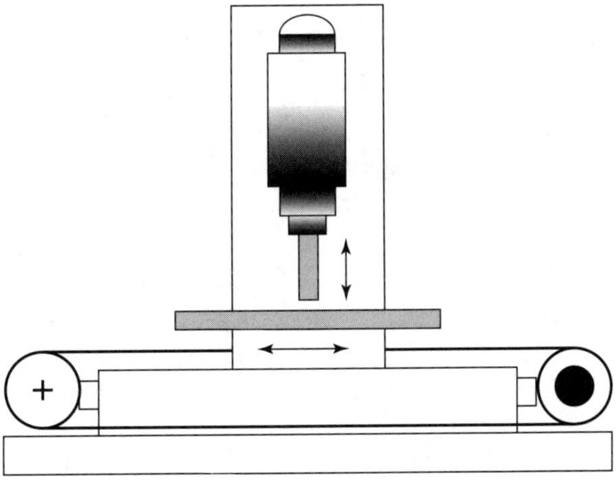

Figure 14.23 Case study: position control system.

Drive Mechanism

The slide table is attached to a toothed belt running in a continuous loop over two pulleys. One of the pulleys is directly mounted on the motor shaft. A gearbox is not used in order to avoid errors due to backlash found in reduction gears. The motor has sufficient torque to handle the load inertia. The belt tension is adjusted properly so that the error during position reversal is minimum.

$$\text{motor + slide + pulley time constant } \tau_M = 0.08 \text{ s}$$

$$\text{motor no-load speed (full voltage)} = 3600 \text{ rpm}$$

$$\text{motor full-scale voltage} = 24 \text{ V}$$

$$\text{power amplifier gain} = 3 \text{ V/V}$$

$$\text{motor pulley diameter} = 10 \text{ cm}$$

Position Transducer

The slide position is measured by an angular potentiometer, which is directly coupled to the other pulley. The potentiometer has a total movement of 270°. The potentiometer pulley is identical to the motor pulley. The pulley diameters are so chosen that the entire linear movement of the slide (end to end) generates just under a 270° rotation of the potentiometer pulley.

$$\text{pulley diameter} = 10 \text{ cm}$$

$$\text{rotational angle (end to end)} = 270^\circ$$

$$\text{applied voltage} = 10 \text{ V}$$

Determine a suitable controller configuration, which will allow the part to be positioned quickly at hole locations and with less than 25% overshoot. (The time specification is descriptive in nature, and for a proper design it should be quantified. At the same time, it is far too dangerous to use a fictitious value, which may be near impossible to achieve with the existing setup. The specifications can be modified after initial testing and design, or can be based upon performance of similar equipment.)

Solution

The complete control system is represented by the block diagram shown in Figure 14.24. Individual transfer functions G_2, G_3, G_4, and H need to be calculated before a closed-loop transfer function can be determined. Because the controller configuration is likely to change, the overall block diagram can first be simplified by combining the G_2, G_3, and G_4 blocks.

G_2: The power amplifier accepts the low-level signal from the controller and generates an output with a capacity to supply a much larger current (power). This power

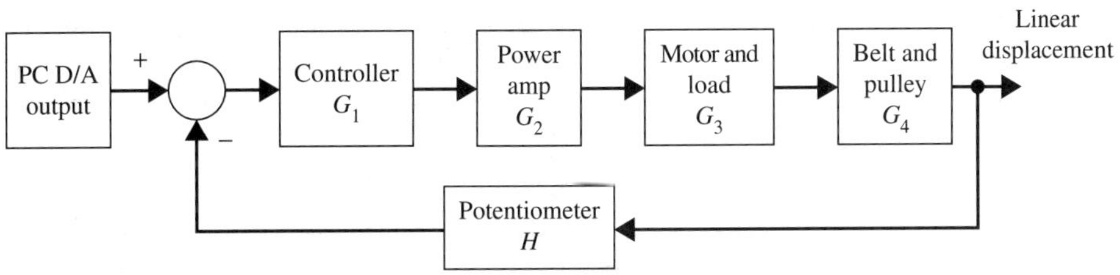

Figure 14.24 Block diagram for case study.

amplifier also amplifies the voltage and has a gain of 3. Because the motor transfer function is based upon the voltage input, the transfer function of the power amplifier is simply equal to its voltage gain. Thus,

$$G_2 = 3 \text{ V/V}$$

***G*$_3$:** A simplified first-order model of motor, belt pulley, and load inertia is going to be assumed. The other s term in the denominator indicates the integration of rotational speed into the angle of rotation.

$$G_3 = \frac{K_M}{s(\tau_M s + 1)}$$

where

$$K_M = \text{motor constant}$$
$$\tau_M = \text{motor + load time constant}$$

The motor constant is

$$K_M = \frac{\text{no load speed}}{\text{maximum motor voltage}} = \frac{(3600 \times 2\pi)/60 \text{ rad/s}}{24 \text{ V}} = 15.7 \text{ rad/s/V}$$

Therefore, the motor (and load) transfer function is

$$G_3 = \frac{15.7}{s(0.08s + 1)} \text{ rad/V}$$

***G_4*:** The belt and pulley simply transfer the rotary motion to angular motion. For every single rotation of the motor shaft, the belt-and-pulley combination will result in a linear motion equal to the circumference of the pulley. Therefore, the transfer function is

$$\begin{aligned} G_4 &= \frac{\pi \times \text{pulley diameter}}{2 \cdot \pi} \\ &= \text{pulley diameter}/2 \\ &= 5\ \text{cm/rad} \end{aligned}$$

***H*:** The belt and potentiometer pulley combination converts the linear distance to angular rotation. It has the same ratio as the motor pulley and belt, because the two pulleys are identical. The potentiometer is connected to a 10-V supply, and this change in output voltage occurs for 270° of rotation of the potentiometer shaft. Therefore, the resultant transfer function is

$$\begin{aligned} H &= \frac{\text{output voltage}}{\text{linear distance for } 270^\circ} \\ &= \frac{10\ \text{V} \times 360^\circ}{\pi \times 10\ \text{cm} \times 270^\circ} \\ &= 0.4244\ \text{V/cm} \end{aligned}$$

The G_2, G_3, and G_4 blocks can now be combined into a single block, G_P.

$$\begin{aligned} G_P &= G_2 G_3 G_4 \\ &= 3 \times \frac{15.7}{s(0.08s + 1)} \times 5\ \text{(cm/V)} \\ &= \frac{235.5}{0.08s^2 + s} \end{aligned}$$

Figure 14.25 shows the modified block diagram.

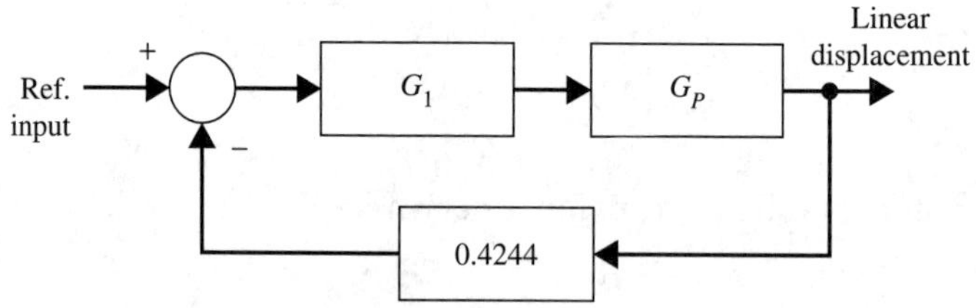

Figure 14.25 Modified block diagram.

G_1**:** The controller is capable of being operated in full PID mode. In the beginning, a proportional-only control mode with a proportional gain of 1 will be assumed. The transient response for this value of gain can first be determined, and based upon the result, the controller configuration or gain can be modified.

Case I Proportional controller, $K_P = 1$: It should be mentioned that a proportional controller with a gain of 1 is the same as if no controller was used and the error signal was directly fed to the power amplifier block. Therefore,

$$G_1 = 1$$

The overall closed-loop transfer function is then

$$\frac{G_1 G_P}{1 + G_1 G_P H} = \frac{1 \times \dfrac{235.5}{0.08s^2 + s}}{1 + 1 \times \dfrac{235.5}{0.08s^2 + s} \times 0.4244}$$

or

$$\text{CLTF} = \frac{235.5}{0.08s^2 + s + 99.95}$$

Step response can be obtained through MATLAB as

```
num = 235.5
den = [0.08 1 99.95]
step(num,den)
```

The resultant step response is shown in Figure 14.26. The response settles down in about 1 s. The amount of overshoot, though, is quite significant (> 50%).

Case II Proportional controller, $K_P = 10$: It may be interesting to see the effect of an increase in K_P on the step response. Increasing the proportional gain to 10,

$$G_1 = 10$$

The overall closed-loop transfer function is then

$$\frac{10 \times \dfrac{235.5}{0.08s^2 + s}}{1 + 10 \times \dfrac{235.5}{0.08s^2 + s} \times 0.4244}$$

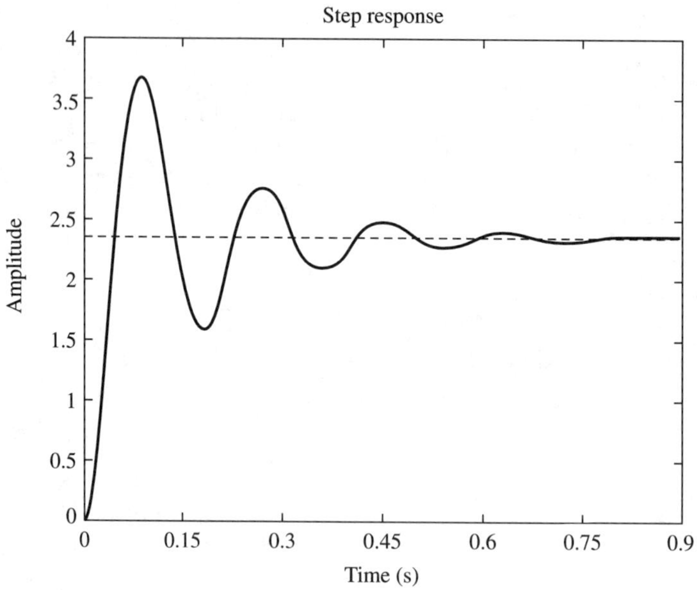

Figure 14.26 Step response for $K_p = 1$.

or

$$\text{CLTF} = \frac{2355}{0.08s^2 + s + 999.5}$$

Obtaining the step response through MATLAB,

```
num = 2355
den = [0.08 1 999.5]
step(num,den)
```

The new step response is shown in Figure 14.27. The response indicates an improvement in the response time of the system, but there is no significant change in the settling time. On the other hand, the amount of overshoot has increased considerably.

Case III P + D controller, $K_P = 10$ and $K_D = 0.01$: Increasing the proportional gain has resulted in an increase in the response time. Although the target position is reached very quickly, it is accompanied by a large overshoot. As a result, a considerable amount of hunting takes place about the steady-state (target) position. There is no point in increasing the proportional gain any further, because the amount of oscillations will not be reduced. The steady-state error does not seem to be much of an issue with this response and as a result the integral mode need not be added. Furthermore,

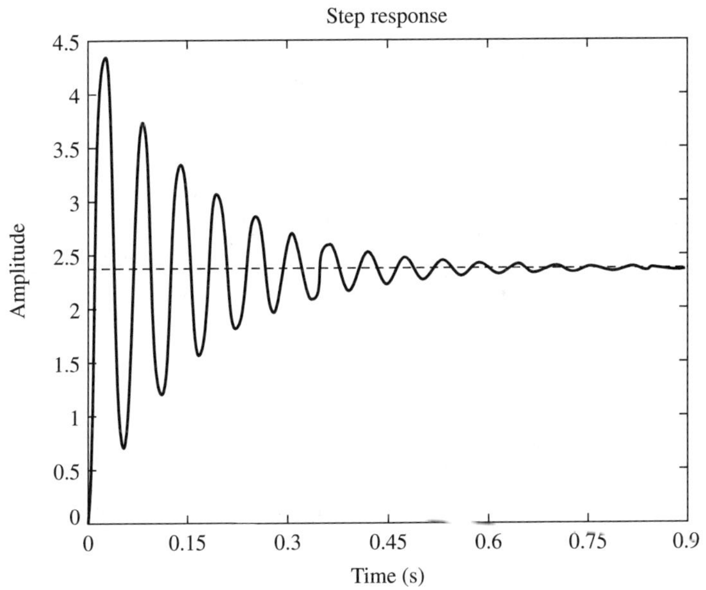

Figure 14.27 Step response for $K_P = 10$.

an integral mode tends to reduce system stability and should be avoided here. The system may be a perfect candidate for the P + D mode of control. The derivative mode tends to add damping to the system and reduces oscillations. Keeping the proportional gain to 10 and adding a small derivative gain of 0.01, the controller transfer function becomes

$$G_1 = K_D s + K_P$$

Substituting the values gives

$$G_1 = 0.01s + 10$$

The overall closed-loop transfer function is then

$$\frac{(0.01s + 10) \times \dfrac{235.5}{0.08s^2 + s}}{1 + (0.01s + 10) \times \dfrac{235.5}{0.08s^2 + s} \times 0.4244}$$

or

$$\text{CLTF} = \frac{2.355s + 2355}{0.08s^2 + 1.9995s + 999.5}$$

Rounding,

$$\text{CLTF} = \frac{2.36s + 2355}{0.08s^2 + 2s + 1000}$$

Obtaining the step response through MATLAB,

```
num = [2.36 2355]
den = [0.08 2 1000]
step(num,den)
```

The new step response is shown in Figure 14.28. The response indicates an improvement in settling time. There is an increased damping in the system, and the oscillations tend to subside somewhat more quickly.

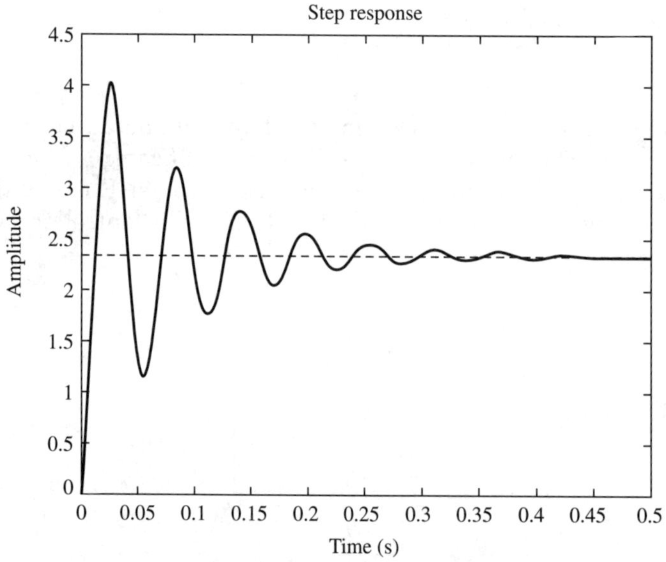

Figure 14.28 Step response for $K_p = 10$ and $K_D = 0.01$.

Case IV P + D controller, $K_P = 10$ and $K_D = 0.1$: The use of the derivative mode has improved the system performance, but the desired performance criteria have not yet been met. There are a number of different steps that can be taken to further analyze the system, but to keep the matter simple the amount of derivative gain will be increased to 0.1. This should add more damping to the system. The proportional gain will still be kept at 10. The controller transfer function is now given as

$$G_1 = K_D s + K_P$$
$$G_1 = 0.1s + 10$$

The overall closed-loop transfer function is

$$\frac{(0.1s+10) \times \dfrac{235.5}{0.08s^2+s}}{1+(0.1s+10) \times \dfrac{235.5}{0.08s^2+s} \times 0.4244}$$

or

$$\text{CLTF} = \frac{23.55s + 2355}{0.08s^2 + 10.99s + 999.5}$$

Rounding gives

$$\text{CLTF} = \frac{23.55s + 2355}{0.08s^2 + 11s + 1000}$$

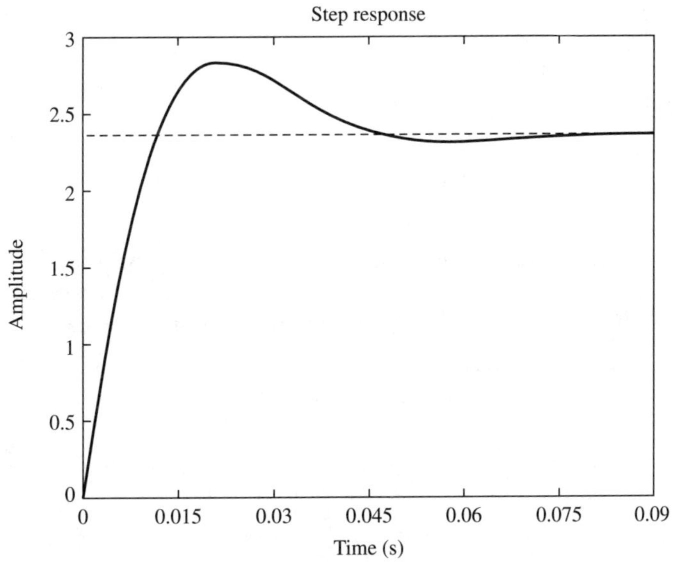

Figure 14.29 Final step response for $K_p = 10$ and $K_D = 0.1$.

Obtaining the step response through MATLAB,

```
num = [23.55 2355]
den = [0.08 11 1000]
step(num,den)
```

The new step response is shown in Figure 14.29. The response indicates a marked improvement in settling time as well as a reduction in hunting. The overshoot has been reduced to less than 20%, which more than meets the design requirements. This concludes the design problem.

Caution: The computer analysis has indicated an excellent step response of the closed-loop system with the selected values of controller gain coefficients. In field-testing, this performance level is not going to be attained, not because of the computer program but because of simplistic and inaccurate modeling of various components. The major culprit in this system is the power amplifier, modeled as a simple proportional element with a gain of 3. As a matter of fact, during a control-loop operation, the amplifier will reach saturation level and will not be able to supply an unlimited voltage level to drive the motor faster. This is not a disadvantage, but to protect against motor burnout it is imperative that excessive voltage levels not be applied to the motor. The output from a power amplifier cannot exceed the supply voltage, and to protect the drive it is sufficient simply to limit the power supply voltage. A more realistic result can be obtained if the nonlinear behavior of these components is included in modeling, but the field tests will still provide the best answers.

PROBLEMS

1. State the reasons for controller tuning.
2. An analog PD controller is being used in a closed loop for positioning control of a machine tool. During one shift, the controller proportional gain was changed from 10 to 11.5. Should this be considered as controller tuning?
3. A closed-loop control system has recently been commissioned, and it seems to be performing satisfactorily. Explain, if it will ever be necessary to perform tuning on it.
4. Does the term *tuning* carry the same meaning for a digital controller as it does for an analog controller?
5. The following results were recorded from step response testing on a DC motor. Estimate its transfer function, assuming a first-order lag with a dead-time model.

$$\text{change in process input} = 2.5\text{ V}$$
$$\text{change in process output} = 400\text{ rpm}$$
$$\text{time } T_{28\%} = 2.5\text{ ms}$$
$$\text{time } T_{63\%} = 12\text{ ms}$$

6. The following diagram shows the frequency response of a process. Estimate the process transfer function.

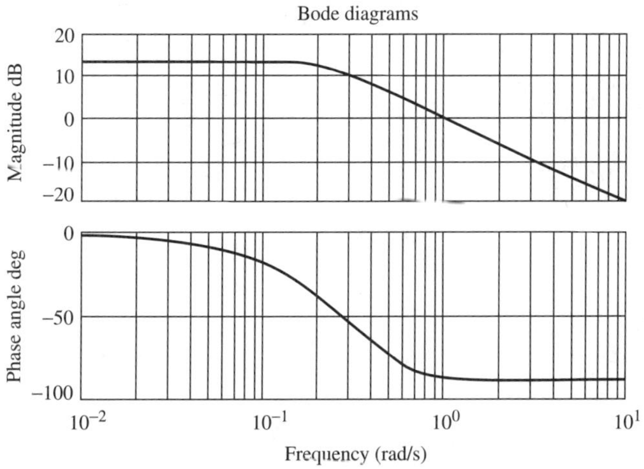

7. The following data were extracted from the reaction curve on a process.

$$\text{process gain} = 8.5$$
$$\text{time } T_D = 1.5\text{ s}$$
$$\text{process time constant } \tau = 6\text{ s}$$

Determine the value of controller coefficients for each of the following modes.
a. Proportional mode
b. Proportional + integral mode
c. Proportional + integral + derivative mode

8. A process being modeled (first-order lag with dead time) is estimated to have a time constant of about 1.25 s. Explain which of the following data-acquisition systems should be considered.
a. 8-bit resolution with a sample rate of 10,000 samples/s
b. 12-bit resolution with a sample rate of 1,000 samples/s
c. 16-bit resolution with a sample rate of 10 samples/s

9. Estimate the process transfer function from the following frequency response curve.

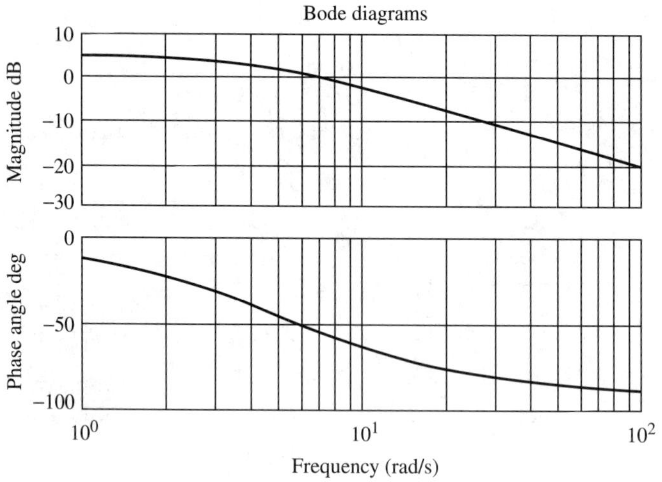

15

ADDITIONAL CONTROL TECHNIQUES

INTRODUCTION

The discussion so far has been limited to single-input and single-output (SISO) linear time invariant (LTI) control systems. Real-life control systems are far from ideal and cannot always be characterized by simple linear models. In this chapter attention is drawn to a number of methods of controls, such as cascade, ratio, and feedforward controls. Modern control theory, which is not limited to SISO control systems, is based upon state-space techniques, so a brief introduction to this topic is included.

FEEDFORWARD CONTROL

A conventional negative-feedback controller works primarily to reduce error in the system. In a proportional controller, there is no controller action if there is no error. In a system working perfectly with zero error, there is no change in controller action unless an error creeps in the system. Ideally speaking, a control system should operate with zero (minimal amount) error, and corrective action should be taken before error can build up in the system.

Consider the following example of driving an automobile at a constant speed.

EXAMPLE 15.1

It is desired to have a car traveling on a road at a constant speed. It can be assumed that the road is absolutely straight without any winding curves and is free of traffic, but it does entail small hills and valleys.

Case I Driver control (feedforward): An experienced driver is able to maintain the speed of a car at a desired value. During a level course, a constant pressure on the accelerator keeps the car rolling at a constant speed. As the driver sees an upward slope (hill), he or she can provide the required increase in engine output by increasing the accelerator pressure, thus keeping the speed to a constant value. Similarly, during a downward slope, a constant speed can be maintained by releasing the pressure on the accelerator pedal. Here the driver takes a visual cue from the change in road conditions and takes an action in advance (*feedforward*), before any drop in speed occurs. Obviously, to be very effective the driver needs to be very familiar with the car as well as the road conditions.

Case II Cruise control (feedback): A car equipped with cruise control can achieve a similar performance. Once set, the cruise control retains the set speed in its memory and constantly compares it with the actual speed of the car. During the level stretch of highway, the cruise control can easily maintain the desired speed. When an upslope is encountered, car speed drops slightly, because the engine is being fed with the same amount of fuel. The cruise controller senses the drop in speed (error) and increases the fuel intake to the engine, thus restoring the car speed. On reaching the top of a hill, car speed increases as load is reduced, but the cruise controller is still supplying an increased supply of fuel to the engine. Again, an error signal (*feedback*) is generated, which causes the cruise controller to reduce the fuel flow and bring the car speed to the desired set value. Road conditions dictate how closely speed can be maintained by the cruise control. The effect is not noticeable for small changes in slope, but it is significant for steep hills or valleys. Because there is no visual indication available to the cruise controller, it relies on only the feedback information and takes an action only after error has set in.

Disturbance Signal

The operation of a plant or a process is affected by other systems operating upstream as well as downstream of this plant. Sometimes environmental conditions also affect plant operation; an excellent example is the home-heating system. The thermal load on the heating system is continually being influenced by wind gusts, sunlight, and rain or snow. All these signals, classified as *disturbance signals*, tend to upset the steady-state plant operation. If these disturbances could be measured, then it would be possible to adjust the process operating condition before a significant error can develop. Feedforward control systems contain a mechanism to monitor the disturbance signal and forward this information to manipulate the plant operation ahead of the feedback.

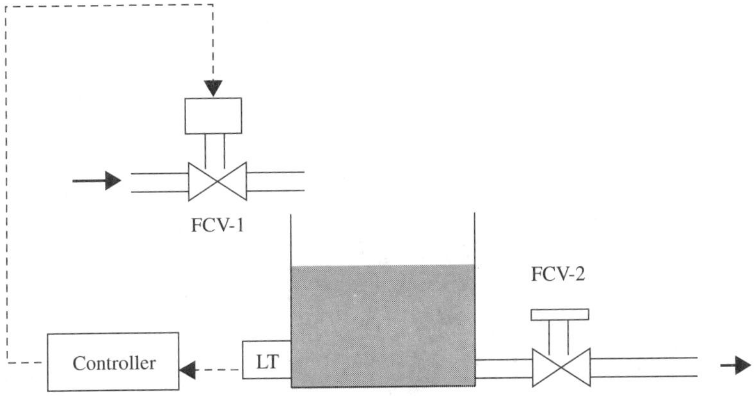

Figure 15.1 Tank level control.

Consider the case of a level control system (Figure 15.1). The level in the tank is maintained by the upstream flow control valve, FCV-1. The flow through this valve is adjusted to match the outflow from the tank. Outflow is dictated by the demand from the downstream system.

If the downstream system requires a larger flow rate, it opens the flow control valve FCV-2, thus increasing the outflow. Because the inlet flow rate through the flow control valve FCV-1 has not yet changed, the tank level starts to drop. The drop in level is detected by the (feedback) level controller through the reduced output from the level transmitter (LT). The level controller opens the flow control valve FCV-1 some more in order to compensate for a drop in tank level.

It is not possible to predict the demand from the downstream system, and as a result the feedback level controller can take an action only after an error has developed in the system. The operation can be improved significantly if a flow transducer (FT) is incorporated in the outflow line. This flow transducer will generate a signal in proportion to the outflow from the tank and can be used to manipulate the opening of flow control valve FCV-1 in conjunction with the level transmitter signal.

Figure 15.2 shows the modified level control system incorporating this modification. The flow transducer measures the disturbance in the outflow rate and feeds it to a controller, appropriately called a feedforward controller. This controller generates an appropriate correction signal, which is added to the output from the feedback controller and sent to the final control element, flow control valve FCV-1. It should be noted that the feedback controller is still finally responsible for error correction, and the feedforward controller is used only to modify the manipulated output signal, based on the disturbance signal. In a system where the external disturbance is the main source of error, the feedforward controller tends to play a bigger role, and the feedback controller simply acts to remove the remaining error.

A simplified block diagram of this system is shown in Figure 15.3. If the mathematical model of the process is precisely known, then it is possible to provide an exact correction through the feedforward mechanism. But in any case, the feedforward

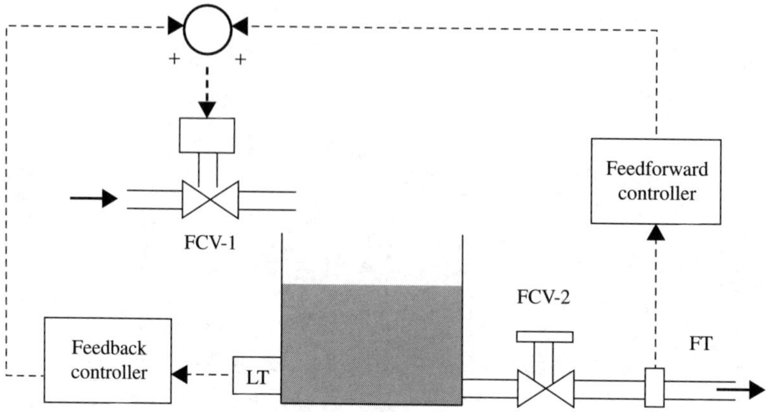

Figure 15.2 Feedforward-feedback tank-level control.

controller does not have to provide a precise compensation for a disturbance signal. An approximate correction is satisfactory in practical systems, because the feedback controller takes care of the remaining error in the system.

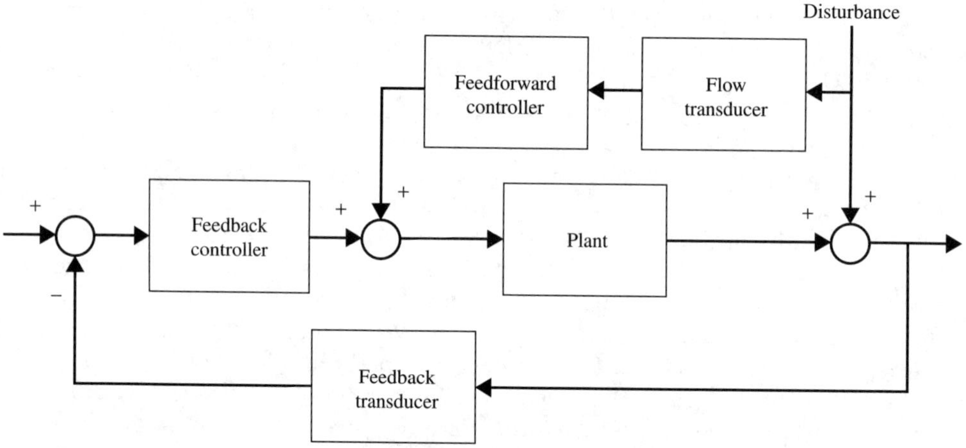

Figure 15.3 Block diagram: feedforward-feedback level control.

EXAMPLE 15.2

Driving a car on a winding road is an excellent example of feedforward control used by every single driver. Based upon the visual indication of an upcoming turn, the driver takes appropriate action of slowing down and adjusting the steering wheel even before the car has approached the turn. If feedback was the only recourse, any action taken by the driver will be too late.

RATIO CONTROL

In some process industries, especially in blending and mixing, there is often a requirement that two material streams maintain a predefined ratio with respect to each other. One example of such a control (or lack of it) is hot water and cold water mixing in a shower. The two water-flow rates should maintain a certain ratio so that a set water temperature is obtained. Figure 15.4 shows a blending process, where two fluid streams, A and B, are mixed in a certain ratio to generate fluid stream C.

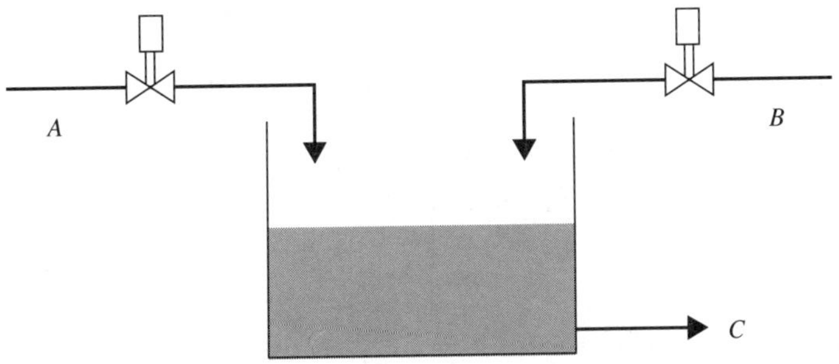

Figure 15.4 Ratio control: mixing process.

Ratio control requires two independent controllers, one for each flow stream. A third controller (a computing unit) acts as a ratio controller. One of the two flow streams is considered the primary flow stream, and its flow rate is controlled by process requirements. The ratio controller gets a feedback signal from the flow transducer monitoring the primary flow stream and generates an output (desired flow rate) for the secondary flow controller. The secondary controller, in turn, ensures that the flow rate of the secondary stream remains at this desired value. If the primary flow rate changes, the required change in the secondary flow stream is passed on to the secondary controller by the ratio unit, and the final flow (product) consistency is maintained.

CASCADE CONTROL

Consider the level control system shown in Figure 15.1. The feedback controller in this case strives to maintain a desired level in the tank, irrespective of the outflow. The main disturbance to level control arrives in the form of demand from the downstream process. Assuming that the outflow from the tank is fairly sturdy, the feedback controller should be able to maintain the level very close to the desired value. Level control systems typically have large time constants, and if an error is set in the system, it may take a while before the level can be stabilized again.

The flow rate of control valves such as FCV-1 at the inlet flow is significantly influenced by the pressure differential across the valve. Line pressure downstream of a flow control valve is fairly constant, but the upstream line pressure can be adversely affected by flow demand originating from other processes connected to the main supply. Hence, the pressure differential across a flow control valve—and, in turn, the flow rate through the valve—is affected by upstream line pressure. Even though the feedback controller may be maintaining the tank level to the desired set point, a change in upstream line pressure will change the control valve flow rate and upset the level. If the system requires very tight level control, then some considerations need to be made for changes in line pressure. One such approach is known as *cascade control*. In this particular application, it may be more sensible to monitor the outflow from flow control valve FCV-1 (inflow to tank) and manipulate the valve in order to maintain a consistent flow rate (Figure 15.5).

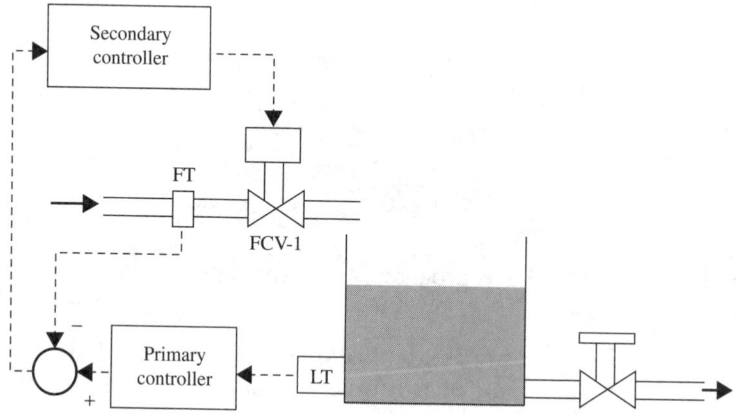

Figure 15.5 Cascade control system.

A secondary controller is set up that controls the valve flow rate. The secondary controller takes its set-point input from the primary level controller, which is working on feedback from the tank level and is primarily responsible for maintaining the level. The inner secondary loop is set up to have a faster response than the outer primary loop (Figure 15.6).

Cascade Control

Cascade control incorporates two feedback controllers arranged in such a fashion that the output of one feedback controller becomes an input to the second controller.

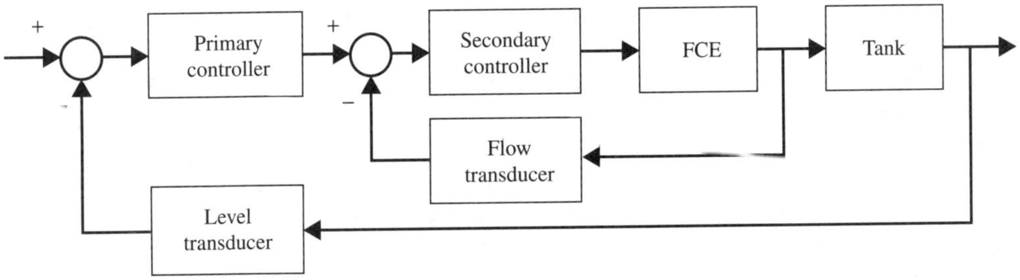

Figure 15.6 Block diagram for cascade level control.

EXAMPLE 15.3

Cascade control can be applied to certain biological/chemical processes, requiring precise control of temperature. In these processes electric heaters are used to supply heat. The heaters are controlled by SCRs (silicon controlled rectifiers), whose firing angle is varied to change the amount of heat that is generated.

Electrical heat so generated is proportional to the square of voltage and is very sensitive to the power supply voltage. If power supply voltage fluctuates (brownout or overvoltage) the amount of heat generated will vary significantly. The temperature controller will not be able to detect a change in heat generation before it is too late and as a result process temperature will change.

A secondary controller is incorporated, which accepts the amount of heat signal (set point) from the main controller and controls the SCR firing angle. If the power supply voltage changes, the secondary controller adjusts the SCR firing angle and maintains the amount of heat generated.

AUTOTUNING CONTROLLER

With the addition of a digital controller, autotuning of controllers is steadily gaining popularity. An autotuning controller can automatically select tuning parameters that are a much better combination then that selected by an inexperienced control practitioner. Control personnel can also make judicious use of an autotuner by letting it come up with starting values and then modifying these later. This is especially significant during a plant start-up, with several tens (sometimes hundreds) of control loops to be set up. A controller with a good autotuning algorithm can also track the subtle changes in process parameters and adjust the control loop operation accordingly. But an autotuning controller is no substitute for experienced control personnel and should be used with caution.

The operating principle of an autotuner is quite simple. The derivation of control coefficients is based upon one or more tests carried out on the plant (similar to those discussed in Chapter 13). Instead of using a person to carry out the test, the computer in an autotuner is delegated the responsibility of carrying it out. In a simple autotuner, the test may be as simple as a step response test. The input and output signals are digitized by an onboard analog-to-digital converter and stored in internal memory. Once the test is complete, a modeler program (part of the computer firmware) scans the recorded data and develops a mathematical model. Once a model is identified, suitable coefficients are selected from internal look-up tables and/or formulas.

Advanced autotuners do not need to generate an upset signal for step response testing but can monitor the process input and output and refine the developed model. Obviously, for new start-ups, some initial coefficient values may need to be specified or the autotuner is allowed to perform a step response (or even a frequency response) test.

STATE-SPACE TECHNIQUES

Classical control system theory allows the control of a process with only a single input and a single output (SISO). In a number of control systems this is quite satisfactory, but an increasing number of modern control systems involve multiple inputs and multiple outputs (MIMO). Besides, some systems cannot be described adequately as linear and time-invariant systems. Such systems require a different approach for control system analysis and design. State-space techniques allow these control systems to be dealt with effectively. These techniques are equally applicable to linear, nonlinear, time-invariant, and time-variant systems, where system characteristics are not constant but are changing with time. Furthermore, space-state techniques allow for better integration with computer analysis, and with the availability of powerful computer programs, such as MATLAB, these are gaining widespread acceptance.

State

The **state** of a system refers to the property (condition) of a system.

State Variables

State variables form the basis of state-space analysis. These make up the minimum set of quantities for a given system necessary to be known in order to completely describe the future response to any given input. The selection of state variables for any given system is not unique. The choice of a state variable is generally (but not necessarily) related to the energy-storage elements of the system.

State Equations

State equations are the set of first-order differential equations incorporating state variables and completely describing system behavior. First-order differential equations can be easily set up as matrix-vector differential equations and solved using computer programs.

EXAMPLE 15.4

Convert the following second-order differential equation to state variable form.

$$\frac{d^2x}{dt^2} + 5\frac{dx}{dt} + 2x = 10y$$

Solution

A second-order differential equation typically requires two state variables. Let the state variables be x_1 and x_2, and defined as follows:

$$x_1 = x$$
$$x_2 = \frac{dx_1}{dt} = \dot{x}_1$$

Their respective first-order derivative terms are

$$\frac{dx}{dt} = \frac{dx_1}{dt} = \dot{x}_1$$
$$\frac{d^2x}{dt^2} = \frac{d}{dt}\left(\frac{dx_1}{dt}\right) = \frac{d}{dt}(x_2) = \dot{x}_2$$

Manipulating the differential equation gives

$$\frac{d^2x}{dt^2} = -5\frac{dx}{dt} - 2x + 10y$$

Substituting the state variables,

$$\dot{x}_2 = -5x_2 - 2x_1 + 10y$$

Thus the two state equations are

$$\dot{x}_1 = x_2$$
$$\dot{x}_2 = -5x_2 - 2x_1 + 10y$$

These equations can be expressed in matrix form:

$$\begin{bmatrix} \dot{x}_1 \\ \dot{x}_2 \end{bmatrix} = \begin{bmatrix} 0 & 1 \\ -5 & -2 \end{bmatrix} \begin{bmatrix} x_1 \\ x_2 \end{bmatrix} + \begin{bmatrix} 0 \\ 10 \end{bmatrix} y$$

$$\dot{\mathbf{X}} = \mathbf{AX} + \mathbf{BY}$$

where **A** is the coefficient matrix and **B** is the input vector.

EXAMPLE 15.5

Express the spring mass damper system (Figure 15.7) in state-space form.

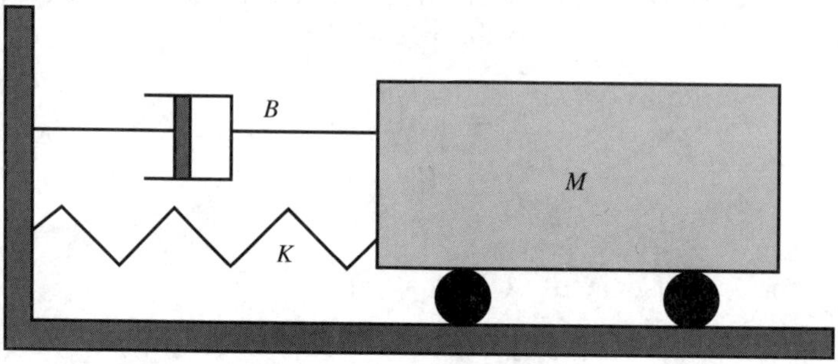

Figure 15.7 Spring-mass damper system for Example 15.5.

Solution

The differential equation describing the movement $x(t)$ of the mass, when subjected to an external force $f(t)$, is given as

$$M\frac{d^2x}{dt^2} + B\frac{dx}{dt} + Kx = f(t)$$

The displacement of mass and its velocity can be chosen as the state variables x_1 and x_2.

$$x_1 = x$$

$$x_2 = \frac{dx}{dt} = \dot{x}_1$$

Rewriting the differential equation gives

$$M\frac{d^2x}{dt^2} = -B\frac{dx}{dt} - Kx + f(t)$$

Substituting the state variables,

$$M\dot{x}_2 = -Bx_2 - Kx_1 - f(t)$$

or

$$\dot{x}_2 = -\frac{B}{M}x_2 - \frac{K}{M}x_1 - \frac{1}{M}f(t)$$

Rearranging and grouping the equations gives

$$\dot{x}_1 = x_2$$

$$\dot{x}_2 = -\frac{K}{M}x_1 - \frac{B}{M}x_2 - \frac{1}{M}f(t)$$

In matrix form,

$$\begin{bmatrix} \ddot{x}_1 \\ \dot{x}_2 \end{bmatrix} = \begin{bmatrix} 0 & 1 \\ -\frac{K}{M} & -\frac{B}{M} \end{bmatrix} \begin{bmatrix} x_1 \\ x_2 \end{bmatrix} + \begin{bmatrix} 0 \\ -\frac{1}{M} \end{bmatrix} f(t)$$

Transfer Function to State-Space Representation

A state-space form is quite suitable for computer simulation and analysis. If a system transfer function is available, it can be converted to state-space form. A common approach is to choose state variables such that each subsequent state variable is the derivative of the previous state variable. State variables such chosen are known as *phase variables*. The following procedure can be adopted.

- Cross multiply the transfer function.
- Obtain the differential equation through the inverse Laplace transformation and assuming zero initial conditions.
- Select state variables.

EXAMPLE 15.6

Represent the following transfer function in state-space form.

$$\frac{C(s)}{R(s)} = \frac{2.5}{2s^3 + 3s^2 + s + 4}$$

Solution

Cross multiplying gives

$$C(s)(2s^3 + 3s^2 + s + 4) = 2.5R(s)$$

Taking the inverse Laplace transformation,

$$2\frac{d^3c}{dt^3} + 3\frac{d^2c}{dt^2} + \frac{dc}{dt} + 4c(t) = 2.5r(t)$$

The differential equation is third-order; hence, three state variables are needed. The output variable is $c(t)$, so it will be chosen as the first state variable:

$$\begin{aligned} x_1 &= c \\ x_2 &= \dot{c} = \dot{x}_1 \\ x_3 &= \ddot{c} = \dot{x}_2 \end{aligned}$$

Rearranging the differential equation gives

$$2\frac{d^3c}{dt^3} = -3\frac{d^2c}{dt^2} - \frac{dc}{dt} - 4c(t) + 2.5r(t)$$

Dividing both sides by 2,

$$\frac{d^3c}{dt^3} = -1.5\frac{d^2c}{dt^2} - 0.5\frac{dc}{dt} - 2c(t) + 1.25r(t)$$

Rearranging the terms,

$$\frac{d^3c}{dt^3} = -2c(t) - 0.5\frac{dc}{dt} - 1.5\frac{d^2c}{dt^2} + 1.25r(t)$$

Replacing the terms with state variables gives

$$x_3 = -2x_1 - 0.5x_2 - 1.5x_3 + 1.25r(t)$$

Arranging all state variables in order, we have

$$\begin{aligned} \dot{x}_1 &= x_2 \\ \dot{x}_2 &= x_3 \\ \dot{x}_3 &= -2x_1 - 0.5x_2 - 1.5x_3 + 1.25r(t) \end{aligned}$$

The output equation is

$$c = x_1$$

In matrix form, it is

$$\begin{bmatrix} \ddot{x}_1 \\ \dot{x}_2 \\ \dot{x}_3 \end{bmatrix} = \begin{bmatrix} 0 & 1 & 0 \\ 0 & 0 & 1 \\ -2 & -0.5 & -1.5 \end{bmatrix} \begin{bmatrix} x_1 \\ x_2 \\ x_3 \end{bmatrix} + \begin{bmatrix} 0 \\ 0 \\ 1.25 \end{bmatrix} r(t)$$

$$c = \begin{bmatrix} 1 & 0 & 0 \end{bmatrix} \begin{bmatrix} x_1 \\ x_2 \\ x_3 \end{bmatrix}$$

PROBLEMS

1. Describe a control system where feedforward control is used.
2. Can all control systems benefit from some form of feedforward control? Explain.
3. Explain if a ratio control system falls under the SISO type of control.
4. It is required to represent the following transfer functions in state-space form. Determine the number of state variables required to fully describe each system.

   a. $\dfrac{Y(s)}{E(s)} = \dfrac{1.4}{s^3 + 5s^2 + 4s + 10}$

   b. $\dfrac{C(s)}{R(s)} = \dfrac{2.73}{s^3 + 3s + 7}$

   c. $\dfrac{C(s)}{R(s)} = \dfrac{1}{0.5s^4 + 8s}$

5. Use the data in Problem 4. Following the phase variable approach, identify the state variables in each case.
6. Each of the following differential equations represents a system model. Represent each system in state-space form.

   a. $\dfrac{d^2x}{dt^2} + 6\dfrac{dx}{dt} + 12x(t) = y(t)$

   b. $\dfrac{d^3x}{dt^3} + 3.8\dfrac{d^2x}{dt^2} + \dfrac{dx}{dt} + 12x(t) = 2.5y(t)$

   c. $\dfrac{d^2x}{dt^2} + 1.2x(t) = y(t)$

BIBLIOGRAPHY

1. Bateson, R. N. 1999. *Introduction to Control System Technology*. Upper Saddle River, N.J.: Prentice Hall.
2. Bennet, S. 1998. *Real-Time Computer Control*. Upper Saddle River, N.J.: Prentice Hall.
3. Bibbero, R. J. 1997. *Microprocessors in Instruments and Control*. New York: John Wiley.
4. Bogart, T. F., Jr. 1982. *Laplace Transforms and Control System Theory for Technology*. New York: John Wiley.
5. Bolton, W. 1998. *Control Engineering*. Addison Wesley Longman Ltd.
6. Cannon, R. H., Jr. *Dynamics of Physical Systems*. New York: McGraw-Hill.
7. Glass, C. M. 1976. *Linear Systems with Applications and Discrete Analysis*. St. Paul, Minn.: West.
8. *M68HC11RM/AD HC11 Reference Manual*. 1991. Motorola. Phoenix, www.motorola.com.
9. The Math Works, Inc. 1999. *Using MATLAB*, version 5.
10. The Math Works, Inc. 1995. *The Student Edition of MATLAB*, version 4. Upper Saddle River, N.J.: Prentice Hall.
11. *MOTU-1 Module*. 1999. Forth Microsystems. London, Ont. www.mcjournal.com/forthmicro.
12. *MPLAB*. 1996. Microchip Technology. Chandler, www.microchip.com.
13. Ogata, K. 1997. *Modern Control Engineering*. Upper Saddle River, N.J.: Prentice Hall.
14. *PIC-2 Module*. 1999. Forth Microsystems. London, ON. www.mcjournal.com/forthmicro.
15. *PIC C Compiler*. 1998. Custom Control Systems, Brookfield www.execpc.com/www.ecsinfo.com.
16. Saadat, Hadi. 1993. *Computational Aids in Control Systems using MATLAB*. New York: McGraw-Hill.
17. Sinha, N. K. 1986. *Control Systems*. New York: Holt Rinehart and Winston.
18. Skrodder, J. C. 1997. *Using the M68HC11 Microcontroller*. Upper Saddle River, N.J.: Prentice Hall.
19. Spasov, P. 1999. *Microcontroller Technology The 68HC11*. Upper Saddle River, N.J.: Prentice Hall.
20. Stefano, D. III, A. R. Stubberud, and I. J. Williams. 1967. *Schaum's Outline Series Theory and Problems of Feedback and Control Systems*. New York: McGraw-Hill.
21. Weyrick, R. C. 1975. *Fundamentals of Automatic Control*. New York: McGraw-Hill.

A

MATLAB

INTRODUCTION

MATLAB (an abbreviation of MATrix LABoratory) is a computer program developed by The Math Works Inc. It is widely used to solve computational-intensive problems in control system analysis and design, signal processing, and other science and engineering fields. This chapter provides an introduction to MATLAB, with the intention of providing a way to solve the problems in controls as covered in this book. Some of the instructions for using MATLAB have been provided in the text with relevant topics, but they are repeated here for clarity. It is assumed here that a Windows version of MATLAB is being used, although the syntax is very similar for the DOS version.

CONVENTIONS USED

The following conventions are used in describing the MATLAB commands.

- MATLAB commands are shown in *italics*.
  - Every command has to be followed by a carriage return <cr> (enter key), which is not indicated.
  - MATLAB commands are case sensitive.
- MATLAB output is shown in **bold**.
- Instructions and description are provided in normal case.

VARIABLES

MATLAB allows free-form variable use. Any single alphabetic character or a combination of characters can be used. The first character in the variable name cannot be a numerical digit. The following are valid variable forms.

```
n,    num,    N,    Denominator,    d,    D,    den,    d123
```

- Variables need not be defined before they are used in an expression. For example, *num = 10.235* means to define a variable named num and set it equal to 10.235.
- A variable contains the most recently defined or calculated value.
- A list of all variables in use can be obtained by the command *who*.
- A variable can be cleared from memory by typing the command *clear* followed by the variable name. For example,

```
x = 10       (Command entry.)
x = 10       (MATLAB output)
clear x      (Clear x variable)
x            (Find the value of x)
Undefined function or variable 'x'
             (MATLAB does not recognize the name x)
```

- All variables can be removed from memory by the command *clear*.

MATHEMATICAL OPERATIONS

MATLAB can be used as a glorified calculator, if so desired. The expression is entered and is processed in order (left to right).

```
2*5 + (2^2 - 5) -5    (Enter an expression)
ans = 4               (MATLAB output)
```

Vector

A vector is a row or column of numbers. A vector is represented by enclosing the numbers in brackets. Numbers separated by spaces or commas indicate a row vector, and numbers separated by semicolons indicate a column vector.

```
row = [1 2 3]      (Row vector)
col = [1; 2; 3]    (Column vector)
```

Polynomial Expression

MATLAB is being intensely applied in control system analysis and design, where Laplace transform expressions are used to express transfer functions. A transfer function contains polynomial expressions in s. When entering the data in MATLAB, the s terms are not entered; instead, only their coefficients are entered.

```
num = [3 1]          (Set num = 3s + 1)
den = [2 1.5 4]      (Set den = 2s^2 + 1.5s + 4)
n = [4 0]            (Set n = 4s + 0 → 4s)
y = [3 0 5]          (Set y = 3s^2 + 0s + 5 → 3s^2 + 5)
```

- The terms are entered in descending order.
- The first term is the coefficient of the highest-power term.
- The last term is the coefficient of the zero-power term (literal).
- It is necessary to enter the coefficients of all the terms.

Polynomial Product

The product of two polynomial expressions can be obtained by the *conv* command. This command is used to multiply two vectors or polynomials.

```
x = [1 1]          (Set x = s + 1)
y = [1 -1]         (Set y = s - 1)
z = conv(x,y)      (z = x · y)
z = [1 0 -1]       (z = s^2 + 0s - 1 → s^2 - 1)
```

Range of a Variable

If a number (vector) has to take multiple values, the values can be specified individually or in a more compact arrangement.

```
Kp = [1, 2, 3, 4, 5]     (Kp ranges from 1 to 5 in steps of 1)
Kp = 1:5                 (Kp ranges from 1 to 5 in steps of 1)
Kd = 0:0.1:10            (KD ranges from 0 to 10 in steps of 0.1)
```

Having given this much understanding (barely enough), some of the tools available in MATLAB for control system analysis can now be applied.

FREQUENCY RESPONSE

The frequency response of a closed-loop control system is plotted using its open-loop transfer function. The program automatically determines a suitable frequency range and draws the magnitude and phase-angle plots. The command to be used is *bode (numerator,denominator)*. The *bode* command can be used to draw an open-loop or a closed-loop frequency response, depending upon the transfer function specified.

EXAMPLE A.1

Draw a frequency response of a unity feedback closed-loop system with a forward gain of $s/(10s^2 + s + 1)$.

Solution

$$\text{open-loop transfer function} = \frac{s}{10s^2 + s + 1}$$

Enter the numerator:

```
num = [1 0]
```

Enter the denominator:

```
den = [10 1 1]
```

Obtain the plot:

```
bode(num,den)
```

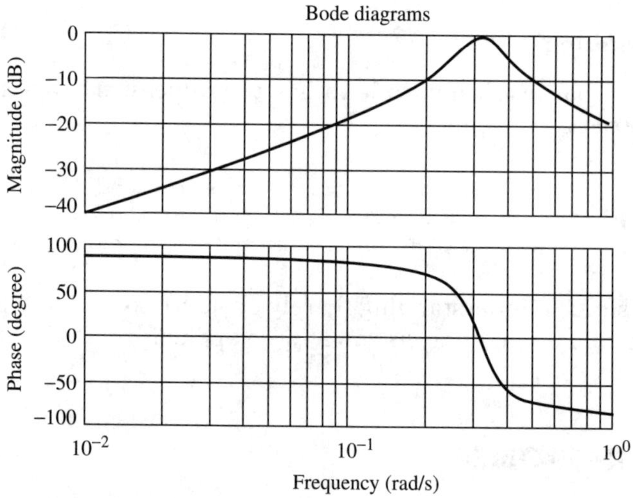

Specifying the Frequency Range

Even though MATLAB automatically determines a suitable frequency range, sometimes it is desirable to select a different range. Range is specified using the *logspace (MIN, MAX)* command, where *MIN* represents the minimum frequency of 10^{MIN} and *MAX* represents the maximum frequency of 10^{MAX}.

EXAMPLE A.2

Specify a frequency range of 0.1 to 1000 rad/s.

Solution

$$\text{MIN} = -1 \qquad (0.1 = 10^{-1})$$
$$\text{MIN} = 3 \qquad (1000 = 10^{3})$$

Use a variable *w* to specify the frequency range:

```
w = logspace(-1, 3)
```

Obtain the frequency plot over the new range of frequency,

```
bode(num,den,w)
```

Tabular Frequency Response Data

It is also possible to obtain the numerical values of gain and phase angle instead of graphical output.

EXAMPLE A.3

```
[gn, ph] = bode(num,den)          (Vectors gn and ph will contain gain and
                                  phase angle data.)
[gn, ph, w] = bode(num,den)       (Obtain gain, phase angle, and corresponding
                                  frequency information.)
[gn, ph, w]                       (View data in a table with three columns.)
```

TRANSIENT RESPONSE

The transient response of a closed-loop control system can be obtained using the command *step(num,den)*, where *num* and *den* are the numerator and denominator part of the closed-loop transfer function. The range of the time interval is automatically selected, but it can be specified if so desired.

EXAMPLE A.4

Obtain the step response of a closed-loop control system with the transfer function $\text{CLTF} = s/(10s^2 + 2s + 1)$.

Solution

Specify the numerator:

```
num = [1 0]
```

Specify the denominator:

```
den = [10 2 1]
```

Obtain the step response (graphical output):

```
step(num,den)
```

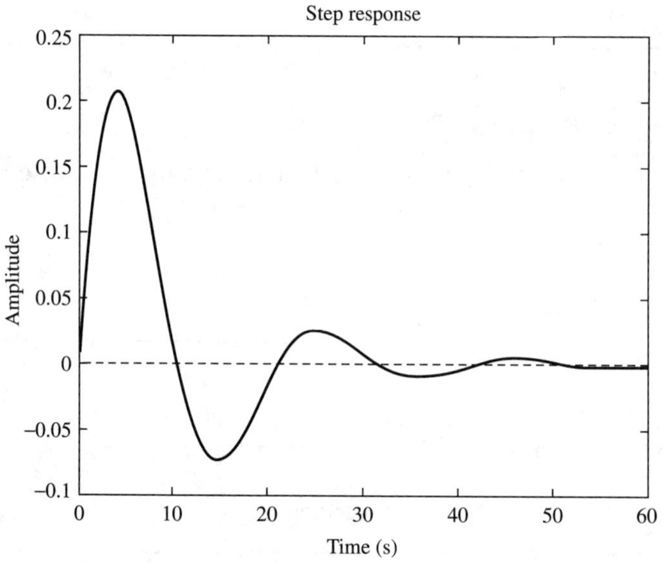

Specifying the Time Interval

The time interval is automatically selected by the program but can be specified to allow viewing of the transient response over a narrow or wider time interval. The step response plot displays time on a linear scale. The points in time over which the response is to be calculated can be specified individually or as a range using the *linspace (Start, End)* command. *Start* represents the starting point in time and *End* represents the final time value. This interval is divided into 100 linearly spaced time values.

EXAMPLE A.5

Specify a time range of 0 to 10 s.

Solution

Use a variable t to specify the time vector,

```
t = linspace(0,10)
```

Obtain the step response over this time interval.

```
step(num,den,t)
```

Tabular Transient Response Data

It is also possible to obtain the numerical values of amplitude and time instead of graphical output.

EXAMPLE A.6

```
t = 0:10              (Set up the time interval of 0 to 10 in steps of 0.1 s)
g = step(num,den)     (Vector g will contain amplitude data)
g = bode(num,den,t)   (g contains amplitude data over the specified time)
[g,t']                (View data in a tabular form, two columns)
```

Root Locus

The root locus of a closed system is plotted with a knowledge of its open-loop transfer function. The command used is *rlocus (num,den)*. Here the *num* and *den* are the open-loop transfer function, which has been expressed in the following form.

$$1 + K\frac{\text{num}(s)}{\text{den}(s)}$$

The root locus is the path of poles of the closed-loop transfer function for the entire range of values (0 to ∞) for the system parameter K.

EXAMPLE A.7

Obtain the root locus for the following system.

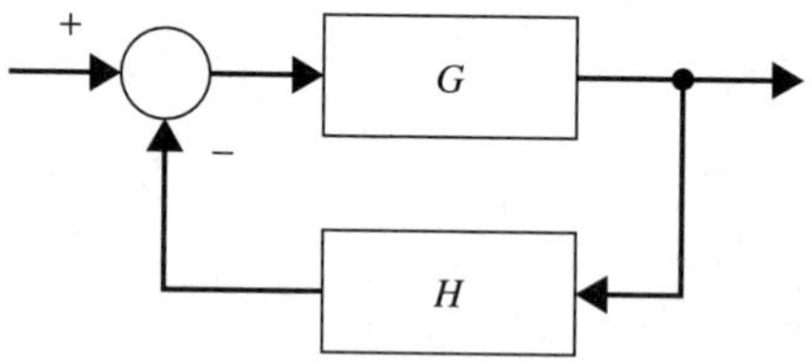

The forward and feedback transfer functions are as follows:

$$G(s) = \frac{K(s+10)}{10s^2+s+1}, \qquad H(s) = \frac{1}{s+1}$$

Solution

The open-loop transfer function is

$$G(s)H(s) = \frac{K(s+10)}{(10s^2+s+1)(s+1)}$$

$$G(s)H(s) = K\frac{(s+10)}{(10s^3+11s^2+2s+1)}$$

Enter the open-loop transfer function:

```
num = [1 10]
den = [10 11 2 1]
```

Obtain the root locus plot:

```
rlocus(num,den)
```

The following root locus plot is obtained:

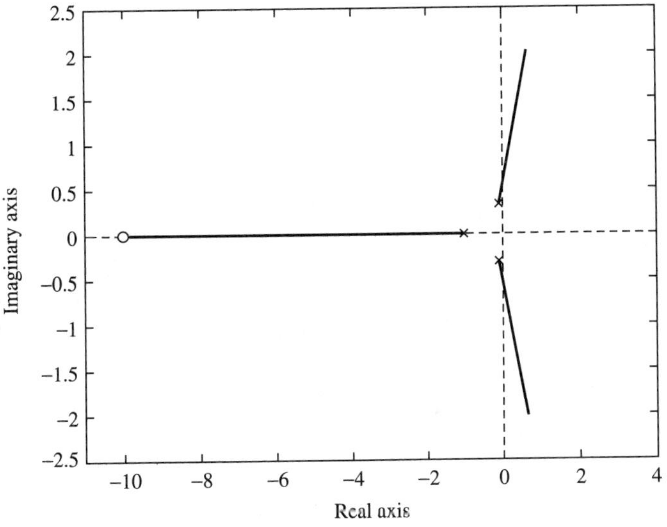

Specifying the Range of Factor *K*

MATLAB plots the root locus for the entire range of values for K. Sometimes it is desired to obtain the plot for specific values of K. As with step response and frequency response, the range of values of K can be specified.

EXAMPLE A.8

```
k = 0: 4              (k varies from 0 to 4)
rlocus(num,den,k)     (Obtain the root locus)
```

The resultant root locus is shown next. The plot shows five values of K: $0, 1, 2, 3,$ and 4.

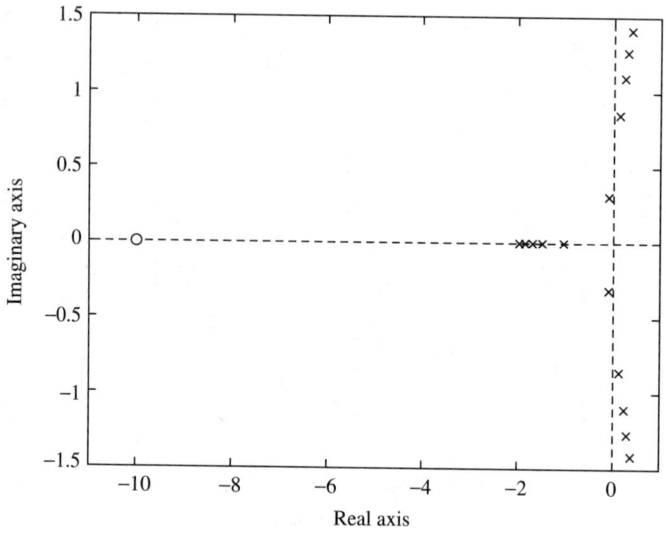

Copying a Plot

Once a plot has been obtained, it normally needs to be integrated in some sort of text document. The easiest method is to click on the Edit menu on the plot screen and select Copy Figure. This action copies the diagram to the Window clipboard, from where it can be pasted to the required document.

This concludes the brief introduction to MATLAB. Again, the emphasis has been on being able to apply the program for solving some of the problems in the text. It is strongly recommended that the on-line help available with MATLAB be utilized to view the syntax as well as to learn more about the capability of the MATLAB program. It is indeed a very powerful program and is changing the ways in which control systems are being taught and analyzed.

GLOSSARY

automatic	self-action without any human intervention
automatic control system	a control system that is self-regulating, without any human intervention
block	set of elements that can be grouped together, with overall characteristics described by an input/output relationship
cascade control	two feedback controllers arranged in such a fashion that the output of one feedback controller becomes an input to the second controller
characteristic equation	an equation obtained by equating the denominator of the transfer function to zero
closed-loop control system	a control system in which the control (regulating action) is influenced by the output
control	regulate, direct, or govern
control system	a group of physical components arranged to control themselves or another system
crisp value	a value that can be specified numerically
crisp variable	a physical variable that can be measured through instruments and can be assigned a numerical (crisp) value
damped natural frequency	the frequency at which the system oscillates before settling down
dead band	see *neutral zone*
defuzzification	the process of taking a fuzzy value and converting it into a numerical (crisp) value
degree of membership	the degree to which a value (crisp variable) belongs to a fuzzy set (variable, characteristic, or specification; expressed either as a fractional value ranging

	from 0 to 1.0 or as a percentage ranging from 0% to 100%
delayed step function	a function of time ($F(t - a)$) that has a zero magnitude before $t = a$ and a constant amplitude after that
element (component)	smallest part of a system that can be treated as a whole (entity)
error band	see *neutral zone*
frequency response	the response of a system, when it is subjected to sinusoidal signal, over a range of frequencies
fuzzification	the process of taking the numerical (crisp) variable and relating it to a fuzzy set through a membership function
fuzzy value (linguistic term)	a value that cannot be expressed using a numerical value but is expressed in linguistic terms
fuzzy (linguistic) variable	a variable that cannot be specified as a numerical (crisp) value
gain crossover frequency	the frequency at which the open-loop gain drops to 0 dB (gain of 1)
gain margin	the additional gain that can be added to open-loop gain before a closed-loop system becomes unstable in operation
heat capacitance	the capacity of an object to store heat
knowledge-based rule	knowledge or experience expressed as a rule, typically taking the form "If___, then___"
linear system	a system where the input/output relationships may be represented by a linear differential equation
manual control system	a control system regulated through human intervention
marginally stable system	a closed-loop control system in which roots of the characteristic equation lie on the imaginary axis; for all practical purposes, an unstable system
mathematical model	an equation or set of equations that define the relationship between the input and output (variables)
membership function	a function that is normally expressed graphically and that tends to illustrate how completely a crisp variable belongs to a fuzzy set
multiple-input, multiple-output (MIMO) system	a system where several parameters may be entered as input and output is represented by multiple variables

neutral zone — the region of error over which the controller does not change its output; also known as dead band or error band

open-loop control system — a control system in which the control (regulating action) is independent of the output

peak time — the amount of time t_p taken to reach the peak amplitude, measured from the instant the step input is applied

peak value — the maximum value of the output, reached after application of the unit step input after time t_p

percentage overshoot (P. O.) — the difference between the peak value and the final steady-state value, expressed as a percentage of the final value

phase crossover frequency — the frequency at which the open-loop phase angle drops to $-180°$

phase margin — the additional drop in phase angle that can be added to a system before it becomes unstable

polar plot — the loci of vectors drawn for a frequency range extending from zero to infinity, with each vector representing system gain and the phase angle at a certain frequency

pole of a transfer function — the root (solution) of the (characteristic) equation obtained by setting the denominator polynomial of the transfer function equal to zero; the value of s that makes (the value of) the transfer function approach infinity (hence the term *pole* (rising to infinity)); complex poles always appear as complex conjugate pairs

pole-zero map — an s-plane diagram showing the location of poles and zeros of a system transfer function

proportional band — the maximum percent error that will cause a change in controller output from minimum (0%) to maximum (100%)

pulse function — the difference between a step function and a delayed step function

ramp function — a function whose amplitude increases linearly with time

regulator system — a control system where the reference (input) is normally fixed

rise time — the time taken for the output amplitude to rise from 10% of final value to 90% of final value (the first time)

root locus — the path of the poles of a closed-loop control system as one of the system parameters, usually the system gain factor K, is varied over a certain range

sample time — the time taken for a computer to complete all the tasks in a loop, or the time interval in which the computer will repeat a task

servomechanism (follower system) — a control system where the reference (input) varies continuously and the system operates so that the output follows the input

settling time — the time taken for the system to settle down to within a certain percentage of the final value

single-input, single-output (SISO) system — a system where only one parameter is entered as input and only one parameter represents the output

singleton — the single crisp value being used to represent a set

sinusoidal function — a function of time, which is periodically changing and is described by

$$f(t) = 0 \qquad t < 0$$
$$f(t) = A \sin \omega t \qquad t \geq 0$$

where A = peak amplitude, $\omega = 2\pi f$, and f = frequency

stable closed-loop system — A system in which the open-loop gain is less than 0 dB at a frequency at which the phase angle has reached $-180°$

state — the property (condition) of a system

steady-state response — the system response after the transients have died and output has settled (time response after transient response)

step function — a function of time, which has a zero value before $t = 0$ and has a constant value for all time $t \geq 0$

system — a collection, set, or arrangement of elements (subsystems)

subsystem — part of a system

thermal capacitance — the constant of proportionality relating the change in temperature to the heat-flow rate

thermal resistance — the change in temperature for a unit change in heat-flow rate

time-invariant system — a system described by a differential equation with constant coefficients

time response — the output response (in time) of a system when subjected to an input

time-variant system — a system described by a differential equation with variable coefficients

total response — the response of a system from the time of application of an input to the point when time approaches infinity

transient response — the response of a system from the time of application of input to the time when output has reached steady state

unit step function — a function of time that has zero magnitude before time $t = 0$ and unit magnitude after that

unstable system — a closed-loop control system in which one or more roots of the characteristic equation lie in the RHP (*R*ight-*H*and side of the s-*P*lane)

zero of a transfer function — the root (solution) of the equation obtained by setting the numerator polynomial of the transfer function equal to 0; the value of s that makes (the value of) the transfer function equal to zero (hence the term *zero*)

INDEX

E

F

P

Q

R

S

T

U

V

Z